Exploring Planetary Climate

This book chronicles the history of climate science and planetary exploration, focusing on our ever-expanding knowledge of Earth's climate, and the parallel research underway on some of our nearest neighbors: Mars, Venus and Titan. From early telescopic observation of clouds and ice caps on planetary bodies in the seventeenth century, to the dawn of the space age and the first robotic planetary explorers, the book presents a comprehensive chronological overview of planetary climate research, right up to the dramatic recent developments in detecting and characterizing exoplanets. Meanwhile, the book also documents the discoveries about our own climate on Earth, not only about how it works today, but also how profoundly different it has been in the past. Highly topical and written in an accessible and engaging narrative style, this book provides invaluable historical context for students, researchers, professional scientists, and those with a general interest in planetary climate research.

RALPH D. LORENZ is a planetary scientist at the Johns Hopkins Applied Physics Laboratory. He has worked for the European Space Agency on the Huygens probe to Titan, and has been involved in many NASA and international space projects, including Cassini, Mars Polar Lander and the Japanese Venus Climate Orbiter. He enjoys visiting exotic locations on Earth – from the Arabian Desert and Alaska to Vanuatu and New Zealand – to learn about processes on other worlds, notably dust devils, sand dunes and volcanoes.

Exploring Planetary Climate

A History of Scientific Discovery on Earth, Mars, Venus and Titan

RALPH D. LORENZ

Johns Hopkins Applied Physics Laboratory

CAMBRIDGE
UNIVERSITY PRESS

CAMBRIDGE
UNIVERSITY PRESS

University Printing House, Cambridge CB2 8BS, United Kingdom

One Liberty Plaza, 20th Floor, New York, NY 10006, USA

477 Williamstown Road, Port Melbourne, VIC 3207, Australia

314–321, 3rd Floor, Plot 3, Splendor Forum, Jasola District Centre, New Delhi – 110025, India

79 Anson Road, #06–04/06, Singapore 079906

Cambridge University Press is part of the University of Cambridge.

It furthers the University's mission by disseminating knowledge in the pursuit of education, learning, and research at the highest international levels of excellence.

www.cambridge.org
Information on this title: www.cambridge.org/9781108471541
DOI: 10.1017/9781108677691

First published 2019

Printed in the United Kingdom by TJ International Ltd. Padstow Cornwall

A catalogue record for this publication is available from the British Library.

ISBN 978-1-108-47154-1 Hardback

Contents

Foreword

Having worked with Ralph on mission proposals to Saturn's moon Titan as well as on the Cassini RADAR team, I am quite familiar with his exceptionally broad expertise, and his unusual ability to distill complex concepts into straightforward explanations. Anyone who knows him is also always prepared to hear historical factoids which illustrate that not only does he understand an engineering or scientific problem, he also knows the history behind how the answer has been puzzled out over time (although he has a suspicious ability to tie most work back to Scottish scientists and engineers!). In this book, Ralph turns these talents to the history of planetary climate, to the benefit of us all.

To the general public, discussions of climate change seem to have started around the time of Al Gore's 1992 book *Earth in Balance*, and therefore seem to have some sort of murky, political origin. As Ralph ably lays out in this book, the history of trying to understand how Earth's climate works stretches back well over a thousand years, with our understanding accelerating as we obtained the capability to make more sophisticated and precise measurements. The potentially damaging effects of increasing carbon dioxide in our atmosphere through the burning of fossil fuels was argued within the scientific community for many decades before the current scientific consensus was reached. This long history of inquiry by hundreds of individuals over centuries makes for fascinating reading, but also serves to help the reader understand why scientists feel confident, and very alarmed, about the pace of climate change, its current and very visible effects on this planet, and the model predictions of what will come if we do not decarbonize the world's economies.

The book also puts the Earth in context: part of the reason we understand our own climate so well is that we have been able to study the climates of Venus, Mars and Titan. These other solar system bodies have helped us to understand the forcings on planetary climates, including the role of greenhouse gases. These other worlds have helped us understand that planetary climates can change radically over time, which helps inform our search for life beyond Earth. Early Mars and Venus may both have supported life, with extensive surface water oceans, before they became too cold or too hot, respectively. As we begin our study of planets around other stars, searching for Earth 2.0, our understanding of planetary climates will undoubtedly continue to evolve. Ralph's excellent summary of what we know based on our own solar system illustrates our strong foundation as we look further outward, and also our strong basis for concern as we look homeward.

Ellen R. Stofan
Former Chief Scientist, NASA

Preface

I was 11 years old. It was 1980, my first week at the King Edward VI Grammar School in Stratford-upon-Avon, England – the same institution at which William Shakespeare had some centuries before learned his small Latin and less Greek. Indeed I had a Latin class that week, but also an hour of Physics. Our Physics teacher had the class – all boys, in uniform suits and ties and required to wear caps in town – copy down from the blackboard into our exercise books a table of the (nine!) planets, with their size, mass, distance from the Sun, length of their year, and their surface temperature (in some cases only recently measured). Our attention was drawn to the systematic order of things, how the temperatures declined with distance from the Sun. This all made sense to me. "I want to be a physicist", I told Mr Hurlbut as we trooped out of the class.

In fact, I didn't become a physicist exactly. I knew, especially after the Voyager encounter with Uranus, the Challenger disaster (which brought some reality to aspirations of becoming an astronaut) and especially the daring close encounter of the European Giotto spacecraft with comet Halley – all of which happened in the first three months of 1986 – that I wanted to work in planetary exploration. I ended up studying for a degree in Aerospace Systems Engineering, and that got me, with exceptional good fortune, into my first job, working as a young engineer for the European Space Agency. This was at the beginning of the Huygens project, a joint endeavor with NASA to send a probe to Saturn's moon Titan. I subsequently became more of a physicist after all (but with some geology,[1] chemistry and other topics thrown in – when I am asked my profession, these days I say "planetary scientist"). Frequently I have been confronted with

---

[1] Actually, I have never sat in a formal geology class. I have, however, been on many field trips with the planetary sciences graduate students at the Lunar and Planetary Lab, at the University of Arizona. These gave me a love of desert landscapes, and an ability to at least fake some understanding of geological processes.

engineering questions of what the spacecraft or instrument I am working on is going to observe, or how it will be affected by the planetary environment it is in.[2] These basic-sounding questions, such as "What is the probability the Huygens probe will be struck by lightning?" or "How strong is the wind on Mars?", turn out to be very, very difficult to answer, but enormously interesting to consider scientifically. Many of these questions led me to explore planetary climate in one way or another.

I had written a couple of books about my work on Titan, and chatting with Ingrid Gnerlich, my editor at Princeton University Press, she suggested that planetary climate would be a good subject for a book. I heartily agreed – although I then got substantially distracted with other projects for several years. That delay, however, has made the subject all the more interesting.

First, Titan is now rather better understood as Cassini's observations have seen seasonal change, seen clouds and rain come and go, and showed a dramatic difference between north and south in the extent of its lakes and seas – a fundamentally climate-related question.

Second, the growing number of extrasolar planets shows that there is a vast realm of planetary possibilities, with different sizes, masses, atmospheres, stars and orbital configurations. All, however, are subject to the same laws of physics, which let us explore which of these worlds might be habitable, motivating study of climate more generally.

And third, our own climate continues to evolve. As I worked on the book and carbon dioxide concentrations climbed above 400 ppm, an international agreement was reached in Paris to attempt to limit temperature rise to 2 K.

This book largely confines itself to the four principal climates in our solar system – those of Earth, Mars, Venus and Titan. The last of these, formally speaking, is not a planet, but it acts like one and has more interesting processes going on than many actual planets. I largely avoid the giant planets Jupiter and Saturn, and the ice giants Uranus and Neptune, for a couple of reasons. First, they don't have accessible surfaces, so there is nothing really for a climate to affect – no sand dunes, no river valleys, etc. Second, rather than having "no climate", in a sense they each have an infinite number of climates – since there is no surface, one level in their thick atmosphere is as significant as any other. If you feel too cold, descend into the warm depths. Lacking infinite patience, but mostly for the first reason, I have not covered them except in passing, although I recognize they are very interesting from a purely fluid-mechanics standpoint.

---

[2] My ruminations on the difference in outlook between the scientific and engineering profession were published recently: R. Lorenz, Engineers are Dogs, Scientists are Cats, *Aerospace America*, pp. 40–43, June 2018.

There are several moons and an ex-planet that have borderline climates, in particular Triton and Pluto. Their atmospheres are tens of thousands of times thinner than that of Earth. While some layers of haze are present on Pluto, and wind streaks and even plumes seen on Triton, their climates are of limited practical significance. In fact, Pluto turned out to be a lot more interesting than I had expected, but the data from New Horizons' encounter will take years to digest. Similarly, relatively little can be said about individual exoplanets. While their collective study is a burgeoning field, it is not one in which I am regularly immersed, and so my review of these developments is really only a cursory guide to the state-of-the-art, but should serve to provide some literacy and context for further reading.

One of the most enriching pleasures of writing a book is that it sets a framework and motivation for learning things. In fact, I had planned to write something of a textbook about planetary climate itself, but as I wrote what was going to be an introductory chapter or two for historical context, I found the history itself to be a rich and interesting story. My instinct is that if I find something interesting, hopefully readers may do also, and so the history itself became the focus of the book. Other books have since emerged (see Further Reading) that amply fill the textbook need, and so I have indulged my instincts to concentrate on how our ideas about climate have evolved and how our knowledge of planetary environments and our own past climate has been won.

As well as telling a story, I hope the book will serve as a springboard for further study, and I have attempted to be generous in documenting sources, resources and directions for further reading with numbered references, although no book of this length could hope to cite every relevant work on a topic this wide. The attentive reader will have already noted I also have a propensity to make parenthetical amplifications – included as footnotes. My excuse is that reality is not perfectly linear, so the text describing it cannot be only one-dimensional – the notes are literary fracticality. Also, in describing parallel threads of endeavor, it is impossible to be strictly chronological – the text is obliged to hop back and forth in time somewhat in order to retain at least a punctuated equilibrium of topic.

I should caution that, while the book is written essentially as a history, I am not trained as a historian. I hope I have applied the necessary degree of scholarship in researching events long before my time. I have known at least some of the scientists involved in the era of planetary exploration (Chapter 5 onwards) and am somewhat familiar with the literature. Since 1990 or so (Chapter 7 onwards) I have become a protagonist myself, writing some of the papers, and reviewing others, so the account thereafter cannot be considered totally objective. Inevitably, my own contributions receive more attention than others, but I hope the story is balanced overall. In trying to tell a story that spans centuries, I have likely failed to do full justice to the large body of work that has emerged in the past two decades about the planet Mars, but again, hope that I have provided most of the highlights.

I have been fortunate to have worked with many talented individuals over the years on topics related to planetary climate. They are too many to list here, although many will become obvious on examining the notes. Reviewers for Princeton University Press and Cambridge University Press provided important comments that improved the text, Margaret Patterson's attentive copyediting made for a much more polished product, and the enthusiasm of Emma Kiddle at Cambridge is much appreciated. A few colleagues deserve mention, however, for kindly commenting on early chapters (Laura Kerber, Jani Radebaugh and Mark Bullock), and especially Conor Nixon for bravely subjecting himself to a draft of the whole text. Kevin McGouldrick provided helpful remarks on the first four chapters (published in abridged form as "Planetary Climate before the Space Age" on Createspace). Their stabilizing feedback prevented some runaway text. Any errors in what remains, however, are mine alone.

Special acknowledgement is due to Elizabeth Turtle, who not only provided encouragement and critical comment on the text, but has patiently tolerated her husband littering the living room with higher piles of books and papers than usual, only to substitute complaints about "never getting time to write books" with complaints about how long the book was taking to write. Her love brings warmth to a cold universe.

1

The Age of Wonder: Learning the Earth, Oceans and Sky

440 BC to ~1760

440 BC – Herodotus considers Nile flood lagging seasons, notes sea-shells on hills and salty soil indicating Egypt was once underwater, recognizing terrestrial conditions vary with time

350 BC – Aristotle writes *Meteorology*, a work that discusses a range of phenomena in the sky, including meteors as well as weather and climate

c. AD 80 – Chinese philosopher Wang Chong states that rain is evaporated from water on the earth into the air and forms clouds and rain

1021 – Alhazen investigates refraction of light and twilight, and deduces an effective height of the Earth's atmosphere as about 20 miles

1450 – Leone Battista Alberti develops a swinging-plate gauge, considered the first anemometer

1543 – Nicolaus Copernicus proposes a heliocentric architecture of the solar system

1604 – Cornelis Drebbel writes a *Treatise on the Elements*, noting the tremendous expansion of water into vapor. Drebbel goes on to invent a submarine, and a self-regulating oven with a thermostat

1607 – Galileo Galilei constructs a thermoscope, and defines heat as a distinct property of matter, rather than as one of the classical four elements (Fire, Water, Air and Earth). Observes mountains on the Moon, phases of Venus

1643 – Evangelista Torricelli invents the mercury barometer

1648 – Blaise Pascal rediscovers that atmospheric pressure decreases with height, and deduces that there is a vacuum above the atmosphere

1650–1690s – Christiaan Huygens discovers Titan, contemplates conditions on other worlds, recognizes atmosphere and fluid may be made of different substances, different densities etc. Calculates how long a bullet would take to reach the planets

1660s – Cassini observes polar caps on Mars

1686 – Edmond Halley makes a systematic study of the trade winds and monsoons and identifies solar heating as the cause of atmospheric motions. Defines the relationship between barometric pressure and height above sea level

1724 – Gabriel Fahrenheit creates reliable scale for measuring temperature with a mercury-type thermometer

1729 – Pierre Bouger pioneers light measurement, and develops the law of attenuation of light by atmospheric absorption (later called the Beer–Lambert law)

1735 – The first idealized explanation of global circulation is the study of the trade winds by George Hadley

1742 – Anders Celsius, a Swedish astronomer, proposes the centigrade temperature scale, which led to the current Celsius scale

1743 – Benjamin Franklin is prevented from seeing a lunar eclipse by a hurricane, he decides that cyclones move in a contrary manner to the winds at their periphery

1755 – Tobias Mayer calculates dependence of temperatures on latitude and altitude

1761 – Joseph Black discovers that ice absorbs heat without changing its temperature when melting

Climate is a subject that concerns everyone, and considerations of climate problems enter some of the first intellectual records in existence. For example, Herodotus' *Histories*, written (or at least told) around the 440s BC, document aspects of geography as well as history. In particular, Herodotus notes the non-intuitive behavior of the Nile river, that its level begins to rise around the summer solstice and does so for a hundred days, then the level falls throughout winter. Herodotus considers whether winds could be a factor (the Nile's usefulness as a transport artery arises from the generally northerly winds, allowing boats to sail south upriver, they could then furl their sails and float back down). In fact summer snowmelt is responsible for the changing flow but Herodotus (incorrectly) dismisses snow, since it was known that the further south one got, the hotter and drier it was, so how could there be snow near the torrid equator?

Interestingly, Herodotus also ventures some speculations on the origins and past of Egypt, noting that the soil is dark, as one might expect alluvial (water-delivered) deposits to be, in contrast to the windblown sands to the west in Libya. Furthermore, he notes that shells characteristic of aquatic animals are found on hilltops and that there is salt in the soil, and ventures that much of Egypt was once underwater. Herodotus recognized (Bishop Ussher not yet having set an age of the Earth of only four millenia before his time!) that these changes may have taken tens of thousands of years.

A more focussed work, and generally acknowledged as the first serious work on the topic (~350 BC), is the *Meteorologia* compiled by Aristotle. This treatise covers a wide range of geophysical and astronomical topics, such as the saltiness of the sea and the nature of comets as well as the causes of wind and rain (the separation of the study of meteors and meteorites into disciplines distinct from "meteorology" came later, as science became more specialized). Naturally, since the concepts of heat and energy were unknown, and the description of matter was limited to four elements (Earth, Air, Fire and Water),[1] Aristotle's explanations are flawed, but he does a fair job of laying out some interesting problems (for example, why should hailstones, made of ice and obviously associated with cold, be more common during summer than winter?). He captures the essence of the hydrological cycle as follows:

Now the Sun, moving as it does, sets up processes of change and becoming and decay, and by its agency the finest and sweetest water is every day carried up and is dissolved into vapor and rises to the upper region, where it is condensed again by the cold and so returns to the earth.

---

[1] Four or five elements feature in much ancient thought. The "classical" four were originally proposed by Empedocles, and Aristotle popularized them, adding a fifth, "aether", for the uncorruptible heavens.

Other Greek philosphers of the time contemplated that there might be many worlds: Metrodorus (a disciple of Epicurus) suggested in the fourth century BC, *"To consider the Earth as the only populated world in infinite space is as absurd as to assert that in an entire field of millet, only one grain will grow."*

After Aristotle's compilation, there was little substantial progress in meteorology in the Western world. Yet, as in other fields, many Muslim philosophers and scientists made steps towards understanding the natural world, notably from the tenth through fifteenth centuries. The scholar Ibn al-Haytham ("Alhazen"), in what is now Iraq, used geometric arguments to deduce the thickness of the sensible atmosphere.[2] Knowing the diameter of the Earth, and the fact that twilight began or ended with the Sun 19 degrees below the horizon, he deduced that the effective scattering height of the atmosphere could be no more than 52,000 paces. Indeed, at such heights – not physically reached for almost another thousand years – the air density is only a fraction of one percent of what it is at the ground.

Similarly, in China, some relevant ideas were emerging. In 1074, Shen Kuo[3] reasoned – like Herodotus – that a belt of bivalve fossil shells in mountains inland implied that the terrain there must have once been a seashore, and that mountains must be eroded and sometimes uplifted (in the West, as we see later in this chapter, James Hutton, seven centuries later, reached similar conclusions and took them further). He also recognized that petrified bamboo, revealed by a landslide in a place where bamboo does not grow today, implied that the climate had been different in the past.

The pivotal role of the Sun in controlling climate was obvious even to the ancients, but systematic consideration of the climate of Earth and of other worlds relied on correctly laying out the architecture of the solar system. While some Greek scientists got this right, it was not until the Copernican revolution beginning in 1543 that progress could really be made. An additional early challenge to understanding climate and weather is that diurnal changes of wind are in near synchrony with tidal changes in the sea, and thus the role of the Moon in influencing winds was initially thought to be significant.

With the Copernican layout of the universe established with the Sun at the center, backed by Kepler's laws, it became possible in the 1600s to consider the other planets in their proper place (e.g. Figure 1.1). English astronomer Thomas Digges realized that the stars may stretch into infinite space,[4] and the scholar Giordano Bruno even

---

[2] For a modern version of this measurement, see M. Beech [1].

[3] I'd never learned of Shen Kuo until I started writing this book. Making discoveries is one of the main joys of writing.

[4] Digges attempted to measure the parallax of Tycho Brahe's supernova, but could only establish a lower limit, determining that the star had to be beyond the Moon. Digges is a relatively poorly-known astronomer and mathematician, typically overshadowed by Giordano Bruno whom he may have influenced in England in the 1580s; see Ref. [2].

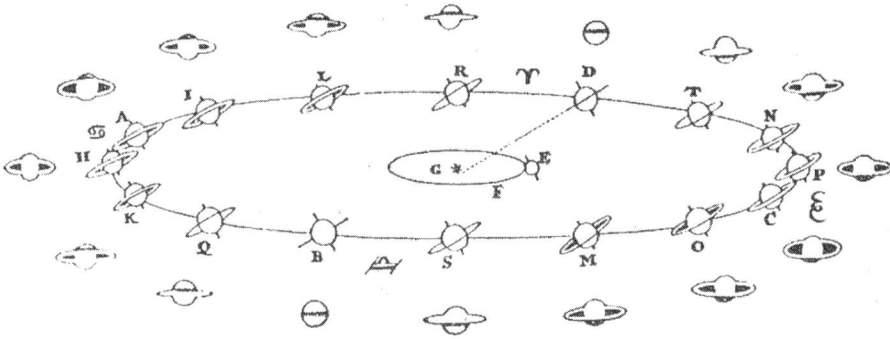

Figure 1.1 Seasons on another planet. This admirable graphic from Huygens' 1659 book *Systema Saturnium* shows not only Saturn's relationship to the Sun and Earth, but also how Saturn's appearance from Earth changes due to the fixed orientation of Saturn's ring and pole with respect to space.

speculated that stars were other suns with their own planets, and that such planets might be inhabited. The invention of the telescope and Galileo's observation of crescent phases of Venus made it clear that Venus was closer to the Sun than is Earth.[5] On the other hand, Mars, Jupiter and Saturn (e.g. Figure 1.1) always appeared almost completely round, implying they were further away.

The development of instruments not only augmented astronomers' vision, but also enabled the quantitative measurement of the world. Instruments to first visualize, and later measure, temperature were also first developed around this time. Galileo developed a thermoscope (only with the addition of a numerical scale does it become a thermometer), while an ingenious Dutchman working in England in the early 1600s, Cornelis Drebbel, devised an oven with a mercury tube that not only displayed a measure of heat, but regulated the air flow into it – a primitive thermostat and perhaps the first example of a feedback control system. Drebbel also ground lenses for optical instruments, showing some to his visiting countryman, the diplomat Constantyn Huygens. Remarkably, Drebbel is also reported to have invented a submarine, and conveyed King James I under the surface of the River Thames in it: the depth of the submersible was indicated with a tube of mercury, in effect a barometer [3].

Evangelista Torricelli in Tuscany, confronting the problem that suction pumps could only lift water 10 meters or so, showed in 1643 that a column of mercury in a tube closed at one end was limited to only 76 cm long – the same weight

---

[5] Although Galileo is sometimes credited with the invention of the telescope, it was likely invented in the Netherlands in 1608, but quickly spread. Thomas Harriot in England observed the sky with it in 1609. Galileo was an early and enthusiastic adopter of the new invention, and quickly built his own and improved the design, and, importantly, wrote down his findings.

per unit area as 10 meters of water – and thus that this had to be the same as the weight per area, or pressure, of the atmosphere. Thus began the systematic, and eventually quantitative, understanding of the physics of atmospheres: he famously remarked, "*We live submerged at the bottom of an ocean of air.*"

Galileo's use of the telescope in 1610, revealing that Jupiter had moons, and that our own Moon is a world with mountains and craters, stimulated thinking about planets as places, and thus about their conditions. The Polish astronomer Hevelius documented the topography of the Moon in detail in 1647. This widening perspective was perhaps most enthusiastically embraced by the Dutch astronomer Christiaan Huygens (Constantyn's son). Huygens ground his own lenses, and was the first to measure the rotation rate of the planet Mars, finding that the length of its day was rather similar to our own, and in 1655 he discovered Titan. He correctly imagined how seasons and shadow would work on other worlds (Figure 1.2). In the mid-1660s, Jean-Dominique Cassini in Paris, observed polar caps on Mars, and Huygens made the first sketch of them in 1672.

Huygens considered that other worlds might have seas composed of fluids with properties different from water [4]: "*Every Planet therefore must have its Waters of such a temper, as to be proportioned to its Heat: Jupiter's and Saturn's must be of such a Nature as not to be liable to Frost…*", a remark that rather nicely anticipates methane rain on Titan. He even speculates that there is a lot of room in between

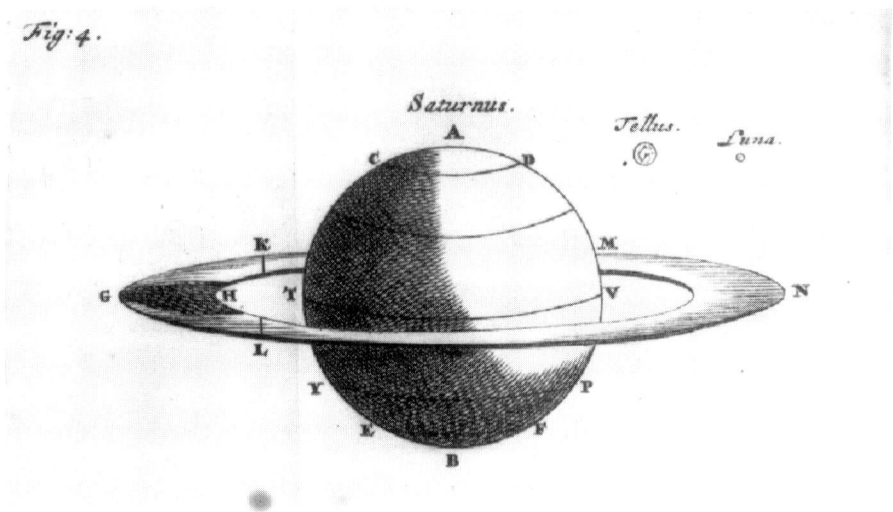

Figure 1.2 A billion-mile leap of the imagination. Not only does Huygens discover Titan and ascertain correctly the architecture of the rings, he expresses his knowledge in a picture, showing a view that cannot be obtained from Earth. He mentally transports himself out to Saturn, to visualize how in summer Saturn would cast a shadow on its rings, but that shadow is generally hidden from our view.

the densities of air and water on Earth for fluids on other planets: "*The Sea perhaps may have such a fluid lying on it, which tho' ten times lighter than Water, may be a hundred Times heavier than Air*", and notes that on worlds with denser atmospheres it would be easier for birds to fly. Indeed, this is not too far off what Titan is like today: Titan's air is four times denser than ours, and its seas about half as dense. So Titan's seas are indeed about 100 times denser than its air, rather than the ~800 times for Earth. If birds – or even people – could breathe and not freeze, they would find that Titan's low gravity makes it a very easy place in which to fly.

Huygens closes his book with breathless happiness:

> *What a wonderful and amazing Scheme have we here of the magnificent Vastness of the Universe! So many Suns, so many Earths, and every one of them stock'd with so many Herbs, Trees and Animals, and adorn'd with so many Seas and Mountains! And how must our Wonder and Admiration be increased when we consider the prodigious Distance and Multitude of the Stars?*

Huygens notes, in an appealingly self-consistent if slightly optimistic argument, that the astronomers on hot Mercury, so close to the Sun, would think that the Earth would be inhospitably cold. And yet here we are; and therefore it follows that, while we might think Jupiter or Saturn too cold to support life, it might nonetheless be present, albeit in somewhat adapted form. Huygens draws on the fact that plants and animals had been found in another new world – America – that were quite different from those known in Europe, and wonders if the differences between lifeforms on different planets might actually be less than the differences between those from widely separated parts of the Earth.

Huygens outlook, one we might call an extreme NeoCopernican, was not limited to astrobiologists on other worlds, but also sailors who probably used similar sails and anchors, pullies and rudders on other seas.[6] He even expresses some jealousy of "*the great Advantages Jupiter and Saturn have for Sailing, in having so many Moons to direct their Course, by whose Guidance they may attain easily to the Knowledge that we are not Masters of, of the Longitude of Places*" (the problem of determining longitude at sea was not to be practically solved for another several decades[7]). Yet, while his imagination ranged across the solar system, Huygens

---

[6] A French author of the time, Bernard de Fontenelle, lauded planetology and astrobiology, writing, "What can more concern us, than to know how this world which we inhabit, is made; and whether there be any other worlds like it, which are also inhabited as this is?" (1688 translation of *La Pluralite des Mondes*).

[7] Galileo had suggested the arrangement of the Jovian moons could be used as a sort of universal clock (for terrestrial, rather than Jovian, sailors to calculate longitude) but this relied on telescopic observations from a heaving ship, which had severe challenges. Halley confronted the problem in various ways, considering occultations of stars by the Moon, or

was well-grounded enough to recognize the practical challenges of traveling to other worlds. He noted that a bullet (the fastest object with which he was familiar – which traversed a hundred fathoms in a heartbeat) would take 250 years to reach Saturn.

Huygens was not alone in his age – an English clergyman and co-founder of the Royal Society, John Wilkins [5], imagined the Moon might be inhabited, and while recognizing that the upper air was cold and thin and that there would be no inns en route to offer victuals and shelter,[8] affirmed "*it possible to make a Flying Chariot, in which a Man may sit, and give such a Motion unto it, as shall convey him through the Air. And this perhaps might be made large enough to carry diverse Men at the same time . . .*". He went on to make the remarkable prediction "*Supposing a Man could fly, or by any other means, raise himself twenty miles upwards, or thereabouts, it were possible for him to come unto the Moon*" – three centuries later these two landmarks in aviation were in fact met only thirteen years apart.[9]

Daniel Defoe briefly considered the climate on the Moon in his 1705 *The Consolidator*,[10] where "*the Elasticity of the air is quite different, and where the pressure*

the use of the dip of the Earth's magnetic field. Ultimately it was the development in 1761 of the marine chronometer, a clock that maintained sufficiently accurate time despite temperature changes and the ship's motion, that allowed practical navigation.

[8] Indeed, this logistical aspect of spaceflight was also pointed out by the *Spokesman-Review* newspaper in 1920, on the publication of Robert Goddard's proposals for rocket-powered spaceflight "THAT FLIGHT TO PLANET MARS – Would take a year and the traveler would be hungry". See Ref. [6].

[9] No human reached 20 miles (105,600 ft or 32 km) upward until the 1950s (38.5 km in the Bell X-2 rocket plane in 1956, 13 years before Apollo 11). Although ballooning began long before heavier-than-air aviation, the 20-mile threshold was only breached (113,740 ft) in the StratoLab High V balloon in 1961, shortly after Kittinger's famous parachute jump (102,800 ft, 31 km) from the Excelsior III balloon in August 1960. See Table 10.1 in [7]. The StratoLab V flight of over 9 hours gave its two aeronauts the honour of being the highest-flying Americans for exactly one day, that title passing to Alan Shepard on a suborbital flight on a Mercury rocket (though of course Yuri Gagarin had flown in orbit a month before).

[10] The work, *The Consolidator; or, Memoirs of Sundry Transactions from the World in the Moon* (London, 1705), is actually a satirical commentary, using the device of people on the Moon viewing the Earth from afar to show the absurdity of aspects of human society. The "Consolidator" is the name of the feathered machine ("Engine"), which has wings about 50 yards in breadth, and cavities "*filled with an Ambient Flame . . . order'd as to move about such springs and wheels as kept the Wings in a most exact and regular Motion*". This seems somewhat inspired by Wilkins' idea, and indeed the book mentions him. Upon attaining a height where "*gravity having passed a certain line*" (Newton's *Principia* was published only 7 years before), the wings became more applied to controlling the engine's descent down to the Moon than to raising it, a scenario that nicely anticipates Jules Verne's *From the Earth to the Moon*. Defoe was of course most famous for his *Robinson Crusoe*, written in 1722. The literary device of extraterrestrial visitors commenting on Western culture was

of the Atmosphere has for want of Vapor no Force" (in spite of which the narrator is able to travel to the Moon in a feathered machine!). The Moon has "*very seldom any clouds at all, and consequently no extraordinary storms*". The previous year Defoe had written *The Storm*, an early journalistic work compiling personal accounts of the Great Storm of 1703, a cyclone that felled millions of trees and killed thousands.

The New World too was reported to be meteorologically challenging – even just the title of one account, by James MacSparran, says it all: "*America Dissected, Being a full and true account of the American Colonies, shewing the intemperance of the climates, excessive heat and cold, and sudden violent changes of weather, terrible and mischievous thunder and lightning, bad and unwholesome air, destructive to human bodies, etc.*" That said, these were exciting times – as Robert Hooke noted, "*new Lands, new Seas are daily found out, and fresh descriptions of unknown Countreys still from both brought in; so that we are forced to alter our maps, and make anew the Geography of both again.*"

An understanding of the Earth's climate at the planetary scale first requires a realization that the Earth is round. While this was of course understood by many prior to the age of exploration, the quantitative implications in terms of the amount of sunlight deposited at different latitudes requires trigonometry, and the amount was first calculated by astronomer Edmond Halley [8]. In a paper published in 1693 he showed by geometric arguments[11] how the combination of the declination of the Sun and the varying length of day can conspire to deliver more sunlight (averaged over a 24-hour period) to the pole at midsummer than to the equator (Figure 1.3). However, the nature of heat was not yet fully understood, and so the exact implications for temperature had to wait another century or two.

Halley may also have been the first to quantitatively consider the hydrological cycle. More particularly, he adopted the thoroughly modern approach of obtaining experimental data, and applying it to a global-scale model [9]. Specifically, he warmed an eight-inch pan of water on a set of scales to the temperature of a hot summer day, regulating the temperature by holding coals nearby, and observed that in the course of two hours the pan lost half a Troy ounce of liquid.

This value of 1/10 of an inch per day is rather close to the terrestrial average flux of evaporation and precipitation, of about one meter per year. That said,

---

also employed by Voltaire in *Micromegas*, where a 37-km tall visitor from a planet around Sirius enounters a 1.8-km tall member of the Academy of Saturn with whom he travels to Earth, and is astonished to find that the tiny beings on that planet are actually intelligent. The notion of a planet around Sirius appears to have raised no more eyebrows than the rest of the book.

[11] The result is expressed in rather arcane terms: "*to give the proportional degree of Heat or the sum of all the Sines of the Sun's Altitude while he is above the Horizon in any oblique sphere, by reducing it to the finding of the Curve surface of a cylindrick Hoof, or of a given part thereof.*"

Lat.	Sun in ♈ ♎	Sun in ♋	Sun in ♑
0	20000	18341	18341
10	19696	20290	15834
20	18794	21737	13166
30	17321	22651	10124
40	15321	23048	6944
50	12855	22991	3798
60	10000	22773	1075
70	6840	23543	000
80	3473	24673	000
90	0000	25055	000

Figure 1.3 Halley's calculation of the relative amounts of sunshine over 24 hours as a function of latitude and season. The rows are latitude, and the columns refer to spring equinox, summer solstice and winter solstice, respectively. Note the surprising result that there is more sunshine in total in polar summer (25,055 units) than at the equator; Halley is essentially using integer arithmetic, hence the numbers are expressed relative to the equator at equinox (20,000).

the agreement is perhaps fortuitous – although Halley notes that he made the water salty like the sea, he does not record the weather conditions such as wind or humidity, which we now know would substantially affect such an experiment.

Halley then bravely upscales this measurement to the Mediterranean Sea, which he briskly approximates as a 4×40 degree box (one degree being 69 English miles) to determine that the Mediterranean should lose 5280 million tons of water per day.[12] He estimates the water influx via large rivers (scaling from the Thames, whose depth and flowspeed were known to him by personal experience, having experimented with diving bells for salvage) and concludes that the Mediterranean must experience a deficit of water, restored by inflow from the Atlantic through the Straits of Gibraltar. This is probably the earliest example of a mass balance calculation.

Halley also considers the salt cycle on the Earth. Anticipating the nineteenth-century geology view that "the present is the key to the past", he notes that the

---

[12] I made similar sweeping approximations myself in performing essentially the same calculation for methane evaporation from Titan's seas [10]. I think Halley would approve.

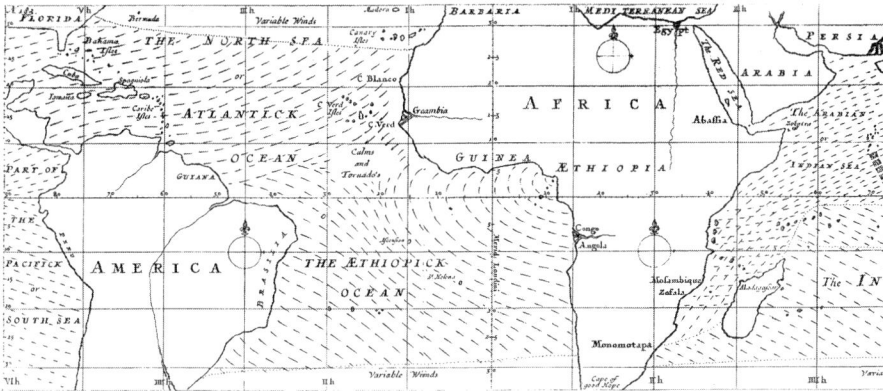

Figure 1.4 Part of Halley's vector map of the trade winds, a first planetary-scale depiction of atmospheric circulation.

seas are salty and that the rivers are less so [11]. But while the water transported into the seas is evaporated, to complete the cycle by raining back onto the land and driving the rivers, the salt must accumulate and there should be a progressive increase in the saltiness of the sea[13] (indeed, one might cheekily suggest such a trend as "secular" – in this paper Halley grapples with the Age of the Earth inferred from scripture, noting the difficulty of taking the "seven days" indicated in Holy writ literally on this question). Halley then proposes systematic measurements of the saltiness of lakes and seas. Recognizing that detecting a measurable increase might demand an interval much longer than his own lifetime, he proposes that at the least such a survey could act as a baseline for a future comparison, rueing that the Ancient Greeks failed to record such quantitative data for the benefit of such a calculation!

Halley also made innovations in the presentation of data. He compiled information from a variety of sources into a map of the global trade winds – perhaps the first vector map [12] (Figure 1.4). He similarly mapped the declination of the compass needle; the resultant graphic is considered the first contour plot in the scientific literature. While we take these portrayals for granted today, we should consider that it is only once real measurements are put on a map in this way that the workings of our world on a planetary scale can be understood.

As an island nation, England in the seventeenth (and Britain as a whole in the eighteenth) century relied for her prosperity and security upon seapower.

---

[13] A rather similar scenario was considered for Titan, where it was assumed that methane may steadily be converted by sunlight into ethane. Thus, by measuring the ethane: methane ratio in Titan's ocean (at the time it was assumed there was a single global ocean), then its age might be constrained.

Geophysics and meteorology (at the time, of course, both disciplines that fell under the umbrella of "natural philosophy") were therefore of strategic interest in connection with navigation at sea. Innovations in compasses, telescopes, and famously in clocks, were all stimulated by seafaring.

Halley's synthesis of reports from as many sources as he could find into a unified picture of the trade winds is probably his most well-known meteorological contribution. Interestingly, Halley recognizes that his big picture is just that, an idealized approximation of the sea winds at some distance from the land. He recognizes that terrain can cause substantial perturbations to the winds:

> Upon and near the shores, the Land and Sea Breezes are almost everywhere sensible; and the great Variety which happens in their Periods, Force and Direction, from the situation of the Mountains, Vallies and Woods, and from the various texture of the Soil, more or less capable of retaining and reflecting Heat, and of exhaling or condensing Vapours is such, that it were an endless task, to endeavour to account for them.

Of course, the modern approach to climate prediction and weather forecasting is to do exactly this, to divide the world up into tiny boxes and assign estimates for just these effects. In modern terms, Halley is saying a good climate model should include fields of topography, albedo and thermal inertia, well established for the Earth and now Mars, although we are still catching up on these for Titan and Venus.

A successor of Halley, George Hadley, offered a dynamical explanation of why the trade winds should blow the way they do [13, 14], recognizing that warm air rises at the equator, where the solar heat is strongest, and so surface air from higher latitudes moves in to replace it.[14] However, the eastwards motion of this air is less than that of the Earth's surface at the equator (which is furthest from the Earth's axis of rotation) and so the wind appears to come from the northwest or southwest rather than purely north or south. Hadley even remarks in his paper that navigation would be "tedious" without the Earth's rotation, and that winds must balance out over the surface of the globe, otherwise the rotation of the Earth would be affected.

Contemplation of other worlds began to move on, with astronomer Giacomo Maraldi (a nephew of Cassini) observing the poles of Mars in 1704 and suggesting, in 1719, that they are ice caps. A rather curious coincidence of fiction and fact is the speculation by Jonathan Swift in the 1727 adventure *Gulliver's Travels* that Mars has two moons (which it has – although they'd be impossible to see with the telescopes of the time).

---

[14] The averaged flow in a plane normal to the equatorial one is formally termed the Mean Meridional Motion, but is often referred to informally as "Hadley circulation". The most equatorward part of the circulation on Earth is termed the Hadley cell (see also about Ferrel in the next chapter).

Around this time, the measurement of heat became systematized with the adoption of standard scales of temperature, by Gabriel Fahrenheit (1724) and Anders Celsius (1742). Once standard temperature scales began to be used, it became possible to compare temperatures recorded in different places. The German astronomer Tobias Mayer developed what might be called the first parametric model of climate around this time, devising mathematical expressions to suggest what the temperature of a given place should be as a function of its latitude and elevation.[15] He even, in essence a methodology like that used by astronomers to predict tides, suggested a simple sinusoidal function to predict how much later the warmest day should be after the summer solstice.

The French geophysicist Jean-Jacques de Mairan gathered as many temperature measurements as he could – in mines as well as on the surface – and, noting that water from springs could be much hotter than the air, deduced that the interior of the Earth must be warm.[16]

Mairan's colleague Buffon attempted to calculate how long the Earth might have taken to cool from an initially hot state by extrapolating from an experiment [15]: he placed iron spheres of different sizes (up to 5 inches) in a fire and observed how long they took to get cold enough to touch, and extrapolated to the size of the Earth, yielding an impressively precise estimate of 96,670 years. Flawed though the calculation was, and indeed the assumption on which it was founded, it was at least a very long time, longer than many had contemplated at the time.

Mairan wondered at how the varying obliquity of light in summer and winter controlled temperature, much as Halley had. Even though there is no direct sunlight at all at the polar regions in winter, they do not become unimaginably cold. Mairan reasoned that perhaps heat from the Earth's interior moderated the temperatures, a paradigm also considered in a more quantitative way by Fourier a century later. Mairan admitted that he could not quantify the effect of the atmosphere, although his colleague Pierre Bouger in 1729 did derive the mathematical law (usually called today the Beer–Lambert law) describing the reduction of light by an absorbing medium.[17]

Some of the first attempts to measure the intensity of sunlight came around this time. Horace-Benedict de Saussure, of Geneva, a geologist and physicist, loved the mountains and enthusiastically and systematically made measurements with barometers (in part to determine the most reliable conditions or

---

[15] Mayer's work, from 1755, lay in some obscurity until published posthumously as *Opera Inedita* by Lichtenberg in 1771.

[16] Mairan made some early studies of the aurora borealis, as well as phosphorescent glows, but his most significant work was in biological clocks. He noted that certain heliotrope plants open and closed their leaves on a regular cycle even when in a dark room.

[17] Bouger made a number of contributions in astronomy, and sailed to Peru on an expedition to measure the meridian arc. A measure of gravity variations, the Bouger anomaly, is named after him. Most of his later contributions were in fact in naval architecture.

time of day to use a barometer to determine altitude) and thermometers, spending some 17 days at the summit of Col de Geant (3371 m) in 1788 recording two-hourly readings. He also devised a reasonably reliable approach to measuring humidity, or the water vapor content of the air: his instrument (a hygrometer) used a human hair as its sensing element. De Saussure experimented with using thermometers in all kinds of different ways, noting the need to ventilate the bulb and shield from sunlight to get a reliable air temperature measurement. He also placed thermometers in glass vessels with a long response time, so that they could be dangled down to the bottom of lakes and then hauled up quickly enough to retain a "memory" of the lake temperature – he found that lake bottoms consistently had a temperature of 4 degrees Celsius.

De Saussure also invented, in 1767, the first practical solar oven – an insulated box with three layers of glass to trap the heat. He found that the maximum temperature he could attain (230 °F) was more or less the same at the summit of Mt. Crammont as it was on the valley floor, 4852 feet lower down, in sharp contrast to the air temperature, which was appreciably warmer at lower altitudes. These experiments provided the first insights into radiative heat transfer in atmospheres, the foundation of modern climate models.

The late 1700s also saw advances in chemistry, with elements such as oxygen and sulfur being recognized by Lavoisier. Also of note in this period are developments due to polymath Benjamin Franklin, who not only explored the electrical nature of storms (famously inventing the lightning conductor to protect buildings), but engaged in a wide range of meteorological investigations. In 1755 he wrote a vivid account of an encounter with a dust devil in Maryland, describing how he rode alongside it and attempted an early experiment in weather modification, seeing if his whip would break up the whirlwind (it didn't!) [16].

As US Postmaster, Franklin investigated the great difference in trip times of mail ships that took weeks longer to cross the Atlantic coming from England than going to it. Assembling accounts from whalers and ship captains, he assembled in 1770 a chart (Figure 1.5) of the current responsible, which he named the Gulf Stream. This current plays a large role in moderating the climate of northern Europe, transporting heat from lower latitudes. Franklin also suggested a possible link between the 1783 eruption of the Laki volcano, and the harsh winter in Europe the following year.

In fact, Franklin was not the first prominent American to discuss changes in the weather and climate, and indeed the possible influence of human activity on climate. Thomas Jefferson, who as a farmer took a close interest in the land and climate, kept diligent records of weather, and reduced his records into averages. He noted that conditions were warming:

Figure 1.5 Benjamin Franklin's chart of the Gulf Stream.

A change in our climate however is taking place very sensibly. Both heats and colds are become much more moderate within the memory even of the middle-aged. Snows are less frequent and less deep. They do not often lie, below the mountains, more than one, two, or three days, and very rarely a week. They are remembered to have been formerly frequent, deep, and of long continuance. The elderly inform me the earth used to be covered with snow about three months in every year. The rivers, which then seldom failed to freeze over in the course of the winter, scarcely ever do so now. This change has produced an unfortunate fluctuation between heat and cold, in the spring of the year, which is very fatal to fruits. [17, 18]

Jefferson speculated that clearing of the East coast forests by settlement now allowed the sea breezes to penetrate further inland, moderating the climate.[18]

---

[18] As Fleming's book [18] notes, the role of trees in introducing moisture into the air was a noted effect at the time, of which even Christopher Columbus two centuries before had been aware. The possibility that land-clearing had changed the climate in Europe had been speculated upon (see e.g. Ref. [19]), but the Americas offered the opportunity of a stronger experiment on this question.

Jefferson also makes an observation recognizing the difference between air temperatures and ground temperatures:

> *The access of frost in autumn, and its recess in the spring, do not seem to depend merely on the degree of cold; much less on the air's being at the freezing point. White frosts are frequent when the thermometer is at 47 degrees, have killed young plants of Indian corn at 48 degrees, and have been known at 54 degrees.*

Jefferson even speculated, in what must be one of the most early instances of contemplated planetary engineering, what the change in climate might be if the Spanish were to cut a channel in the isthmus of Panama wide enough to allow tropical ocean currents to pass, perhaps eliminating the fog banks off Newfoundland.

A remarkable compilation of climate information (*An Estimate of the Temperature of Different Latitudes*) was published in 1787 by Richard Kirwan, an Irish geologist and chemist.[19] He reviews the ideas of Mayer and de Mairan (having spent 19 years in London, he maintained a vigorous correspondence with many scientific luminaries of the time) and tabulates Mayer's calculations of temperature versus latitude. More usefully, he also compiles the measurements he obtained from his contacts, from Kamchatka and Algiers, to Labrador and Quito, the Cape of Good Hope and Tibet. Although the exercise is descriptive rather than predictive, Kirwan assesses various factors influencing the weather (e.g. whether a cool summer preceded a cold winter, and the role of Siberian winds) and notes that journals implied that winds measured in America were subsequently encountered in Europe, an early correlation study. He also confronts the puzzle of interannual variability, yet offers hope that the scientific method may eventually crack the problem:

> *Since the astronomical source of heat is permanent, and the local causes of its modification undergo no annual variation, and yet the temperature of no two succeeding years is perfectly alike, it is evident that this annual variation proceeds from causes equally variable. Of these, there may be many, but at present we know of none, that have a demonstrable influence on the weather, but winds, and since winds themselves, however uncertain in appearance, are like all the other phenomena of nature, governed by fixed and determinate laws, they deserve the most serious investigation, for which we are at present, tolerably well-prepared.*

---

[19] Published by P. Elmsly, London, 1787. This, and many of the other wonderful old publications cited in this chapter, is available online.

2

Planets and Greenhouses

1760 to ~1870

1761 – Lomonosov observes transit of Venus, infers presence of atmosphere

1767 – Horace-Benedict de Saussure devises a solar oven, and finds solar heat independent of air temperature

1777 – Antoine Lavoisier discovers oxygen and develops an explanation for combustion

1780 – Charles Theodor, the Elector of Palatine, charters the first international network of meteorological observers, lasting until 1795

1783 – Antoine Lavoisier proposes a caloric theory of heat

1783 – Montgolfier brothers and Jacques Charles fly hot-air and light-gas balloons

1783 – Benjamin Franklin notes Laki eruption as a possible cause for cold winters

1783 – First hair hygrometer demonstrated by Horace-Bénédict de Saussure

1785 – James Hutton describes geological cycles, argues that Earth is much older than generally thought

1787 – Thomas Jefferson maintains weather records, suggests US climate has changed due to land clearing

1787 – Richard Kirwan compiles a catalog of world temperatures

1780s – Alexander von Humboldt's explorations

1796 – William Herschel reports a catalog of stellar variability, noting that stars may be a window into variations on the Sun

1801 – Herschel notes a correlation between sunspots and the price of wheat, suggesting a direct influence of solar variability on Earth's weather

1804 – John Leslie observes that a matte black surface radiates heat more effectively than a polished surface, suggesting the importance of black body radiation

1817 – Alexander von Humboldt publishes a global map of average temperatures

1820 – Daniell develops dew-point hygrometer

1822 – Fourier describes paradigm of climate as balance of heat from Sun and interior versus loss to space, noting the resistance of the atmosphere to these fluxes

1824 – Sadi Carnot analyzes the efficiency of steam engines and lays the foundations of thermodynamics, describes the weather as a heat engine

1837 – Agassiz proposes ice ages, Adhemar considers astronomical change as cause

1837 – Samuel Morse develops electrical telegraph system

1847 – Ebelmen outlines the carbon cycle

1847 – Francis Ronalds and William Radcliffe Birt describe a stable kite to make observations at altitude using self-recording instruments

1848 – William Thomson extends the concept of absolute zero from gases to all substances

1849 – Smithsonian Institution begins to establish an observation network across the United States, with 150 observers, via telegraph, under the leadership of Joseph Henry

1853 – The first International Meteorological Conference held in Brussels at the initiative of Matthew Fontaine Maury, US Navy, recommending standard observing formats for weather reports

1856 – William Ferrel publishes his essay on the winds and the currents of the oceans

1856 – Eunice Foote notes absorption of sunlight by water vapor and carbon dioxide

1856 – Piazzi Smyth suggests correlation of rock temperatures with sunspots, and makes high-altitude measurements of sunlight

1859 – John Tyndall measures thermal absorption of carbon dioxide and other gases, the fundamental mechanism of the greenhouse effect

1864 – James Croll calculates heat flux in ocean circulation, combines geological and astronomical analyses to develop astronomical theory of ice ages, proposes ice–albedo feedback as amplification mechanism

1873 – Challenger expedition maps deep ocean temperatures

Developments in planetary astronomy moved somewhat slowly in this period. However, remarkable but rare geometric events, specifically the transits of Venus across the face of the Sun in 1761 and 1769, acted as important stimuli to international collaboration in science. Simultaneous widely spaced observations of a transit could refine the estimate of the astronomical unit, the fundamental scale of the solar system. French, British and Austrian astronomers fanned out across the world, observing the 1761 event from South Africa, Siberia, Madagascar and Newfoundland. The results of observations in the American colonies were published in the very first volume of the American Philosophical Society's Transactions, and in fact the Declaration of Independence in 1776 was read from a platform originally built as a temporary observatory for the 1769 transit [6].

The 1761 transit yielded an important result about Venus – the first strong indication that our sister planet had an atmosphere. The Russian astronomer Mikhail Lomonosov observed the passage of Venus across the disk of the Sun using a small refractor telescope with a smoked glass filter to suppress the light. He noticed that the "tidy" edge of the Sun was distorted by Venus' passage, and particularly that a bulge in the solar disk was caused by refraction of the Sun's light by an atmosphere.[1]

This observation incidentally underscores that much of what we know about planetary atmospheres is not from light reflected by them, but rather light refracted or absorbed through them. Such occultation or absorption measurements are generally much more sensitive than observation of reflected light: as we discuss in later chapters, they were used (with radio waves from spacecraft) to make definitive measurements of the thickness of the Mars and Titan atmospheres in the late twentieth century, and with light spectra to study the composition of atmospheres of planets around other stars in the early twenty-first century.

Similar observations a quarter century later showed to the German/British astronomer William Herschel that the Martian atmosphere was different from ours, in his 1784 paper entitled "On the remarkable appearances at the polar regions on the planet Mars, the inclination of its axis, the position of its poles, and its spheroidal figure". When two faint stars passed close to Mars without showing variation in brightness (i.e. what would later be called a stellar occultation experiment), Herschel inferred that the Martian atmosphere had to be thin. Herschel also noticed that the south polar cap on Mars wasn't quite

---

[1] Lomonosov also argued that the Earth must have a southern polar continent (i.e. Antarctica) because icebergs seen in the southern oceans must have formed on land.

centered on the geometric pole, which he saw to be inclined to Mars' orbit around the Sun much the same as the Earth's, meaning Mars had seasons much like Earth.

One of Herschel's most important discoveries, in 1800, was not made through a telescope, however. He passed sunlight through a glass prism to make a rainbow, and placed blackened thermometers in different colors to measure their temperature, in effect measuring the power in each wavelength of light. He found that the temperatures increased from the violet to the red end of the spectrum,[2] but noticed that a position just beyond red (where there was no visible light at all) had the highest temperature. He inferred there must be invisible radiation, which he called "calorific rays", which was reflected, refracted and absorbed just like light, but with a longer wavelength: what we now call the infrared.

Herschel also observed the Sun, and sunspots in particular, which had been noted by Meton of Athens in the fifth century BC, as well as by Galileo and many others after. Strikingly, in 1801 he noted [20] that the price of wheat varied (as documented in Scottish economist Adam Smith's *Wealth of Nations*) over the previous century in a manner that seemed to relate to the number of sunspots (which he assumed might be associated with a drop in solar output). His discussion highlights the fundamental challenge in climate change studies, that of attribution: correlation does not guarantee causation:

> *The result of this review of the foregoing five periods is, that, from the price of wheat, it seems probable that some temporary scarcity or defect of vegetation has generally taken place, when the Sun has been without those appearances which we surmise to be symptoms of copious emission of light and heat. In order, however, to make this an argument in favour of our hypothesis, even if the reality of a defective vegetation of grain were sufficiently established by its enhanced price, it would still be necessary to shew that a deficiency of the solar beams had been the occasion of it.*

This work embraces three important ideas[3] – that the climate changes, that the Sun changes (contrary to Kirwan's assumption), and that the Sun might affect climate. Herschel ventures to suggest on the basis of the sunspot cycle that harvests in the coming years should be good, but cautions: "*the subject, however, being so new, it will be proper to conclude, by adding, that this prediction ought not to be relied on by any one.*"

---

[2] Note that individual blue photons have a higher energy than red ones, but there are a lot more red photons in the solar spectrum reaching the thermometer through the atmosphere and the prism.

[3] These ideas are discussed more fully in, for example, Ref. [21]

About this time, a Scottish farmer and scholar, James Hutton, had developed a perspective of the Earth as undergoing constant change. His "Theory of the Earth; or an Investigation of the Laws observable in the Composition, Dissolution, and Restoration of Land upon the Globe" was read to the Royal Society of Edinburgh in 1785. He recognized that much of the geological record was shaped by processes that can be observed in action today, and thus that these processes must have acted for a considerable period to form the landscape we see today, with layer upon layer of sediments laid down underwater, heated and turned to rock, and sometimes thrust back up as mountains, only to be worn down again. This ongoing cycle saw *"no vestige of a beginning, no prospect of an end"*.

Hutton, a close friend of fellow Scot Joseph Black (who discovered latent heat, and identified carbon dioxide as "fixed air"), also developed a Theory of Rain, reasoning that, since the amount of water vapor that could be held in the air increased nonlinearly with temperature, mixing of warm and moist air with cooler air would prompt condensation and thus rain. The properties of water vapor and heat were of intense interest in the development of the steam engine at this time; Black helped influence, and finance, James Watt's experiments.

Chemistry emerged as a discipline, and scientists began to understand that the Greek element "air" was a mix of different compounds; with careful weighing, the French chemist Antoine Lavoisier deduced that "fixed air" was about 25% by weight carbon and 75% of what had been called "vital air" or "dephlogisticated air", but what he named "oxygen".[4] Plants and animals exchanged these gases: animal respiration in particular was analogous to combustion, drawing in "vital air" and exhaling "fixed air", which dissolved in water to make a weak acid, which Lavoisier called "carbonic acid". We now call the gas (knowing the relative amounts and atomic weights) carbon dioxide – two atoms of oxygen to one of carbon. Carbonic acid had also been called "chalky acid" – indeed, in the laboratory, pouring acid on limestone, chalk or other carbonate rocks was the easiest way to make pure carbon dioxide, just as adding acid to metals such as iron was a way to make the gas "hydrogen".

Another notable French development in this period was that of the balloon – using hot air by the Mongolfier brothers in 1783, and using hydrogen gas that same year by Jacques Charles.[5] These developments paved the way for the

---

[4] It is possible that Cornelis Drebbel may have replenished the oxygen in his submersible by cooking saltpetre.

[5] In fact the Montgolfier brothers did not actually fly in the pioneering flight themselves, which was made by Pilatre de Rozier, who later went on to develop a hybrid balloon (using hot air and light gas, combining advantages of both). Charles' gas balloon flight was viewed by some 400,000 spectators, among them was Benjamin Franklin, then the diplomatic representative to France of the new United States of America.

exploration of the atmosphere, and Joseph Gay-Lussac in 1804 ascended to 7 km, recording temperatures from which he deduced a steady fall of temperature with altitude. Later with Alexander von Humboldt (then resident in Paris) he determined that the composition of air did not change measurably with height, and that water was made of two parts hydrogen and one of oxygen.

Humboldt, a vigorous Prussian naturalist, had explored South America extensively. Among his discoveries were the electric eel in the Amazon, and the cold ocean current off Peru that bears his name. He was an enthusiastic mountain climber, reaching 5800 m on the Andean volcanic peak Chimborazo in Ecuador, a world record at the time. Humboldt wrote as widely as he travelled,[6] and he documented how elevation produced a change in local climate that mirrored changes in latitude (e.g. the climate at sea level at 30° latitude was the same as that at the equator at 1500 m elevation), which he mapped out to develop a picture of the terrestrial climate overall. The power of graphics in scientific communication is once again underscored – although Kirwan, de Mairan and Mayer had all quantified variations with latitude and altitude, it is Humboldt that usually gets the credit for laying these variations out in a readily assimilated way (Figure 2.1). In this period, the major European powers engaged in wide-ranging expeditions to explore the globe (e.g. Figure 2.2).

The invention and standardization of the barometer and thermometer enabled quantitative observations, long-term climate records. Further, in 1803 an English chemist and amateur meteorologist, Luke Howard, proposed a systematic classification of clouds into three principal categories – cumulus, stratus and cirrus – and their combinations [23]. Howard also compiled measurements into *The Climate of London 1818–1820*, a first review of urban climate. He noted "city fog" (or what we would call smog) and that the temperatures in the city at night were several degrees warmer than simultaneous measurements made in the nearby countryside, an important indication of the human impact on climate. Another English chemist at this time, John Daniell, best known for his design of an electric cell, developed a measure of humidity, the dew-point hygrometer: by expressing moisture as a temperature, this method was easy to standardize, in contrast to the catgut and hair instruments used before. In 1824 Daniell also wrote an *Essay on Artificial Climate considered in its Applications to Horticulture*, documenting the importance of enhanced humidity in greenhouses.

---

[6] An excellent biography of this fascinating man is Andrea Wulf's book [22]. Humboldt's acquaintances read like a *Who's Who* of the period: considering just those mentioned elsewhere in the present book, Humboldt stayed at the White House with Thomas Jefferson, gave financial support to Louis Agassiz, and inspired Charles Darwin.

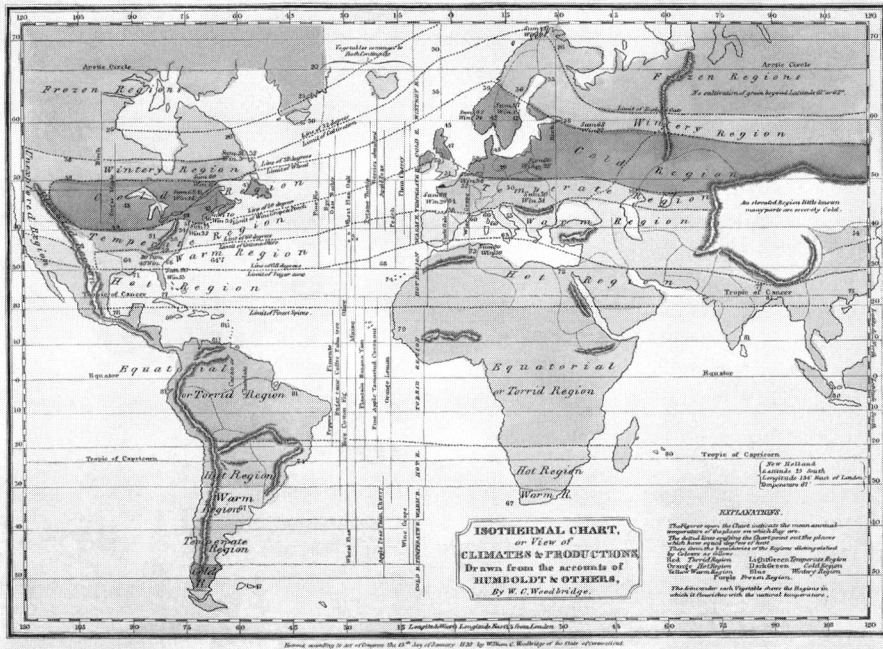

Figure 2.1 Humboldt's assignment of climate regions on the globe, drawn more nicely by W. Woodbridge. Note that, while the zones are generally delineated by parallels of latitude, they are displaced northwards in Europe relative to the Americas, due to the influence of Atlantic ocean currents and winds bringing warmth from lower latitudes. Without this effect, the British Isles should have a wintry climate like Canada.

Back in France, Joseph Fourier in 1822 calculates that the Earth would be far colder if it lacked an atmosphere, and notes (in papers in 1824 and 1827) that underground temperatures do not show diurnal variations, but do show a variation with latitude. He recognizes that this varation with latitude is less than pure geometry of sunlight would produce, but must be moderated by the transport of heat in the atmosphere and oceans. He even recognizes anthropogenic climate change: "*The movements of the air and the waters, the extent of the seas, the elevation and the form of the surface, the effects of human industry and all the accidental changes to the terrestrial surface modify the temperatures in each climate.*"

Much has been written,[7] in connection with the history of understanding the politically charged greenhouse effect, about what Fourier does and does not say

---

[7] First and foremost, it is best to read Fourier's writings yourself, ideally in French. An English translation of his 1824 piece is found in Ref. [24]. Misquotations of Fourier are described in Ref. [25]. A very nice summary of early research on the greenhouse effect and glaciations, fittingly in both French and English, is Ref. [26].

Figure 2.2 The European powers were busy cataloging the world and documenting its features. Here, the French ship *Astrolabe* negotiates icebergs on Dumont D'Urville's expedition in 1838. (Source: NOAA Photo Library: Image ID libr0229.)

in his paper, so let's spend a little time on it here. My impression – emphasizing the planetary perspective – is that he gets a lot of things right, and he acknowledges what is not known for sure.

First, he outlines (see my interpretive framework in Figure 2.3) that the temperature of the Earth depends on three reservoirs of heat – the Sun, the interior of the Earth, and the temperature of planetary space. We'll come back to the last one. Fourier (whose eponymous decomposition of functions into sums of sinusoids is useful in problems of thermal conduction) recognizes that the heat leaking out from the interior of the Earth has a miniscule influence on our present climate.

He recognizes that he cannot quantify the effect of the atmosphere on terrestrial temperatures, but draws attention to measurements by Swiss physicist and alpine geologist Horace-Benedict de Saussure of temperatures in an insulated box with a glass window, which he brought to different mountain elevations. The contemplation of these experiments and similar analogies led to the term "greenhouse effect" – we leave aside here how formally accurate an analogy to an

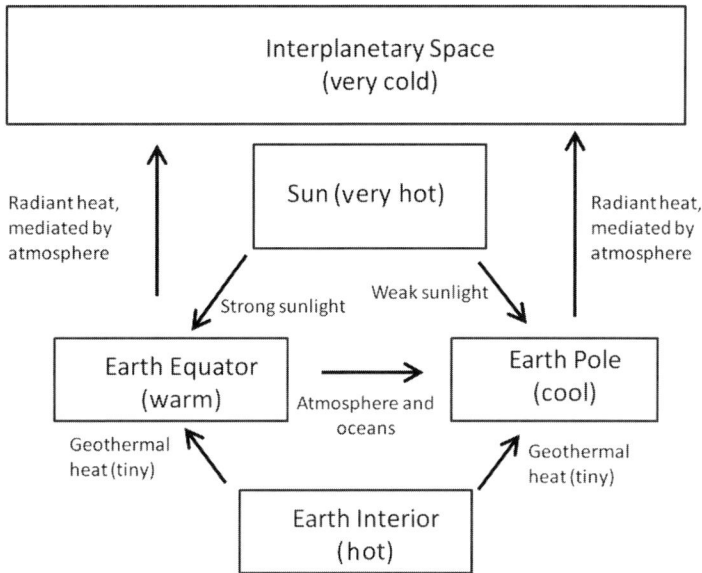

Figure 2.3 Schematic of Fourier's ideas on planetary climate as interpreted by the author. The numbers weren't known, but the framework, wherein heat flows are driven by temperature gradients, is essentially correct.

atmosphere it is, since the suppression of convection and the absorbance of thermal radiation are different things, but it is the warming-by-blocking effect that matters here. He also recognizes the distinction between "luminous" and "non-luminous" heat – in essence what modern climatologists would call visible and thermal radiation, or short-wave and long-wave (infrared) radiation. Further, he understands that materials may allow these two streams of energy to pass in different degrees. So, he certainly sets the stage for understanding the greenhouse effect – as he puts it *"luminous heat flowing in penetrates . . . and non-luminous heat has more difficulty in finding its way out."*

   He doesn't quantify how much heat is moved around the Earth by the atmosphere and oceans, but notes that these motions do attenuate the difference in temperature between equator and pole. Not having an estimate, he then (it is a slightly rambling and self-contradictory paper, worth reading nonetheless) more or less argues that the poles of the Earth should get unimaginably cold, since they get very little heat from the Sun. So they must get some heat from space – arguing that either the temperature of space around the Earth, or the effect of starlight keeps the poles from getting too cold. In effect he imagines space as a heat sink: *"the influence of stars is equivalent to the presence of an immense hollow sphere, with the Earth in the center, the constant temperature of which should be a little below what would be observed in the polar regions.*" He's wrong here,[8] but the picture fits the data. He suggests that the other planets cannot be much colder than our poles, since even though they get less sunlight than we do, they are still immersed in this same heat sink of space.

Impressively, while explaining how he gets a picture mostly consistent with what is observed, Fourier admits (of the other planets) "*Now, they would be entirely different if we were to admit an absolute cold in space*" (which is in fact the case). So Fourier gets the overall picture just about right, although there isn't enough known about atmospheric opacity and air motions transporting heat to distinguish their effects from those of (also unknown) space.

Interestingly, around the same time as Fourier's memoir, his countryman Sadi Carnot, making efforts to understand the fundamentals behind the performance of steam engines, opens his treatise "Reflections on the Motive Power of Heat", with the solar heating/air motion causality in the other direction:

> *To the agency of heat may be ascribed those vast disturbances which we see occurring everywhere on the earth; the movements of the atmosphere, the rising of mists, the fall of rain and other meteors,[9] the streams of water which channel the surface of the earth, of which man has succeeded in utilizing only a small part.*

Carnot saw atmospheres, correctly, as engines. We will return to this useful perspective later. The following decades saw the formulation of the more formal laws of thermodynamics: the concept of entropy by Rudolf Clausius, and he and Benoit Clapeyron laid out a basis for the vapor pressure of materials such as water, of critical importance to climate. William Thomson, later Lord Kelvin (who became wealthy from innovations in telegraphy and was the first scientist to become a member of the House of Lords in the British Parliament) took their work further and devised the absolute scale of temperature that bears his name in 1848.[10]

---

[8] Well, mostly wrong. We are of course all immersed in a universal bath of about 3 K of cosmic background radiation, and neither Earth's poles nor the other planets can get colder than this. So the picture is correct in a sense, but not right for explaining our polar temperatures, which depend mostly on heat transport from lower latitude.

[9] The term hydrometeors is still used in English as a catch-all for hailstones, raindrops, snowflakes etc.; it is in this sense that Carnot's word "meteor" is meant.

[10] Throughout this book I will unashamedly use both Celsius (centigrade) degrees and kelvin, as the occasion demands. Note that, since the determination by the General Convention on Weights and Measures in 1967, the degree symbol (°) is used with Celsius, but not with kelvin. Temperature (kelvin) = temperature (Celsius) + 273.15 K, thus room temperature is about 293 K. The scale defined by Daniel Fahrenheit had its zero defined by a water-ice-salt freezing mixture, and 32 °F at the freezing point. The 180 °F difference

Kelvin, in 1864, armed with a better understanding of heat conduction, and with measurements of the thermal conductivity of rocks and temperature gradient in the Earth's crust, deduced the geothermal heat flow and a cooling age for the Earth of tens of millions of years. Among the underground temperatures bearing on this were a set measured with special thermometers installed in 1837 at depths of 3, 6, 12 and 24 feet in the porphyry rock of Calton Hill in Edinburgh, Scotland. These instruments had long tubes with scales at the surface, and were read once a week for the next 16 years, with a claimed precision of a hundredth of a degree Fahrenheit. Scottish astronomer Charles Piazzi-Smyth [27] examined this formidable dataset in 1856, noting how the thermal wave of summer propagated into the rock, with the 3-foot wave with a 15-degree amplitude peaking in August, but at 24 feet a mere 1.2 degrees, peaking in January.[11] The mean temperatures also showed a consistent increase with depth, of about 1 °F for 21 feet of depth. Piazzi-Smyth then attempted to subtract all these effects out, to see what variation was left, in the hope of detecting some inter-annual variation that might be due to the Sun. He plotted the temperature variation (which was pushing the quality of the data rather hard) against observations of sunspots, but concedes "*no very near approach can be claimed to the temperature curves*".

That same year, Piazzi-Smyth set off on an expedition to explore the question of how much better an astronomical observatory might be by being placed on a mountaintop. He sailed to Tenerife, a volcanic island off the west coast of Africa, armed with a portable telescope (portable, that is, by four mules – see Figure 2.4) and a variety of meteorological instruments. His team set up on the flanks of Guajara, at an elevation of about 9000 feet. Although this height was well above the usual cloud-deck and skies were often beautifully clear, Piazzi-Smyth noted (in his book [28],[12] which is remarkably illustrated with "photo-stereographs")

---

between the freezing and boiling points of water is a nice round number, convenient for integer arithmetic and making divisions on a scale in the 1700s, but seems archaic in a modern world of computers and decimals. It is nonetheless still widely used in the United States, and I have used it where the original source featured it prominently. There is even an absolute version of the Fahrenheit scale suitable for thermodynamic calculations of engine efficiency and the like: the Rankine scale has 0 °R equal to −459.57 °F or 0 K.

[11] In fact, De Saussure had noticed similar delay in propagation of the summer peak temperatures into the ground.

[12] The stereo pairs of photographs give striking 3-D viewing. On p. 97 Piazzi-Smyth describes with some pride his construction of his wife's tent, with a built-in canvas floor that helped prevent it being blown away by wind, and also kept out dust and sand. He also attempted, with a sensitive thermopile, to measure the heat of the Moon, finding that the galvanometer needle twitched about the same from the Moon as it did a candle

Figure 2.4 Photograph from Piazzi-Smyth's Tenerife expedition. In the foreground is the portable telescope installation brought up the mountain (much as Langley's apparatus would be later). On the rock wall is a thermometer being exposed in its wind-shield box to the Sun to measure solar intensity.

that astronomical conditions were sometimes degraded by a thick haze of dust (which had blown in from the Sahara Desert, some 300 miles to the east). When conditions were good, however, they were very good, and the views of Jupiter allowed him to note the calm nature of its poles with cirrostratus clouds, while the tropics had tempestuous cumulostratus: he observed that the "*medial line of calm*" was not exactly coincident with the equator, and wondered, if the causes were the same (greater proportion of land area in the northern hemisphere), why the calm between the northern and southern trade winds was pushed north of the equator on Earth, towards the latitude of Tenerife.

By day, Piazzi-Smyth noted the tough conditions at altitude: the varnish on his barometer (which read 22 inches at the Guajara station) had blistered in the harsh sunlight after only one day. He observed how dry the air was, the dew point having dropped by some 40 degrees in their ascent: "*No wonder we felt our lips cracking, our hair frizzling, our nails becoming brittle, and saw each other's faces scarlet.*"

Piazzi-Smyth's team made measurements with solar thermometers in the thin air, making the observation that would become a familiar feature of the Martian climate, namely dusty days (when the solar thermometer's reading was reduced

15 feet away. The book also discusses the effect of the reduced boiling temperature of water at these high elevations on the quality of tea.

by attenuation of sunlight by the dust) saw the air temperature in the shade being elevated. Piazzi-Smyth even describes being blasted by pebbles and blinded by dust in a dust devil, a whirlwind that came out of nowhere – as they often do on Mars. At the time the generation of hurricanes and other rotating storm systems was not understood, and it was even speculated that electricity might somehow be responsible – Piazzi-Smyth pondered hurricanes after his dust devil encounter, noting that even the dust devil "*required an exertion of a considerable amount of mechanical energy; and the question to be settled, is, which of the two elements present on the occasion, was most capable of producing the effect observed, wind or electricity?*"

P. Baddeley, a surgeon in the (British) Bengal Army, confronted the same question around 1850 [29, 30]: whether electrical phenomena were the cause, or the result, of whirlwinds. In India, he chased dust devils, observing the behavior of a gold-leaf electroscope, and noting that a paper moist with starch and iodide solution became discolored, indicating the presence of ozone. (Ozone itself had only been isolated in 1839, and was determined to be a triatomic molecule of oxygen in 1867.)

The nature of storms was a matter of considerable interest in the 1800s, as it has always been, but getting a wider picture of weather systems was somewhat easier on the expanding continental United States than in fractured Europe, where storms tended to arrive from the vast and unobserved Atlantic. William Redfield, a New England saddle-maker, noticed that the alignment of fallen trees after a major storm in 1821 differed from place to place, and reconstructed the storm's path from newspaper reports and interviews with sailors. He advanced the idea that winds swept circularly around storms – they were large whirlwinds. A contrary view was vigorously espoused by James Espy, who argued that winds converged towards storms, and rose in the center. Of course, each proponent had truth partly on their side, so while the debate was not immediately resolved, it stimulated the acquisition of better data. The rapidly expanding telegraph network allowed weather systems to be reported before they arrived, and Joseph Henry, an electrical engineer and director of the Smithsonian Institution in Washington, D.C., arranged for telegraph operators to provide brief routine weather reports.[13] By 1858, Henry had a large map on display in the lobby of the Smithsonian, displaying weather reports telegraphed that day from as far away as Iowa and Missouri, as near to a real-time global display as the nineteenth century got. Another party in the debate was mathematician William Ferrel, who understood more clearly than others how the Earth's

---

[13] A nice history of meteorology, and of weather maps in particular, is Ref. [31].

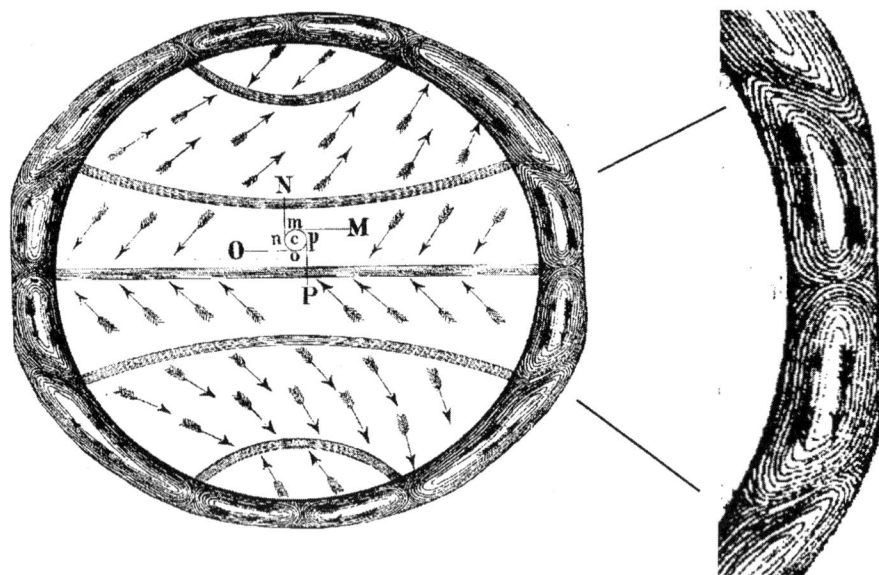

Figure 2.5 Ferrel's sketch of the pattern of trade winds and their relationship to the meridional circulation (shown in cross-section at the edge of the globe, enlarged at right to show the lowest-latitude pair of cells, now called the Hadley cells, separated by the Intertropical Convergence Zone). Ferrel correctly saw that there were three cells in each hemisphere.

rotation influenced air masses, and elaborated on Hadley's ideas.[14] Ferrel also recognized the structure of the vertical and meridional (north–south) circulation of the atmosphere (Figure 2.5), with upwelling at the equator and downwelling near 30 degrees latitude (which is termed the Hadley cell), a midlatitude cell rotating in the opposite sense, named after Ferrel, and a polar cell. The position of the downwelling depends on planetary rotation, and is quite different on other planets.

As the physical nature of heat was becoming understood in the laboratory (and, indeed, the brewery[15]), outdoors the realization was emerging that the

---

[14] The deflection of objects on a rotating sphere was recognized, in part, by Hadley, but there is an additional factor noted by Ferrel. Indeed, the deflection is often referred to as the Coriolis effect, after Gustave Coriolis, but might be more properly attributed at least in part to Ferrel. A general relationship between pressure gradient and wind direction was noted by the Dutchman Buys Ballot in 1857.

[15] James Joule, who quantified the mechanical equivalent of heat and after whom the unit of energy is named, was a brewer. Hevelius, who mapped the lunar mountains, built his observatory with the profits from his brewery. The T-test, a widely used statistical criterion, was published by William Gosset in the early twentieth century under the pseudonym "Student" because his employer, Guinness of Dublin, had a policy against

Earth's climate was not invariant. In particular, suspicions grew, most vigorously propounded by Swiss Louis Agassiz *circa* 1837, that much of Europe had once been covered by a great sheet of ice, perhaps a mile thick. This staggering idea could explain many features of the recent geological record, such as scratches on rocks, and the presence of large erratic boulders hundreds of miles from their source (in fact Hutton had speculated ice might have been their cause). But how could the climate have changed so dramatically? In 1842 Joseph Adhemar recognized that the Earth's elliptical orbit leads to slight asymmetry in the seasons, and he suggested in his book *Revolutions of the Sea* that astronomical change (specifically, the 26,000-year precession of the equinoxes, known even to ancient astronomers) might be responsible.

Meanwhile, the world was becoming smaller. Not only were voyages of exploration, such as Humboldt's and that in 1836 of HMS *Beagle* carrying Charles Darwin, bringing a more complete understanding of the world and its inhabitants, but communications were fast improving. First railway networks, and then the telegraph, meant that information could travel faster than a horse. It now became possible to compare weather conditions at different places simultaneously, to literally develop a picture of the atmosphere.

While observations at different places of a few storms began to be assembled into weather maps (after the fact) in the 1820s and 1830s,[16] it was James Glaisher – appointed in 1839 as the Superintendent of the Magnetic and Meteorological Department at the Royal Observatory in Greenwich, England, who first systematically began assembling such observations (brought by railway) and his consolidated reports of the weather the previous day began to be published in newspapers in 1849.

The offices of an American newspaper, the *New York Sun*, were besieged in 1844 by crowds trying to buy copies that told of a new wonder of the age, the first crossing of the Atlantic by balloon. The account was full of technical detail of the balloon, ellipsoidal of varnished silk, filled with coal gas and propelled by an Archimedean screw. This remarkable article had hurriedly been pieced together from early reports of a narrative of the 75-hour voyage – remarkably from east to west, Wales to South Carolina. Unfortunately, the article – penned by one Edgar Allan Poe – wasn't actually true. But it showed nicely that the public were ready to believe that such planet-girdling aerial voyages were possible.[17]

---

publication by its scientists. Science and beer have a long history together, even ignoring the contributions of Wilhelm Beer in mapping Mars.

[16] A very readable story of these early attempts by William Redfield, Elias Loomis and others, and the meteorological careers of Glaisher and Fitzroy, is found in Ref. [32]

[17] The story, generally referred to as "The Balloon Hoax", ran at 10 a.m. on April 13, 1844, under the headline "ASTOUNDING NEWS! BY EXPRESS VIA NORFOLK: THE ATLANTIC

Matthew Maury, an ambitious oceanographer with the US Navy, saw that to take the next step, to anticipate or forecast the weather, would demand international collaboration, and convened an International Meteorological Conference in Brussels in 1853. In the following year, a dedicated British government department was formed, essentially the forerunner of the modern Met Office, led by the *Beagle*'s former sea captain Robert Fitzroy (France's meteorological service was set up by the astronomer Urban Leverrier at the Paris Observatory the same year). Fitzroy set up stations with standardized instruments, and had them send the readings several times a day via telegraph. This allowed him to issue storm warnings to ports when violent conditions were expected. It was only a small step further to report conditions when no storms were expected, and the first regular weather forecasts began to be published in 1861. Because of its strategic importance, in many countries the weather reporting function was seen as part of the military – for example early US weather maps were published by the Signal Service of the War Department (Figure 2.6).

Although the day-to-day variations in weather were now being tracked and progressively better understood, scientists were still grappling with the factors that controlled the long-term conditions. In particular, the effects of the atmosphere on sunlight and heat had to be quantified. A paradox was the variation of temperature with altitude – how could it get colder when you got closer to the Sun? Glaisher made a famous ascent in a gas balloon in 1862, carrying instruments to measure pressure, temperature and humidity, and found that Gay-Lussac's altitude trend did not persist indefinitely – these balloon flights gave the first inklings of the existence of the stratosphere.[18] In fact, Glaisher lost consciousness during his flight, which inadvertently reached above some 30,000 feet, the lower reaches of the stratosphere – he was only saved by his pilot Henry Coxwell, who managed to pull a gas vent valve with his teeth, his hands rendered useless by frostbite (Figure 2.7).

The role of humidity in the atmosphere turned out to be crucial to the climate question. While the blanketing effect of the atmosphere was long suspected, it

CROSSED IN THREE DAYS! SIGNAL TRIUMPH OF MR. MONCK MASON'S FLYING MACHINE!!!" It contrasts the success of the machine's Archimedean screw against a more conventional windmill propeller, a topical debate in ship propulsion at the time. The credulity of the public on this story mirrors that for H. G. Wells' *War of the Worlds* several decades later, and is interesting to consider in the present era of "fake news". Poe's Balloon Hoax story was likely a significant inspiration for Jules Verne's novel *Five Weeks in a Balloon.*

[18] A wonderful history of early ballooning is found in Ref. [33]. This story covers not only scientific flights such as Glaisher's, but also the role of balloons in the American Civil War, and in the Prussian Seige of Paris in 1870.

Figure 2.6 An early Signal Service weather map on record in NOAA Library: September 1, 1872. Careful compilation and study of these maps led to scientific forecasting. Note the apparent lack of any weather in the thinly populated western half of the country. (From "Daily Bulletin of Weather-Reports … for the Month of September, 1872". Image ID: wea01800, NOAA National Weather Service (NWS) Collection.)

was John Tyndall[19] who demonstrated in 1859, using a thermopile heat detector, the blocking of radiant heat by certain gases, notably water vapor and carbon dioxide. Tyndall, a great communicator as well as experimentalist, suggested that changes in the concentrations of these gases could alter climate with the following analogy: "*As a dam built across a river causes a local deepening of the stream, so our atmosphere, thrown as a barrier across the terrestrial rays, produces a local heightening of the temperature at the Earth's surface.*"

Tyndall's progress arose in part from the exquisite sensitivity of his apparatus (Figure 2.8): he set it up, not to measure the absolute amount of caloric rays transmitted by the gas (held in a metal tube with rock-salt windows – glass is opaque to infrared radiation, so the "greenhouse effect" is not a complete misnomer), but to measure the difference between that and a reference, which he could adjust to be equal to that transmitted by the gas tube evacuated of gas.

---

[19] Tyndall was born in Ireland, both areas of which were part of the UK at the time; Kelvin was born in Northern Ireland, which still is part of the UK.

MR. GLAISHER INSENSIBLE AT THE HEIGHT OF SEVEN MILES

Figure 2.7 Glaisher's near-fatal 1862 ascent above the clouds, Glaisher passes out at the right while Coxwell struggles with the vent cord to save their lives. (Source: NOAA Photo Library: Image ID libr0641.)

Measuring this difference, with a delicate galvanometer,[20] let him detect much smaller absorptions. The Italian Macedonio Melloni used a similar detector, at the focus of a 1-m wide Fresnel lens, to make the first detection of the heat of the Moon in 1845 [34].[21]

Tyndall's experiments showed that the ability of gases to emit infrared radiation was directly related to their ability to absorb it. He also showed that the amount of absorption increased proportionally with the amount of gas – at first – but at larger amounts, the absorption leveled off.

He determined that the water vapor in the atmosphere was the major factor – in his experiment causing 13 times more absorption than the other gases [35]. He concluded: "*every variation of this constituent must produce a change of climate. Similar*

---

[20] The galvanometer was built by an acclaimed instrument-maker in Berlin, but this wasn't good enough, there was an annoying offset in the reading, which Tyndall traced to magnetic impurities in the wire. With some sleuthing, he traced that to some kind of dye in the silk insulation on the wire, which he replaced with white silk.

[21] Melloni used a Fresnel lens in part because it meant the overall thickness of the glass would be less than for a conventional lens, and glass absorbed infrared rays.

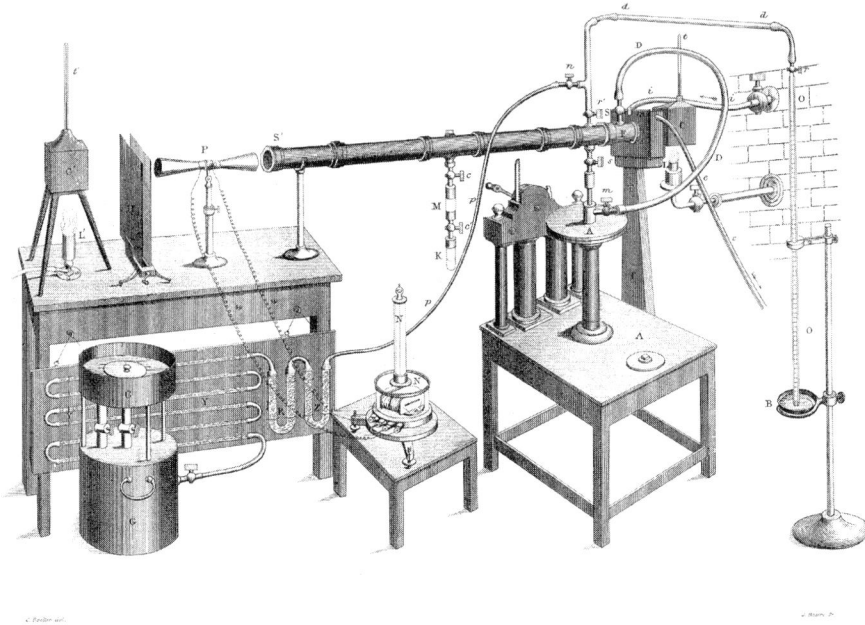

Figure 2.8 Tyndall's apparatus: Victorian equipment has the virtue that one can see how it all works from a picture, rather than a schematic diagram. The key elements are the pair of thermopiles (P) with their horn collimators, and the sensitive galvanometer N on the small table, which allowed the tiny thermopile signal – the difference in flux of caloric rays between the reference Leslie cube (C) and its screens at left and that transmitted from the other cube at right through the tube of gas (S). The cubes are maintained by gas burners at a constant temperature, measured by thermometers (t). The other equipment here is for handling the gas – from supply tank (G) at lower left, through some purifying steps; a vacuum pump A controls the gas pressure, measured with the manometer/barometer O at right.

remarks would apply to the carbonic acid diffused through the air; while an almost inappreciable admixture of any of the hydrocarbon vapours would produce great effects on the terrestrial rays and produce corresponding changes of climate." A little over a century later, planetary scientists would invoke such hydrocarbon vapors to solve climate puzzles, but of course carbonic acid (carbon dioxide) would occupy center stage on Earth.

It was only at this time realized that invisible radiations that caused chemical reactions ("actinic" or what we'd now call ultraviolet), visible light, and thermal radiation (infrared) were the same thing – Melloni, for example realized only in 1842, "*The dark undulations responsible for chemical or calorific actions are perfectly similar to luminous undulations; they only differ from them by wavelength.*" Although it is the blockage of long-wavelength ("thermal", longward of say 3 micrometers) infrared radiation from the Earth's warm surface that is responsible for our greenhouse effect, the distinction between this flavor of invisible infrared light and another (what we'd call near-infrared, between about 0.8 and 3 micrometers, and abundant in sunlight) took some time to be appreciated.

The absorption by gases of the latter radiation was documented in a short paper [36] a couple of years earlier than Tyndall, by Eunice Foote, a notably early American woman scientist.[22] By simply setting out two glass vessels with thermometers in the sunlight, she determined that the warming varied with the density of the air (using a pump to increase the air density in one and decrease it in the other). Furthermore, she measured that the warming was greater in moist air than dry air, and found that the effect was particularly strong with "carbonic acid gas" – strong enough to detect by hand without even using the thermometer. She speculated, "*An atmosphere of that gas would give to our Earth a high temperature.*"

The possibility of variations in the amounts of carbon dioxide and oxygen in the atmosphere was presciently outlined by French chemist and mining engineer Jacques-Joseph Ebelmen, who wrote in the 1840s [34, 35] "*I see in volcanic phenomena the principal cause that restores carbon dioxide to the atmosphere that is removed by the decomposition of rocks*" – his countryman Boussingault in the 1830s had sampled the gas from several volcanoes and found carbon dioxide to be a major component [36], whereas the Swiss Nicolas de Saussure (the son of Horace-Benedict, whose glass-box experiments had inspired Fourier's thoughts on the greenhouse) showed in 1804 that atmospheric air contains only a tiny amount. Meanwhile, Ebelmen found carbon dioxide was removed by reaction with silicate rocks, a process (sometimes referred to in modern times as the "Urey" reaction) usually mediated by the action of plants, and then its ultimate sink was in carbonate mud in the oceans: "*The terrestrially-derived carbonates end up by being deposited or they are taken up by marine animals.*" Without the benefit of Tyndall's measurements, he qualitatively guessed correctly, "*. . . in ancient geologic epochs the atmosphere was denser and richer in CO_2, and perhaps O_2, than at present. To a greater weight of the gaseous envelope should correspond a stronger condensation of solar heat.*" Ebelmen introduced the phrase "carbon rotation",

---

[22] This crisp, straightforward paper appears immediately after a rather muddled five-page article by her husband Elisha Foote. Note that this neat observation is not a discovery of the greenhouse effect – if anything it is the opposite (i.e. an antigreenhouse effect, the absorption of sunlight rather than thermal infrared). The short-wave infrared absorption by carbon dioxide is very important, however, in the climate of Venus. Foote's observation is also hardly the earliest detection of near-infrared absorption – Fizeau and Foucault (of pendulum fame) recorded a solar spectrum in the near-infrared in 1847 which shows absorptions due to the Earth's atmosphere – see Lequeux [34].

noting that carbon dioxide in the air might be taken up by plants, and (as Hutton and many others thereafter recognized) those plants might be turned into coal, the carbon locked up in the ground, or made into the carbonate shells of marine life and perhaps into limestone, and then the carbon might be released again by combustion or volcanic heating of carbonates. The modern carbon cycle emerged.

Meanwhile, an impressively multidisciplinary approach to the question of the ice ages was doggedly made in the spare time of a Scottish janitor, James Croll, who, after several unsuccessful careers in other fields, enjoyed his job at the Andersonian College and Museum in Glasgow because it allowed him use of its library. Croll published a short paper in August of 1864 estimating the changes in the Earth's climate that might be due to changes in orbital eccentricity. His subsequent work on the ice-age problem was to occupy his attention for the next 20 years.

Croll recognized that the changes in sunlight due to these astronomical changes were small, but suggested one means by which they could be amplified. He noted that snow-covered ground (or ice-covered seas) would reflect more sunlight than their bare equivalents, and so slightly colder conditions could allow a bit more terrain to be snow covered for longer, and would absorb a bit less heat as a result, allowing snow to persist a bit longer again. Croll introduced the idea of a climate feedback (specifically, the "ice–albedo" feedback).

Croll was perhaps the first person to calculate how much colder regions at high latitude would be if there were no heat transport from lower latitudes, recognizing that high latitudes receive so little sunlight (as originally calculated by Halley) that the contribution of ocean currents and winds dominates. He calculated that London was 10 degrees warmer than the average temperature at that latitude, and that the latitude was on average 30 degrees warmer than it should be from sunlight considerations alone. He noted that the Gulf Stream provides some 77,479,650,000,000,000,000 foot-pounds of energy to the Atlantic, or roughly one-fourth of the heat received from the Sun.

The oceans were a hot scientific topic at the time, with amazing deep-sea creatures among the 4000 species discovered by HMS *Challenger*. *Challenger*'s epic expedition 1873–6 made hundreds of soundings of temperatures and salinity at depth, as well as sampling seafloor sediments. The tens of thousands of pages of the expedition's findings took some 19 years to be fully documented, and its deep-sea temperature measurements are still referenced today.[23]

Croll's researches were summarized in an 1890 book with the grand title *Climate and Time* (more fully, *Climate and Time in their Geological Relations: A Theory*

[23] This is still a fascinating volume [40]. The *Challenger*'s temperature soundings form an important benchmark on climate change – see [41].

Figure 2.9 Erratic boulders in the Cuillin mountains of Skye, considered as evidence for ice sheets. (Source: Popular Science Monthly, Vol. 4. April 1874.)

of the Secular Changes of the Earth's Climate). Although he got the big picture right, his calculations on temperature were based on the assumption that the temperature was proportional to the supplied heat (both from the Sun, as a function of the changing orbital characteristics and latitude, and from ocean currents), whereas the correct dependence was only at this time being worked out by physicists, primarily in Germany and Austria.

However, Croll not only considered the astronomical aspects of the problem, calculating at length the influence of changes in the Earth's orbital eccentricity and obliquity (i.e. the tilt of its spin axis) on the amount of sunlight at different latitudes and seasons, but he also explored the geological evidence for glaciation (e.g. Figure 2.9). In particular, he investigated the widespread boulder clay, erratic boulders and scratched rocks, and the shells of arctic creatures now extinct in interglacial Scotland. In the century to follow, his multidisciplinary approach would become rare as science became more and more specialized. The end of this epoch, in the last decades of the 1800s, marks the onset of ever-more quantitative methods in science, and, at last, the accurate evaluation of conditions on other worlds.

3

The Age of Numbers: Calculation of Planetary Climates

1870 to 1930

1879 – Stefan determines fourth-power law of heat radiation

1885 – Christiansen in Denmark makes first "modern" calculation of planet temperatures using Stefan law

1880s – Langley and Rosse observe heat from the Moon; Langley measures infrared spectrum

1896 – Arrhenius calculates climate change in response to carbon dioxide doubling

1902 – Richard Assmann and Léon Teisserenc de Bort independently discover the stratosphere with balloons

1907 – Jose Comas-Sola detects limb-darkening on Titan, suggesting an atmosphere

1900s – Lowell popularizes discussion of Mars' climate as clement; contested by Poynting, Douglass, Alfred Russel Wallace, Comas-Sola

1913 – Maunder considers Mars' climate vs latitude; invokes the term "habitable zone"

1920s – Milankovitch refines calculations of astronomical forcing of climate on Earth

1922 – Lewis Fry Richardson organises the first numerical weather prediction experiment

1923 – Pettit and Nicholson measure temperatures on Mars with 100-inch Hooker telescope

1930 – Pavel Molchanov invents and launches the first radiosonde to a height of 7.8 kilometers, measuring temperature there in real time

The late 1800s saw the horizons of popular imagination venturing out to the planets. Jules Verne, of course, in 1865 wrote the novel *From the Earth to the Moon*, wherein the Baltimore Gun Club established (in Florida, presciently enough, not far from the modern Kennedy Space Center) a giant cannon to launch a human crew to the Moon. H.G. Wells' 1898 science fiction *War of the Worlds* imagined interplanetary transportation in the other direction, with Martians invading southern England in cylinders whose gun-launch from Mars was visible in the telescope. Interplanetary gunnery was not especially practical, as even Huygens had noted. But in 1867, a Scottish clergyman, William Leitch, wrote a review [42] of what was known about the solar system,[1] and noted:

> The only machine, independent of the atmosphere, we can conceive of, would be one on the principle of the rocket. The rocket rises in the air, not from the resistance offered by the atmosphere to its fiery stream, but from the internal reaction. The velocity would, indeed, be greater in a vacuum than in the atmosphere[2] . . . we might, with such a machine, transcend the boundaries of our globe, and visit other orbs.

While the temperature and atmosphere on Mars was not yet known, circumstantial evidence against its habitability began to emerge. Although Giovanni Schiaparelli, observing Mars in its "Great Opposition" in 1877, had sketched a map with dark lineaments that he called "canali" (channels – mistranslated as "canals"), he was himself doubtful that they were seas, noting that smooth liquid surfaces would reflect the Sun like a mirror. The fact that these "specular reflections" were not observed in telescopes meant that there were no smooth seas. This point was elaborated by Dennis Taylor [43] in 1894, who suggested that surface waters would at least sometimes be smooth, and the absence of recorded sunglint meant that widespread liquid could not be present. He also suggested that the rapid disappearance of the seasonal caps meant they could only be a few yards thick.

Progress in thermodynamics, and in particular in understanding radiant heat, finally laid the quantitative basis for understanding planetary temperatures.

---

[1] The book imagines a journey through the solar system on a comet ("nature's rocket") and describes what might be seen. This appears to be one of the first mentions of rockets as devices for space travel, and of the superior performance in vacuum in particular. In *Around the Moon*, Jules Verne's 1870 sequel to *From the Earth to the Moon*, Verne imagines not only a barren Moon seen by the protagonists through opera glasses from their bullet-space-ship, but also some small rockets intended to soften the bullet's landing being used for a course correction which sends it back to Earth.

[2] The superior performance of rockets in vacuum was not widely appreciated even 50 years after this – Robert Goddard's 1920 work was panned by the *New York Times*, which claimed incorrectly that rockets would not work in space.

Merkur	Venus	Erde	Mars	Jupiter	Saturn	Uranus	Neptun
$t_1 = 189$	65	15	-40	-147	-180	-207	$-221°$
$t_2 = 210$	57	15	-34	-150	-180	-209	$-221°$

Figure 3.1 Christiansen's evaluation of planetary temperatures in degrees Celsius, using the Stefan fourth-power law. The two rows correspond to slightly different assumptions about planetary reflectivity or albedo. These estimates are really only relevant for surface temperatures on Mars, since the Venus values do not take its formidable greenhouse effect into account, and Mercury has such a large day/night difference that an average temperature isn't very meaningful. The estimates for the giant planets are a rough guide for a level in their deep atmospheres that corresponds to about 1 bar.

Jozef Stefan in Vienna deduced in 1879 from Tyndall's measurements that the heat flux radiated by an object varied as the fourth power of its absolute temperature, and Ludwig Boltzmann derived the result theoretically five years later. Stefan even calculated the temperature of the Sun, using an observation that a metal plate at about 2000 K radiated about 29 times less heat than did the Sun. Guessing that about a third of the Sun's flux was absorbed by the atmosphere, this meant the temperature ratio had to be about $(1.5 \times 29)^{0.25}$ or ~2.6 times that of the plate, or about 5700 K, quite close to the modern value.

It takes some detective work to find exactly who first used the formula to estimate planetary temperatures: in a 1903 paper [44], John Poynting attributes it to Wilhelm Wien *circa* 1901, but in a footnote in a later paper credits the Danish scientist Christian Christiansen with computing a table of planetary temperatures in 1885.[3] Specifically, Christiansen (writing in a Danish journal) notes that the radiated heat varies as the fourth power of absolute temperature, and the radiated heat must balance that received from the Sun, which varies as the inverse square of distance. Thus, the absolute temperature varies as the square root of distance. Christiansen's paper lists the temperatures computed using the mean distances of the planets (see Figure 3.1): Venus would be about 60 °C, while Mars was −40 °C, and Saturn and therefore Titan would be −180 °C.

---

[3] In fact, Arrhenius in his 1915 book attributes Christiansen with the first calculation, perhaps being more aware of work in Nordic countries than others. Christiansen was at the time the only Professor of Physics at the University of Copenhagen in Denmark. He made some advances in optics, and was the Ph.D. advisor of Niels Bohr (after whom the university is now named). Christiansen's original report "Nogle Bemaerkninger angaaende Planeternes Varmegrad" is in Danish, in the *Oversigt over det Kong. Danske Videnskarnernes Selskabs. Forhandlinger*, Kjöbenhavn, 1886, pp. 85–108, but was summarized as "Einige Bemerkungen über die Temperatur der Planeten" later that year in a German journal, *Wiedemann's Beiblatter zu den Annalen der Physik und Chemie*, Band X, 1886.

Figure 3.2 Birr Castle, near Parsonstown, Ireland, painted by Henrietta Compton in 1845. The 3-foot telescope that Rosse used to measure the temperature of the Moon is just to the left of center, and the 6-foot telescope (The Leviathan of Parsonstown) is under construction at right; its 3-ton speculum mirror is being manoeuvred into place. The Leviathan was the largest telescope in the world until the 100-inch Hooker telescope at Mt. Wilson in California in 1917. The 6-foot mirror is on display in the Science Museum, London.

The deep cold of the outer planets, recognized qualitatively by Huygens, is evident in these computations, as is the unpromising chill of Mars – about which much more shortly. Venus entertains, perhaps, some hope of being a torrid kind of habitable – but of course the exceptional greenhouse warming of that planet was not known at the time. Christiansen suggests that Mars' conditions were not unlike Greenland.

Around this time, astronomy also began to progress, not only from a proliferation of ever-larger telescopes at ever-better sites, but by the substitution of instruments and photographic film, rather than the human eye, at their foci. After a number of plausible but unreproduced attempts, Lord Rosse, using the 3-foot reflecting telescope at his estate in Ireland (Figure 3.2),[4] in 1868–9 was

---

[4] Rosse's 6-foot telescope, The Leviathan, was the largest telescope in the world until the 100-inch Hooker telescope at Mt. Wilson in California in 1917. The 3-foot telescope was rather easier to use, which probably accounts for its application in this experiment.

able to reliably detect the heat from the Moon with a thermopile. The long-wave and short-wave components of radiation from the Moon could be separated by interposing a sheet of glass, and Rosse reported that the proportion of long-wave radiation was much higher in moonlight than sunlight. Furthermore, the total brightness varied roughly in proportion with the phase of the Moon – in other words, the Moon absorbed much of the sunlight falling on it, re-radiating it as heat, and the Moon warmed up and cooled down relatively quickly. Because the relative absorptions of short- and long-wave radiation by the atmosphere were still unknown, a precise temperature could not be derived, but Rosse reported that the day/night swing of lunar temperatures was 500 °F [45].[5]

These observations were brought to a higher degree of fidelity by Samuel Pierpont Langley, of the Allegheny Observatory in Pittsburgh, and later Director of the Smithsonian Institution in Washington, D.C. He devised a bolometer detector, more sensitive than the thermopile, and, using a rock salt prism to spread infrared light, was able to measure the thermal spectrum of the Sun and Moon (in which we can recognize the absorptions of major gases – ozone, as well as carbon dioxide and water vapor). In the 1880s he was able to measure the temperatures of different parts of the Moon, and observe during a lunar eclipse – confirming the *"extraordinary rapidity with which the lunar surface parts with its heat"* [46].[6]

Langley not only measured these absorptions in the laboratory, but sought to understand how absorption of sunlight and radiation of heat might vary throughout the atmosphere, and mounted an expedition to Mt. Whitney in California [48], dragging his crew and 2 tons of instruments across the country by train, then (with army escort, the West being still wild) up to 12,000 feet by mules over summer 1881. Langley's attentions later moved to aviation, his steam-powered model planes launched by catapult from a boat on the Potomac showing early promise, but he was beaten out by the Wright brothers in the race for human flight. Langley's program of research at the Smithsonian into the solar constant and atmospheric absorptions was later taken up by Abbott.

In part drawing on Langley's data, the Swedish physicist Svante Arrhenius in 1896 published the first calculation of changing global climate taking into account the warming by carbon dioxide [46]. This exercise demanded tens of thousands of tedious calculations. With glaciation in mind, he noted that if CO_2 levels halved, then the Earth's surface temperature would fall by 4–5 °C, taking

---

[5] His result is sometimes misquoted as saying the temperature of the Moon reached 500 °F, which is not the same thing, since the night-time temperatures were high negative values on the Fahrenheit scale.

[6] A detailed account (170 pages) of Langley's experiments is found in Ref. [47]. Langley's book, *The New Astronomy* (1891), is a very readable description of the state of the art in the late nineteenth century.

into account the feedback from water vapor. Arrhenius noted the flipside too: doubling CO_2 levels would trigger a rise of about 5–6 °C. However, he didn't expect this would happen for thousands of years, and (from his wintry perspective in Sweden) if it did, considered that it might be a pleasant change.

At this point, several factors were recognized as potentially contributing to climate change, and the ice ages in particular. First, the astronomical variations suggested by Adhemar and Croll; a challenge here being that glaciation in the northern hemisphere should alternate with glaciation in the south. Volcanic perturbations to the atmosphere caught wide attention after the 1883 eruption of Krakatoa, which was perhaps the first globally recognized geophysical event, the explosion being picked up by barographs (recording barometers) worldwide. The reflective sulfate aerosols injected into the stratosphere caused notable cooling in the following couple of years, and the dramatic light scattered near sunset may have motivated the bloody skies in Edvard Munch's painting *The Scream*. And then there was the possibility that carbon dioxide levels might have varied due to vegetation or similar changes. Finally, there was the Sun itself, which was now known not to be constant in appearance – the 11-year cycle in the number of sunspots was discovered by Schwabe in 1845. Longer-term variations were noticed by another German astronomer, Gustav Spoerer, in 1889. His findings were shortly thereafter elaborated by E.W. Maunder in England, who noted that there had been a prolonged period 1645–1715 when sunspots were comparatively rare, a period often referred to as the "Maunder Minimum" and coincident with a period of lower temperatures in Europe, sometimes called the "Little Ice Age".

As with many scientific debates, framed in the sense of one cause versus another, all of these factors may be significant. But these purely terrestrial considerations were merely paint on a deeper canvas, that of our Sun. Nobody knew how the Sun worked. Realizing that the Moon shone only because it reflected the Sun's light, didn't help explain why it didn't glow in the first place, nor did the findings of Rosse and Langley that didn't explain why it radiated more heat than it reflected light. It was known that the deep interior of the Earth was warm (as Fourier noted) but that the heat leaking out was tiny compared with the Sun's illumination – but was that true everywhere? The estimates by Christiansen of Jupiter's and Saturn's temperatures acknowledged that perhaps they could be much warmer if they glowed from internal heat as well as reflected sunlight. The Earth's heat, and its apparent age (many millions of years, to judge from the evidence of Hutton and his geologist successors) could be explained by gravity, the potential energy liberated as smaller bodies accumulated and impacted together to form the planet as a whole. Knowing the heat flow in the Earth, from temperature measurements in deep mines and from the conductivity of rocks, Kelvin asserted the Earth could be about 100 million years old (radioactivity had not yet been discovered, so Kelvin's error can perhaps be excused, at least at first). Similarly, the Sun could be hot from the gravitational potential energy of its formation – but that meant that the Sun, too, might only be 100 million years old, and that it was cooling down! That meant that more ice ages might be likely in the future. It is against this perspective that studies of the other planets proceeded.

Other evidence of the age of the Earth emerged from calculations essentially quantifying the ideas of Halley and Hutton. Geologist and engineer Mellard Reade in 1879 calculated that the dissolved calcium and magnesium salts in British rivers corresponded to a dissolution of the rocks denuding the land by one foot in 12,978 years [50].[7] Further, scaling this up to the flow of various rivers worldwide, he calculated it would take 25 million years for the sulfate salts to accumulate in the ocean to their present levels. Reade noted that chloride salts would take even longer – 200 million years – establishing this as a lower limit for the age of the Earth.

Other contemplations of the rocks suggested that perhaps conditions had been very different in the deep past. The American geologist T.C. Chamberlin noted in 1898 that limestone rocks (e.g. Figure 3.3) easily held more than 60 times the amount of carbon dioxide that was present in the atmosphere [51, 52]. This made changes in the accumulation or removal of carbon dioxide a powerful lever to push the climate, and for this reason he advocated that greenhouse changes were responsible for the ice ages [53]. (He played a large role in mapping out glacial deposits in the northern United States, recognizing that there was not one ice age, but several episodes of glaciation. He also founded the *Journal of Geology*, in which his glacial speculations appeared.)

Chamberlin drew attention a few years later to the possibility that changes in ocean circulation could lead to large climate variations, writing in 1906 [54]:

> In an endeavor to find some measure of the rate of the abysmal circulation, it became clear that the agencies which influence the deep-sea movements in opposite phases were very nearly balanced. From this sprang the suggestion that, if their relative values were changed to the extent implied by geological evidence, there might be a reversal of the direction of the deep-sea circulation and that this might throw light on some of the strange climatic phenomena of the past and give us a new means of forecast of climatic states in the future.

---

[7] Similar calculations were made by the Irish physicist John Joly in 1899, which seem to be better known, but are also wrong. As Reade notes, the method can only establish a lower limit on the age, since salts can also be removed from the sea.

Figure 3.3 While the 100-m high White Cliffs of Dover (made of chalk) are by no means the largest limestone deposit locking up carbon dioxide on Earth, they are one of the most visually striking. (Credit: Tom Hall/EyeEm/Getty Images.)

Swedish scientist Nils Ekholm reviewed the state of the art of climate science in 1901 [55]. He noted that the hot surface of the Earth early in its history would have had a thick, steam atmosphere. Unsurprisingly adopting the perspective of his friend and compatriot Arrhenius, he noted that a sufficient quantity of carbonic acid will produce not only a warm climate, but also a uniform one over the whole Earth. He recognized that the "*carbonic acid trade of nature is carried on with very little capital and a very great exchange*".

Ekholm also introduced the idea that climate change might modulate volcanism. He calculated that if global temperatures dropped by 20 °C, then the crust of the Earth would cool by a comparable amount within some millions of years, whereas the interior of the Earth would remain essentially as hot as before. The crustal cooling would lead to a contraction of some 12 km of the circumference of the Earth, inducing "*a great many fissures and subsidences in the regions formerly crumpled, with accompanying volcanic eruptions*" to occur. In fact, a century later, the influence of climate on planetary tectonics would be revisited, in the context of Earth's sister planet, Venus (see Chapter 6).

More useful than this speculation, however, is his analysis of historical Scandinavian temperature records and botanical indicators suggesting warm conditions between 7000 and 10,000 years before the present. He notes that, in fact, variations of the Earth's obliquity can explain the general deterioration of climate since then, calculating the excess in northern polar insolation 9100 years ago, and deducing that summertime temperatures would have been 4 °C or so higher than now. In contrast, 28,000 years earlier, they would have been 7 °C cooler – consistent with the ice sheets being present. Ekholm generously recognizes Croll's priority in developing the astronomically forced model, although he notes that Croll considered principally the changes in the Earth's eccentricity, not its obliquity.

Ekholm's wide-ranging paper also considers more recent changes, looking at historical records of Baltic sea ice and (where they existed) temperature records. He concludes that there was "probably" a change towards a milder, more maritime climate in Northern Europe in the last 1000 years, but recognizes that a cause cannot be determined.

He closes by noting that the cooling of the Earth's interior would result in a decline in volcanism and thus slower replenishment of the CO_2, but that this decline would not be significant for millions of years. On the other hand, he notes that human activity added a tenth of a percent to the atmosphere's CO_2 inventory each year, and "*if this continues for some thousand years it will undoubtedly cause a very obvious rise of the mean temperature of the Earth*". But with the calculated obliquity variations pushing Northern Europe towards a new ice age in the next 10,000 years, he thought that this global warming by CO_2 might be a good thing, and might be done deliberately.

Meanwhile, Gilbert Walker, a British mathematician working for the Indian Meteorological Department, devised techniques for analyzing time series data, and in the early twentieth century determined a statistical basis for teleconnections in the Earth climate system.[8] Specifically (helped by a small army of Indian mathematicians), he identified a seesaw anticorrelation between atmospheric pressure over the Indian Ocean and the Pacific (the "Southern Oscillation" – to distinguish it from another correlation he found, the "North Atlantic Oscillation"), which also related to temperature and rainfall patterns, and in particular the agriculturally vital Indian monsoon, which had failed causing a famine in 1899 [56].[9]

---

[8] Now generally known as the Yule–Walker equations. Walker's mathematical interests were wide, and I had cause to cite his pioneering work on boomerang aerodynamics in my earlier book *Spinning Flight* (Springer, 2006). Udnay Yule was interested in exploring variations in sunspot numbers.

[9] The paper itself is rather heavy going, being a discussion of various correlations. An illuminating review of Walker's contributions is Ref. [57]. The formulation of ENSO as a combination of El Niño and Walker's Southern Oscillation was in fact made by Jacob Bjerknes, son of Wilhelm.

Figure 3.4 The strong El Niño of 2015, 90 years after Walker identified the Southern Oscillation with which it is linked, is visible not only as ocean surface temperatures directly, but even as sea surface elevation – the warmer ocean temperatures in the eastern Pacific lowering the density such that the sea level is tens of centimeters higher there than average, as measured in this satellite radar altimetry map. (Source: NASA JPL-Caltech.)

This variation is now known as the El Niño–Southern Oscillation (ENSO), as it also manifests in warmer winter sea surface temperatures off the coast of Peru (Figure 3.4), where fishermen associated it with the Christmas season, hence El Niño (Spanish for "The Boy"). During the "normal" (La Niña) phase, easterly surface winds pile warm water towards Indonesia (such that the sea level is 60 cm higher than on the South American coast, where waters are 3–5 °C cooler than normal). These easterly winds form part of a cell, termed the Walker circulation, with warm, wet air rising (with the low pressure Walker noticed) at its western end, and descending at the eastern end. This phenomenon, relying in part on heat stored in the ocean, is one of the principal causes of interannual variation in the terrestrial climate system and adds quasi-periodic "noise" to climate variables, making the confident detection of secular changes due to increasing greenhouse gases or other factors more difficult. Walker wrote in 1925: "*It is a natural supposition that there should be in weather free oscillations with fixed natural periods, and that these oscillations should persist except when some external disturbance produces discontinuous changes in phase or amplitude.*" Climate might wobble as well as drift.

The notion that carbon dioxide might be a major constituent of the Martian atmosphere was realized around this period by the Anglo-Irish physicist G. Johnstone Stoney (who also invented the word "electron" to denote the fundamental quantity of electricity). Stoney, who had worked in his undergraduate days around 1850 as an assistant at Rosse's observatory, applied the kinetic theory of gases to planetary atmospheres even in the 1860s, although much of his work was published in the Royal Dublin Society's trans-actions, where it attracted little international attention. In a more widely read paper in 1898, he argued that a few molecules in a gas moved at speeds ten or so times the average speed (which depends on the temperature of the gas and on its molecular weight), and if that speed was greater than the escape velocity of the planet or moon, then that gas could escape and so should not be found [58].[10] With that reasoning (more commonly referred to as Jeans escape, after the English astronomer James Jeans who popularized the process in a book a few years later) it was no surprise that the Moon had no atmosphere, since its escape velocity was so low. Stoney noted that hydrogen is supplied to Earth's atmosphere by volcanoes, but notes that it should be able to escape to space. While he saw – somewhat anticipating James Lovelock's ideas in the 1960s–1970s – that today hydrogen would likely be consumed by combustion with oxygen if it accumulated to any degree, he noted that Earth's early atmosphere before oxygen became as abundant as today might allow hydrogen to persist chemically. But he also saw that the light hydrogen molecules would be able to escape to space, so any such hydrogen-rich episode must have been short lived. And similarly, he noted, water vapor should be able to escape from Mars. The implications were profound: "*Without water, there can be no vegetation upon Mars, at least not such vegetation as we know; and in the absence of vegetation it is not likely that there is much free oxygen. Under these circumstances, the analogy of the Earth suggests that the atmosphere of Mars consists mainly of nitrogen, argon and carbon dioxide.*" As indeed it turned out to be. Stoney furthermore suggested that the frosts and fogs seen on Mars were of carbon dioxide, with that heavy gas concentrating in the lowest places, and even that "*extensive displacements of the vapor, consequent upon its distillation towards two poles alter-nately*" might account for the changes in Mars' appearance through the tele-scope. While Mars' atmosphere is generally better mixed than Stoney perhaps imagined (and argon and nitrogen are only minor constituents), his guess at the composition, and his picture of the carbon dioxide frost cycle, were spot on. But they would not be popular.

Meanwhile, the extent of mixing in the Earth's atmosphere began to be explored. After Glaisher's near-fatal experience in high-altitude ballooning, and indeed the deaths of two French balloonists in 1875, exploration of the upper air

---

[10] In this paper he quotes his own work of some 31 years earlier, presented to the Royal Dublin Society, discussing how lighter gases should reach higher altitudes in planetary atmospheres, and that gas molecules moved fast enough to escape the Moon's gravity.

Figure 3.5 "Dr. Simpson sending up a balloon" in 1911 at McMurdo Sound, Antarctica, on Scott's last expedition. Understanding meteorology was a key part of the flurry of polar expeditions in the late 1800s/early 1900s. (Photo: Herbert Ponting, NOAA At the Ends of the Earth Photo Collection, image ID corp2812.)

had stalled until the 1890s with the technical innovation of the (uncrewed) sounding balloon, originating with Georges Besançon and Gustave Hermite in France. This was a small, lightweight balloon (initially paper, silk or goldbeater's skin – the outer skin of an ox intestine, used in making gold leaf) equipped with instruments that recorded information on paper or metal foil: when the balloon burst, the instrument package descended by parachute and was recovered. Their most successful balloon was named *Aerophile*. Two challenges were the intense cold and the difficulty of measuring the temperature of the air without contamination from heating by direct sunlight, a much bigger problem in thin air at high altitude (or, 75 years later, on Mars) than on the ground. Teisserenc de Bort in Paris, and Richard Assmann in Berlin launched numerous balloons in the late 1890s and early 1900s; de Bort launched some 236 to 11-km altitude, of which 74 reached 14 km.

They became confident (Assmann in particular introduced an aspirated thermometer that tackled the sunlight problem, as well as rubber balloons of the modern type, e.g. Figure 3.5) after repeated observations that in fact the

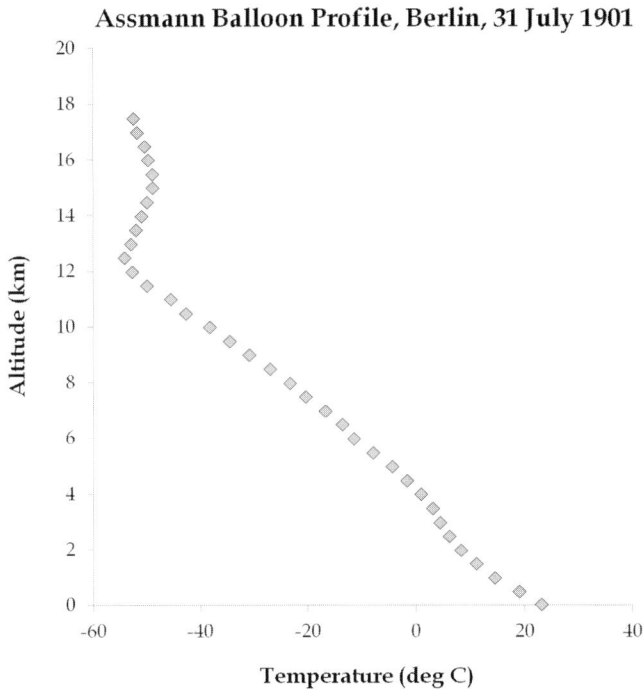

Figure 3.6 The temperature profile of the Earth's atmosphere, measured in 1901 by one of Richard Assmann's recording balloons over Berlin (Assmann's balloons recorded the temperature photographically). The more or less steady drop or lapse rate of -7 K/km is seen here up to 12 km, above which the atmosphere becomes stable and near constant in temperature.

progressive decline in temperatures ("lapse rate") with altitude came to a stop, and perhaps even turned over, above an altitude that varied between 8 and 13 km (Figure 3.6).[11] They had discovered, and de Bort named, the stratosphere.[12] And of course one had to come up with a name for the layer beneath it, the troposphere (and the boundary between, the tropopause). The French term for the sounding balloons, "sonde", was also adopted internationally, and the higher-tech version that used radio to transmit its readings so it didn't need to be recovered was called a "radiosonde", after its development by Pavel Molchanov in Leningrad, Soviet Russia, in 1930.

---

[11] A good summary can be found in Ref. [59]

[12] Sobester has written a very readable book on the challenges of accessing (and of what is now routine passenger transport in) the stratosphere [7].

Meteorological balloons became an established tool in other countries too e.g. [60, 61], and international conventions arose to synchronize the launch of balloons (in the early 1900s the International Aeronautical Commission had set the first Thursday in each month for such observations) to measure the profile of the atmosphere at many locations simultaneously. Meanwhile, the fundamentals of practical space travel began to be laid out. Konstantin Tsiolkovksy in Russia devised the fundamental equation of rocketry in 1897, and his 1903 "Exploration of Outer Space by Means of Rocket Devices" showed that the ~8 km/s speed change needed to enter orbit could be reached using a multistage rocket fueled by liquid oxygen and liquid hydrogen. Robert Goddard in the USA made the first practical experiments in liquid-fueled rocketry in the 1920s [62], and suggested that rocket-propelled vehicles would be a useful means of lofting automatic recording instruments to the upper atmosphere and might one day even photograph the Moon and planets.[13] Goddard later experimented with gyro-stabilization of rockets – automatic control systems would eventually become a key part of spacecraft and rocket launch vehicles, and in fact the first pilotless aircraft guided by Elmer Sperry's gyroscopes flew in 1917–18, only a few years after the Wright brothers.

Goddard's initial work was supported largely by private foundations such as the Smithsonian, the Guggenheim and the Carnegie institutions.[14] Similarly, donations from newly wealthy industrialists in the vibrant American economy were able to support the development of formidable astronomical observatories in that country in the late nineteenth century.

Among the planetary science developments permitted by these new larger telescopes was improved mapping of the planet Mars.[15] However, recording by sketches what the human eye sees is a notoriously subjective means of recording

---

[13] The *New York Times* mocked these ideas at the time, claiming that a rocket had nothing to push against and so would not work in space. It thoughtfully published a retraction in 1969.

[14] A. MacDonald [6] breaks down Goddard's funding by year – it was only in 1942/3, long after Goddard's breakthrough developments, that support from the Navy Bureau of Aeronautics brought government support to levels comparable with those Goddard had received from private sources 1917–41. MacDonald also presents data on the cost-by-analogy of the construction of large astronomical observatories in the USA 1820–1920, and argues the modern equivalent funding amounted to hundreds of millions of dollars a year, mostly from civic institutions and rich industrialists, not vastly different from the funding levels required to support a modern space program. MacDonald's interesting thesis is that the present era of privately funded developments in spaceflight (notably Space-X) may be a reversion to how space interests were supported in the nineteenth and early twentieth centuries, and that the NASA-led Space Age may have been a Cold War aberration.

[15] The story is nicely told by Sheehan [63].

data. Wealthy Percival Lowell, inspired by Flammarion's book on Mars and Schiaparelli's report of "canali", founded a world-class observatory in Flagstaff, Arizona, at about 2000 m elevation, just south of the Grand Canyon. Whereas modern observatories are typically on higher mountains still, in fact for visual astronomy this elevation is actually optimal due to the physiological effects of high altitude on the observer.[16]

Lowell convinced himself he saw a network of dark lines on Mars, and conjured a popular vision of a dying world inhabited by an intelligent civilization, conveying the last dregs of its water from the polar caps to lower latitudes. Yet others were unpersuaded – Maunder in the UK, and even Andrew Douglass, who had scouted locations for Lowell's observatory and later ran it on Lowell's behalf, found that the eye tends to join dots seen on a small disk through a telescope, and suspicions became widespread that in fact the canals were illusory (see Figure 3.7). Douglass was fired, and moved to Tucson in southern Arizona, where he founded not only an observatory, but also a laboratory to study tree rings as an indicator of past climate and solar activity. (Ironically there is today a canal, the Central Arizona Project, which brings water from the Colorado river west of Flagstaff down to parched Tucson –perhaps not unlike Lowell's vision of Mars.)

Another astronomer, in Catalonia, was also gifted with a good telescope, good atmospheric conditions, and good eyesight. At his Fabra Observatory, overlooking Barcelona, Josep Comas Sola could tell with his 37-cm binocular telescope that Mars had no canals [64].[17] He was also able to tell that tiny Titan, whose disk was just visible in the telescope, was like Neptune in that the edge of the disk was subtly darkened, unlike the sharp edge of the Moon. This, he correctly interpreted in 1907, meant that Titan had an atmosphere.

Lowell published an estimate of the climate of Mars, using the fourth-power law [65]. The trick here is that neither the solar constant, nor the effects of the Earth's or Mars' atmosphere, nor the reflectivity of the Earth were known. Lowell was

---

[16] I have astronomer colleagues who describe going outside the telescope dome at 5000 m on Mauna Kea, the site of many powerful telescopes because it is above much of the atmosphere, and reporting that very few stars are visible, because the thin air does not maintain low-light sensitivity in the eye. But when they take a whiff of pure oxygen from a mask (oxygen supplies are on hand in case of altitude sickness), they see an explosion of stars as their retinal function is temporarily restored. An elevation of about 2000 m is roughly optimal between the improved clarity of the atmosphere at high altitudes and the improved sensitivity of the eye at lower altitudes.

[17] The observatory is now overlooked, bizzarely, by a rollercoaster in the Tibidabo amusement park further up the hill. In an odd coincidence, I actually went to kindergarten only a couple of kilometers from this observatory. I had the opportunity to look through the telescope at a conference in Barcelona to celebrate the fifth anniversary of the Huygens probe descent.

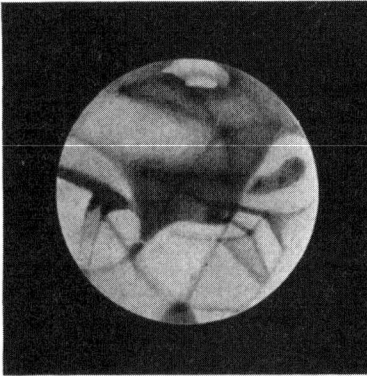

Drawing of Mars
1924, Sept. 2.

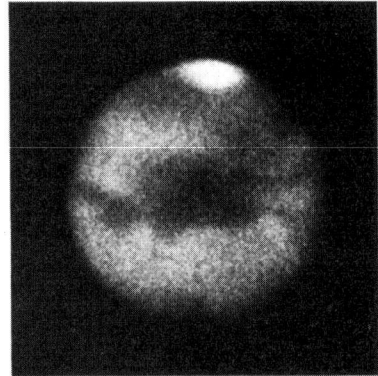

Photograph of Mars
1924, Sept. 2.

Figure 3.7 Mars through the telescope – in this instance the 36-inch refractor at Lick Observatory. Note the polar cap, and the tendency of the drawing to "join the dots", producing linear features that are not evident. (Trumpler, R., 1924. Visual and Photographic Observations of Mars. Publications of the Astronomical Society of the Pacific, 36, pp. 263–269. © *The Astronomical Society of the Pacific. Reproduced by permission of IOP Publishing. All rights reserved.*)

aware (and quotes a calculation by Moulton) that if the Earth and Mars have the same reflectivity, then without atmospheric effects the absolute temperature of Mars should be the fourth root of (4/9) times that of the Earth, since Mars is (3/2) times further away, and thus gets (9/4) times less sunlight. That means about $-30\,°C$. He recognized that the Martian atmosphere is thinner than ours, but has more carbon dioxide, so he considered these factors to balance out. But he argued that the Earth and its cloudy atmosphere had an albedo, or reflectivity, of some 75%, whereas that of dusty Mars is only a third of that value. This factor of 3 in absorption completely compensates for Mars' further distance from the Sun, such that Mars ends up being about the same temperature as the Earth.

It is difficult not to see this calculation as a case of Lowell pushing the numbers to get the answer he was looking for, and indeed a response followed swiftly. Poynting in the UK quickly dissected Lowell's calculations and noted that white cardboard has a reflectivity of nearly 75% but looks much brighter than the sky, so the foundation of Lowell's calculation was likely way off [66]. Lowell was prolific in the popular media as well as in the learned journals, and, among his books, *The Canals of Mars* prompted a fierce book-length rebuttal by Alfred Russel Wallace [67], the co-discoverer with Darwin of the Theory of Evolution by

Natural Selection.[18] Wallace notes that, on Earth at least, temperatures fall with elevation, so that when the atmosphere above becomes thinner, the reduction in the blanketing effect is a stronger effect than the reduction in obscuration of sunlight. Regardless of the apparent delusions of Lowell about the canals, and the deluge of scientific criticism his work drew, Lowell did much to popularize astronomy and contemplation of the habitability of the planets, and the observatory he founded remains a respected and productive institution today.

Arrhenius also judged Mars to be colder than Lowell had advocated, suggesting an average temperature of $-40\,^{\circ}$C, or $-10\,^{\circ}$C at the equator [68]. But he suggested that surface liquids could still exist due to the antifreeze effect of calcium and magnesium salts. He recognized that there must be but little water vapor in the cold atmosphere, with clouds being therefore rare, but imagined (rather presciently) that, in summer, hoar frost on the polar cap would evaporate and, as the relatively humid air swept over hygroscopic salts, these would absorb the water and form local wetness; a process that would hit the headlines a century later. Ironically, Arrhenius failed to consider a greenhouse effect on Venus, calculating its average temperature to be $47\,^{\circ}$C, torrid but not barren. Swirled in clouds, and with the near-surface atmosphere holding "*three times the humidity of the Congo*", he imagined Venus to be dripping wet, covered with luxuriant but primitive vegetation. Interestingly, he suggests that, despite the clouds and moisture, precipitation might not be higher than on Earth, the thick clouds suppressing air currents.

Edward Maunder too considered Mars as well as other aspects of planetary climate in a book *Are the Planets Inhabited?* in 1913. This book ponders briefly the climate of possible planets around double stars, and notes that the shadow of Saturn's rings would profoundly influence the climate of that world, and that the seasons of Uranus (whose poles are tilted almost into the ecliptic plane) would be exceptionally severe. Further, Maunder reasoned that, with its relatively thin atmosphere, the ranges of temperature with season and latitude on Mars would be rather wide, since the atmosphere could not transport much heat to ameliorate the loss of sunlight during polar night. He even drew a diagram (Figure 3.8) to illustrate the calculated temperature range, and concluded that prospects for life were slim: "*Mars is a frozen planet, and the extremes of cold experienced there, not only every year but every night, far transcend the bitterest extremes of our own polar regions.*"

---

[18] The book elaborates on Poynting's objections, as well as more general issues. Interestingly, it tells a fairly plausible story of planetary accretion, and suggests the canals are actually fractures formed by cooling of the planet, much like the cracks in basalt columns such as the Giant's Causeway, Northern Ireland.

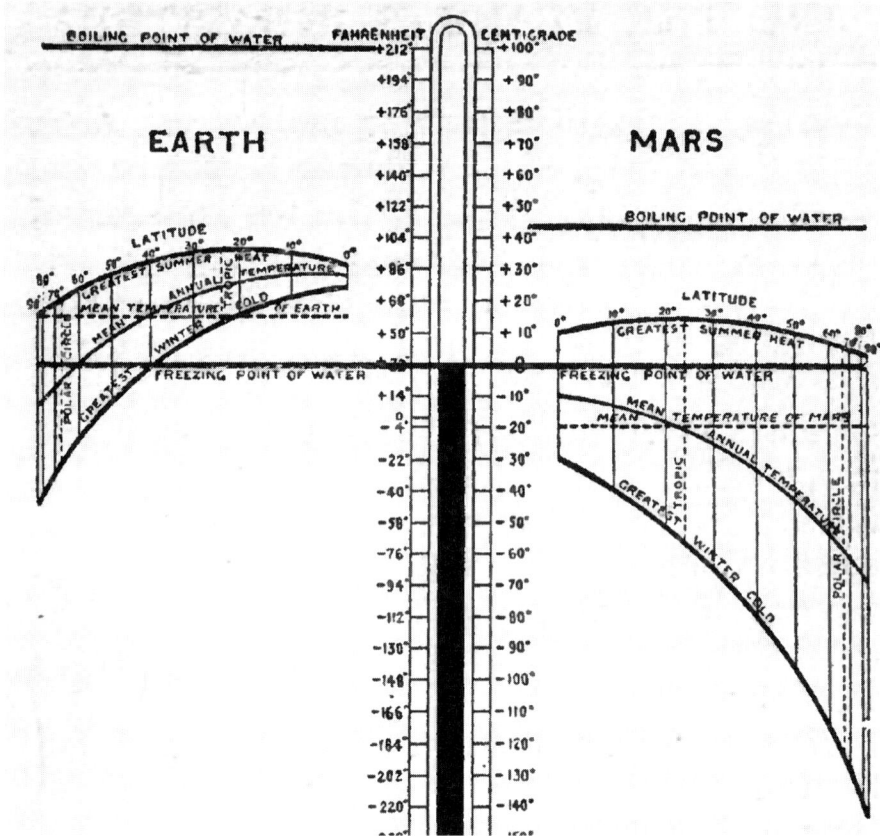

Figure 3.8 Maunder's "Thermographs" of Earth and Mars [69], sketching the seasonal ranges of temperature as a function of latitude (making the assumption Mars' atmosphere was half as effective at transporting heat as Earth's). In fact, this estimate is remarkably close to what we know today. The boiling point of water on Mars assumes that the pressure was higher than it actually turned out to be.

Maunder's book, like that of Arrhenius, aimed at a popular audience, is notable for a couple of other innovations, not (presently) widely appreciated [69].[19] In the book's closing narrative, it ventures a calculation (not quite articulated as an equation, in which form it would be popularized in 1961) as follows:

---

[19] The book also discusses physiological experiments on eyesight as a possible explanation for spurious identification of canals on Mars. It also discusses certain impracticalities of the asserted canals' characteristics for their assumed purpose – a network of uniform canals (a feature Lowell claimed as support for an artificial origin) would not be efficient for irrigation over wide areas.

If we assume that there are a hundred million stars within the ken of our telescopes, we may well believe that not more than one in a hundred of these would fulfil the condition of being a single and stable Sun, such as ours. Of the planets revolving round these million suns – stable and efficient suns – can we expect that in more cases than one in a hundred there will be a planet in the habitable zone fulfilling all the other conditions of habitability, of size, mass, inclination of axis, circular orbit, and rotation? Of these ten thousand earths which may be made fit for the habitation of Man, can we assume that even one in a hundred is now in that epoch . . . when the waters under the heaven have been gathered into one place, and the dry land has appeared, and when the earth and the waters have brought forth life abundantly? Out of a hundred million of planetary systems throughout the depths of space, can we suppose that there are even one hundred worlds that are actually inhabited at the present moment?

Not only does this argument anticipate the product of a number of stars with successive fractions of conditions of habitability and life origin (generally referred to as "the Drake equation"), it may also be the first incarnation of the term "habitable zone".

In this era, the behavior of gas exchanging heat by radiation in a gravity field began to be understood, stimulated in part by the discovery of the stratosphere.[20] Much of this work appeared first in the German language (by Austrian, Swiss and German scientists) and much of it in the astrophysical literature. Beginning with work on the structure of stars by Karl Schwarzschild in 1906, other analyses exploring the fundamentals of how the Earth's stratosphere came about soon followed by E. Gold and by W. Humphreys (in the USA) in 1909; Humphreys rather presciently suggested that absorption of ultraviolet by ozone might be responsible for warming the Earth's stratosphere. Robert Emden's work *Gaskugeln: Anwendungen der mechanischen Wärmetheorie auf kosmologische und meteorologische Probleme* (Gas balls: applications of the mechanical heat theory to cosmological and meteorological problems) sums up the dual applications of this radiative equilibrium theory.[21]

These early works assumed "semi-grey" atmospheres, where the absorption at different wavelengths is the same.[22] Soon, however, some inconsistencies had to be resolved: Hugo Hergesell in Germany noted that the colder upper troposphere

---

[20] An excellent dissection of the work in this period is C. Pekeris' thesis [70], which can be found online. Some discussion is also found in Goody and Yung [71].

[21] Emden, a Swiss working in Germany, in fact married Karl Swarzschild's sister.

[22] In the Sun, light and heat are the same thing, so the meaning of grey atmosphere is clear. The Earth's atmosphere has quite different absorption of light and heat, as recognized since Fourier, and so two different shades of grey are assumed, one for heat and one for light, an approach usually termed "semi-grey".

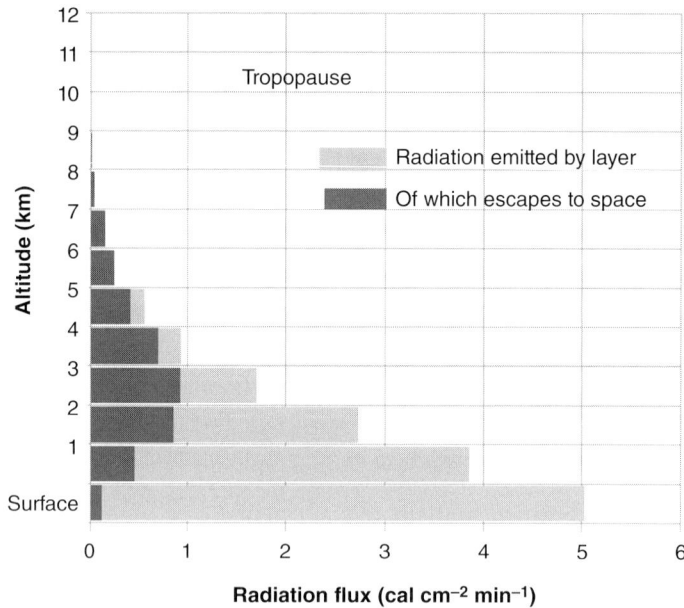

Figure 3.9 Simpson's 1927 calculation (note the units) of emission and transmission of radiant heat from and through the atmosphere – a one-dimensional radiative balance calculation. Simpson assumed a linear profile of temperature, and an amount of water vapor that increased steeply towards the ground. Such calculations form the fundamental basis for modern analyses of terrestrial and planetary climate. (Figure by author.)

couldn't hold enough water vapor for the assumed variation of transparency to hold. Edward Milne in Cambridge and R. Mugge explored how the exchange of radiation should vary with latitude, the latter also calculating more generally what Croll had started to consider, namely how much heat was transported by the atmosphere as a function of latitude.

Some inconsistencies remained, however, and were finally cleared up in a rather important set of theoretical papers by George Simpson, a prominent meteorologist who had accompanied Scott's expedition to the Antarctic (Figure 3.5).[23] In computing heat fluxes (Figure 3.9), assuming that water vapor was the only significant absorber (a prejudice that may have influenced his reception of Callendar's suggestion of a strengthening carbon dioxide greenhouse a decade later), he found the interesting result that there was in fact an

---

[23] "Some studies in terrestrial radiation" by Simpson [72]. He rather unimaginatively followed this up with "Further studies in terrestrial radiation" [73]. The next year's paper bravely broke the mould: "The distribution of terrestrial radiation" [74].

upper limit to how much radiation the atmosphere could export to space. This paradox was the kernel of the "runaway greenhouse" idea that would eventually prove important in understanding Venus.

While the semi-grey methods used by these workers would still be used extensively in the study of climate on other planets, it became recognized at this point that they really were not adequate to explain what had become known about the Earth (in fact Gold had already noted that there had to be part of the spectrum where the Earth's surface could radiate unimpeded). But the calculations began to challenge the ability of scientists to compute by hand.

Meanwhile, the utility of physics for understanding, and more particularly predicting, the short-term weather on Earth began to be recognized. Vilhelm Bjerknes in Sweden in 1907 suggested that weather might be predictable, not from the time-honored expedient of using the weather as a model of itself (i.e. look for when the weather pattern was like today in the past, and then expect tomorrow what happened the day afterwards – an approach dating back to the Greek scientist Theophrastus, and systematically employed by Fitzroy and others) but rather by applying physical laws. In fact, in 1901 the prominent American meteorologist Cleveland Abbe had argued in a long paper in the *Monthly Weather Review* (a journal he founded in 1872) that long-range weather forecasts should use equations of conservation of energy and momentum [75]:[24]

> By thus considering the land and water hemispheres of our globe as the thermal and frictional disturbers of the phenomena that would otherwise pertain to a uniform surface, rapidly rotating ... we shall arrive at the desired result sooner and better by the study of the mechanics of the atmosphere than by the search for elusive empiric periodicities.

The outbreak of World War I interrupted many scientific investigations. Fortunately, the Serbian mathematician Milutin Milankovitch was interned only briefly, and in Budapest contemplated the current climate of inner planets of the solar system [76].[25] In 1916 he published a paper entitled "Investigation of the climate of the planet Mars". Milankovitch calculated that the average temperature in the lower layers of the atmosphere on Mars is $-45\,^\circ$C and the average surface temperature is $-17\,^\circ$C. Also, he concluded:

---

[24] Abbe needed consistent timekeeping to assemble weather reports, and in 1883 persuaded US railroad companies to adopt his system of four time zones across America. In 1903 Abbe wrote to Wilbur Wright to ask him to submit an article on wind currents to the *Monthly Weather Review*.

[25] Milankovitch and Richardson survived the war, not all scientists were so lucky – Karl Schwarzschild perished (of disease) on the Russian Front.

This large temperature difference between the ground and lower layers of the atmosphere is not unexpected. Great transparency for solar radiation makes the climate of Mars very similar to high-altitude climate of our Earth.

In any case, Milankovitch theoretically demonstrated that Mars has an extremely harsh climate.

Milankovitch went on to evaluate with more mathematical rigor the astronomically forced changes in the Earth's climate that led to repeated glaciations [77].[26] While sometimes referred to by his name, these variations – which we now know also to occur on Mars and Titan – were essentially an elaboration on the mechanisms laid out by Croll, and are often termed more appropriately "Croll–Milankovitch cycles".

Elsewhere in the European conflict, Lewis Fry Richardson, a bespectacled British ambulanceman dealing with the horrific casualties in France, attempted – by hand, taking months – to calculate a 6-hour change in weather conditions (starting from 7 a.m. on May 20, 1910 – a day when coordinated balloon measurements had been made across Europe).[27] His result was inaccurate, due to poor quality of his input data, but the mathematical framework was sound, and forms the foundation of modern numerical meteorology. This framework – essentially distilling the state of the atmosphere into a table of numbers by dividing the world up with a rectangular grid (Figure 3.10) and applying equations to process the numbers – was outlined in a book, the manuscript for which was temporarily lost under a heap of coal somewhere behind the front lines where it was sent for safety. Happily it was recovered, and eventually published in 1922 [79].

Richardson envisioned that these calculations would be made by people working like an orchestra. A conductor would be needed to coordinate the calculations, making sure that one part of the grid did not outpace the others. Since we take modern forecasts for granted, it is worth repeating a little of the preface of *Weather Prediction by Numerical Process*:

---

[26] The book – about 400 pages of algebra calculating sunlight variations – was published in Belgrade in 1941, just as Germany invaded Yugoslavia. The story of Milankovitch, and the ultimate confirmation of the idea Adhemar, Croll and he originated, is well told in Ref. [78].

[27] Richardson captured the essence of turbulence in his book with the famous verse "Big whirls have little whirls that feed on their velocity, and little whirls have lesser whirls and so on to viscosity." Richardson also discussed what would later be called "fractals", finding that the measured length of a coastline depends on the length of the ruler one uses.

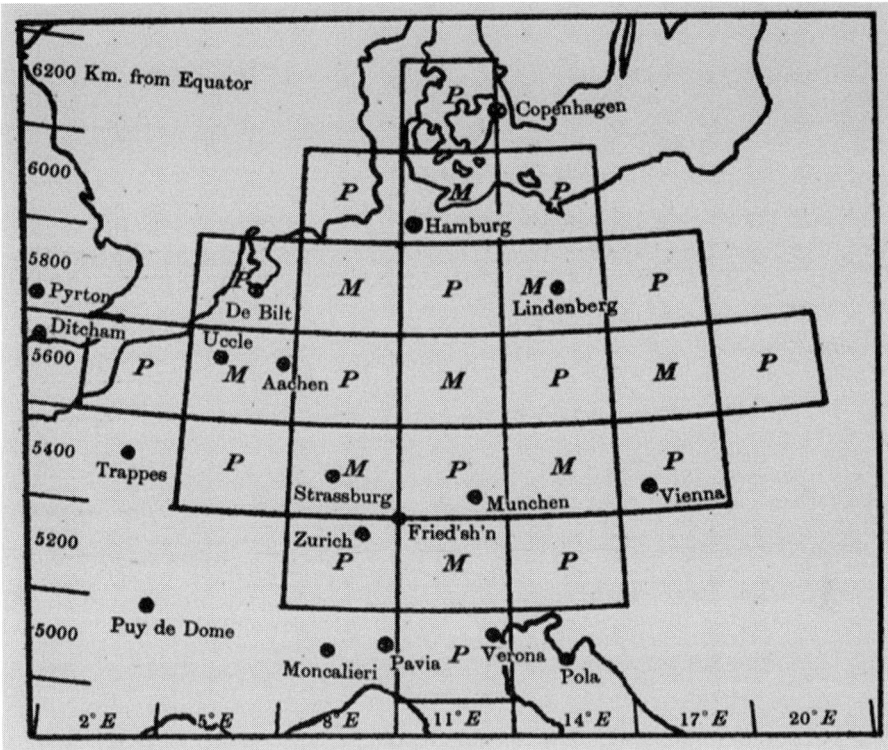

Figure 3.10 Richardson's calculation grid of 2 × 3 degree squares. (L. F. Richardson, *Weather Prediction by Numerical Process.* First edition published by Cambridge University Press, 1922. Used with permission.)

> *In this book, a scheme of weather prediction, which resembles the process by which the* Nautical Almanac *is produced, in so far as it is rounded upon the differential equations and not upon the partial recurrence of phenomena in their ensemble ... The scheme is complicated because the atmosphere is complicated ... Perhaps someday in the dim future it will be possible to advance the computations faster than the weather advances, and at a cost less than the saving to mankind due to the information gained. But that is a dream.*

Richardson, a pacifist, would perhaps be horrified to learn that another world war would see his dream realized, with the development of machines able to perform the needed calculations fast enough, as we describe in the next chapter. Indeed, the computational appetite of weather prediction and climate simulation has advanced to the point where massively parallel computer clusters are now used, and the challenges of orchestrating distributed calculations that Richardson imagined are important issues today.

4

Feeling the Heat

1920 to 1963

1923 – Pettit and Nicholson measure temperature of Mars with 100-inch Mt. Wilson telescope

1923 – J Harlen Bretz identifies catastrophic deglaciation features in US Northwest

1926 – Vernadsky coins the term "Biosphere"

1930s – Dust Bowl

1938 – Callendar proposes CO_2 levels and temperature are rising over several decades, CO_2 rise consistent with human cause, CO_2 rise is cause of temperature increase

1938 – Nuclear reactions in the Sun identified

1940 – Venus greenhouse effect considered by Wildt and Spencer Jones

1944 – German V-2 rockets reach high altitudes en route to London

1944 – Kuiper discovers Titan atmosphere containing methane

1940s – Development of infrared detectors for weapon aiming and tracking. Further measurement of infrared absorption in the atmosphere. Development of electronic computers

1952 – Struve suggests exoplanets might be detectable by radial velocity (Doppler spectroscopy) measurements of stars, or by eclipses (transits)

1952 – Opik makes energy balance model (popularized by Budyko and Sellers ~1970)

1950s – First polar ice cores and electronic numerical weather predictions

1954 – Dollfus ascends with balloon-borne telescope to measure Venus' and Mars' moisture

1956 – Radio measurements of Venus' high temperatures

1957 – International Geophysical Year: Sputnik 1

1957 – Keeler begins systematic CO_2 measurements

1963 – Spinrad and colleagues measure water vapor on Mars at Mt. Wilson

After World War I, bigger telescopes and more sensitive thermopiles opened the way to measuring the temperature of the planets directly, by their infrared emission. Edison Pettit and Seth Nicholson used the 100-inch Hooker Telescope on Mt. Wilson in 1923 to measure the center of the Mars disk (i.e. roughly at noon) as just above freezing, while the polar cap was at $-70\,^{\circ}$C [80].[1] They also found the variation of infrared emission from the limb to the center was like that of the Moon, with no displacement of the warm peak due to rotation – this meant that, like the Moon, Mars' surface warmed and cooled quickly. Photographic methods were also improving, allowing a record of the planet's appearance that did not rely on astronomers' sometimes unreliable vision and memory.

It was still thought at this time that nitrogen and argon might be the principal constituents of the Martian atmosphere, and while likely thinner than Earth's (as Herschel's visual observation of stellar occultation had indicated) astronomers struggled to measure just how thin. Donald Menzel at the Lowell Observatory reported, on the basis of photometric measurements in 1926, that an upper limit might be 66 mbar, while Bernard Lyot in Paris based an upper limit of 24 mbar on measures of the polarization of light. These observational challenges, and perhaps some wishful thinking, allowed scientists to continue to labor under the illusion that the atmosphere might perhaps be as high as 85 mbar, as advocated by Gerard de Vacouleurs (also in France) from the 1930s, in part by estimating how the contrast of Martian surface features varied as the planet's rotation brought them to the edge of the disk.[2] All these methods, however, neglected the role of scattering by dust suspended in the atmosphere, causing the pressure to be overestimated.

Dust became a prominent topic in the 1930s in the USA, as a poor choice of farming methods rendered the ground susceptible to erosion, especially in Texas, Kansas and Oklahoma. Cultivation that had been tolerable in the relatively wet 1920s failed as drought blighted crops and strong winds swept soil up into massive dust storms (Figure 4.1). The agricultural and economic upheaval caused by these impacts – sometimes blotting out the Sun and bringing visibility down to a meter – intensified the suffering of the Great Depression, and precipitated the migration of millions from the plains states. One young scientist in Massachussets, observing the turbidity of airmasses blowing east from the Dust

---

[1] Immediately following their paper is a report by Coblentz and Lampland, using similar methods but using a 40-inch telescope at Lowell. They claim the afternoon side was warmer than the morning, although they don't assign numbers; with six times less collecting area on their telescope, their observation was much more challenging and the results less certain. W. Sinton [81] gives a concise review of these and related endeavors.

[2] The early work on Mars is nicely reviewed in Refs. [63] and [82].

Figure 4.1 A dust storm overruns a farm in Texas in 1935 during the Dust Bowl. (Source: NOAA Photo Library: Image ID theb1365, NOAA's National Weather Service (NWS) Collection.)

Bowl, was Henry Wexler, who would later become a prominent player in US climate and weather, advocating the first meteorological satellites.[3]

Meanwhile, in Egypt, the first systematic survey of dust devil activity was made by W.D. Flower [84], a scientist working for the British Meteorological Office (at the time, part of His Majesty's Air Ministry – as in the USA, weather services began as branches of the military). He documented their numbers, sizes and behavior in several outposts of the Empire: Jordan, Iraq, Palestine and the Sudan as well as Egypt. Flower's careful records have been useful even three quarters of a century later, to identify characteristics that have been compared with dust devils on Mars [85].

In 1928, Richard Courant, a German mathematician, outlined a criterion for the stable convergence of partial differential equations by finite difference methods [86]. While that may sound abstract, it represents a law on numerical climate simulation as inconvenient yet binding as the finite speed of light is on space travel. Essentially the Courant (or Courant–Friedrichs–Lewy) criterion states that when simulating weather or climate dynamics in the way Richardson

---

[3] Wexler's prolific contributions in this and other areas, prior to his death at the age of only 51, are well told in Ref. [83].

Figure 4.2 A giant boulder transported to the middle of nowhere (the Ephrata fan in Washington State) by the Missoula flood. This area was studied as an analog to the Mars Pathfinder landing site in 1997. (Photo by the author.)

had proposed, the time step over which the calculations are marched must be small enough. The failure of Richardson's example attempt was in part due to over-reaching, violating this criterion, which also states that the smaller the grid cell on which the calculations are made, the smaller the time step must be – multiplying the number of calculations even more as a finer grid is proposed. Numerical weather prediction would have to wait.

In the remote northwestern parts of the USA, some indirect but profound impacts of the ice ages, and in particular the process of deglaciation, began to be recognized. Geologist J Harlen Bretz argued that the massive valleys and waterfalls carved into hard basalt rock in what he called the channeled scablands of Washington State could only be made by an immense flow of water [87, 88]. With J.T. Pardee, he recognized that massive Lake Missoula, previously dammed by the ice sheets, had drained suddenly as the ice receded. The deep canyons, falls, submarine dunes and giant strewn boulders attested to massive flow – a megaflood (Figure 4.2). Such catastrophism took some decades to be widely embraced. This region, and the event that shaped it, became a prototype for the outflow channels later to be identified on Mars [89].

Yet more revolutionary than the idea of glaciation and catastrophic floods, perhaps, was the idea that the Earth's continents have moved relative to one another. In 1912, Alfred Wegener, a German geophysicist (with strong interests in meteorology – Wegener and his brother Kurt set a world record for ballooning in 1906 on a scientific flight lasting 52 hours), proposed continental drift. In 1924, Wegener and his father-in-law, Russian meteorologist Wladimir Köppen, who in turn had defined a prominent system of climate classification in the 1880s, wrote a book that lent support to (and drew from) Milankovitch's work, and also embraced continental drift to explain how places such as Svalbard at 78° N could have coal deposits, which required lush vegetation in the deep past [90].[4] Once continental drift is embraced, the story is simple – Svalbard wasn't in the Arctic when the coal-forming vegetation was laid down. At the time, the best evidence in support of continental drift was the fossil record and the match of the coastline shapes of Africa and South America, which surely strikes every child confronted with a map: it was not until magnetized stripes, symmetric about the mid-Atlantic Ridge, were measured in the 1950s demonstrating seafloor spreading that the concept was universally accepted.

The capture of the free energy of sunlight by living things, which have rapidly proliferated across the Earth's surface, and their profound transformation of matter (not least the formation of coal and oil – "fossilized sunshine") was the subject of a work *The Biosphere* by Russian geochemist Vladimir Vernadsky [91].[5] His recognition of the profound role living things have played in chemical cycling in the oceans, atmosphere and even the crust of the Earth was not widely appreciated in the West until the 1960s and Lovelock and Margulis' work on Gaia (see Chapter 5).

Coal, and human combustion of it, entered the climate picture in 1938, when Guy Stewart Callendar, a British steam engineer and amateur meteorologist, proposed that the atmospheric carbon dioxide concentration had measurably increased, from around 290 parts per million (ppm) before 1900 to about 300 in the 1930s, although the uncertainties on these measurements were not well

---

[4] Milutin Milankovitch apparently made significant contributions to the text, but is not named as an author. While usually associated with geology, Wegener wrote a book on the thermodynamics of the atmosphere in 1911. He was also a prominent Arctic explorer, and indeed himself perished on his skis on an expedition in Greenland.

[5] The book was published in Russian in 1926 and in French in 1929, but is not widely known in the English-speaking world (a very abridged English version was published in 1986). Indeed, Lovelock confesses to being unaware of it (and noted that none of his learned colleagues drew attention to the lapse) when he and Margulis in effect retraced some of Vernadsky's steps in 1972. The term "Biosphere" was itself coined in a throwaway sentence in the book *Origin of the Alps* by Austrian geologist Eduard Suess in 1875.

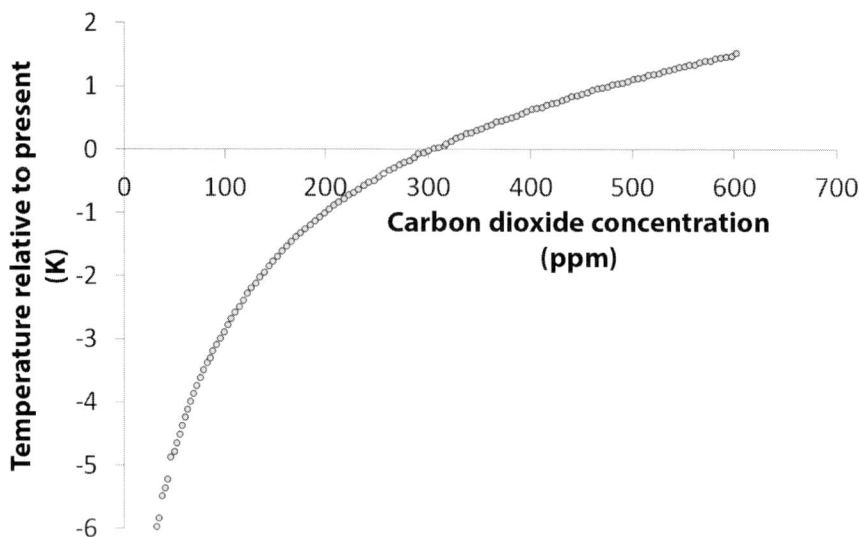

Figure 4.3 Callendar's prediction of warming as a function of carbon dioxide concentration. (Figure by author, from data in Callendar's paper.)

determined [92] (Figure 4.3).[6] He had also carefully compiled temperature records from around the globe, and found a general increase. And then Callendar joined the dots – he not only argued that the carbon dioxide increase was consistent with fossil fuel combustion (and that the extra CO_2 had not been taken up completely by the oceans), but also presented calculations showing that the CO_2 increase could explain the apparent temperature increase (Figure 4.4).

Callendar's unfunded but prescient work received only modest attention at the time, perhaps in part due to a dismissive response from the prominent Sir George Simpson, who praised the amount of work Callendar had put into the paper but observed (my underlining) "*that it was not sufficiently realized by non-meteorologists who came for the first time to help the Society in its study, that it was impossible to solve the problem of the temperature distribution in the atmosphere by working out the radiation*". While Simpson's point that convection plays a role in atmospheric structure was correct, his condescension in referring to Callendar as a newcomer and "non-meteorologist" seems brutal.

[6] Also, J.R. Fleming's account is an excellent story [93]. It fails to notice, however, the strong undertone of dismissal in Simpson's discussion that followed Callendar's presentation of his work at the Royal Meteorological Society, recorded in the pages of the journal after the paper itself. Despite Simpson's eminence, Callendar's response to the comments on his paper was robust, and he went on in further work to improve his estimates in future papers, without changing the conclusions, which of course have been rather strikingly borne out.

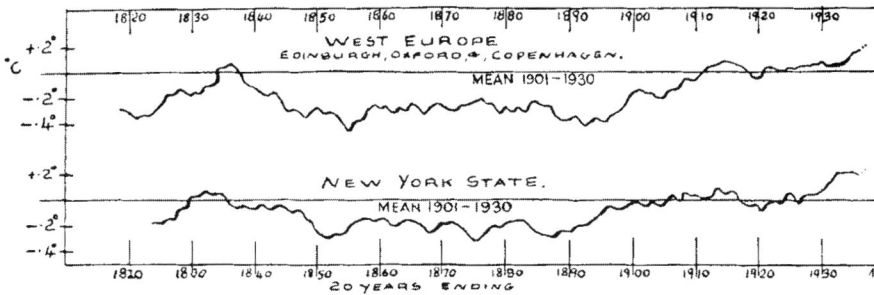

FIG. 3.—The most reliable long period temperature records. Twenty-year moving departures from the mean, 1901-1930.

Figure 4.4 Callendar's compilation of European and American temperatures, showing notable warming since the late 1800s. (From Callendar G. S. 1938. The artificial production of carbon dioxide and its influence on temperature. *Q. J. R. Meteorol. Soc.* **64**: 223–240, with permission from Wiley.)

In fact, Callendar's usual attribution as a "steam engineer", which perhaps conjures an image of a grease-blackened brute with a large wrench, fails to really recognize that his career was all about precision measurement. He had presented scientific measurements on the thermodynamic properties of high-pressure steam at international conferences, so while his work on climate was performed in his spare time, his analysis was by no means amateurish.

Simpson dismissed Callendar's empirical observation of warming as a spurious change: "*the rise in temperature was probably only one phase of one of the peculiar variations which all meteorological elements experienced*". Indeed, global temperatures underwent a dip in the 1960s, which meant Callendar's observations and predictions received relatively little attention. However, recent assessments of the evolution of surface temperatures and CO_2 concentration show that Callendar's predictions were remarkably accurate; see e.g. [94].

About the same time (1940), the idea of a carbon dioxide greenhouse on Venus was advanced by Rupert Wildt, a German astronomer in the USA. Drawing on spectroscopic observations of CO_2 in the Venusian atmosphere (by Adams and Dunham in 1932, using the 100-inch telescope on Mt. Wilson and special new film emulsions sensitive in the near-infrared [95]) and laboratory data on CO_2 opacity, he made the rudimentary calculation that Venus was "*probably warmer than the boiling point of water*" [96].[7] The amount of CO_2 estimated was only, however, that amount near and above the cloud tops, so the strength of the greenhouse and thus the surface temperature were considerably underestimated.

---

[7] His greenhouse calculations were much less elaborate than those of Simpson, Callendar or others. He also made the claim that Venus had clouds of formaldehyde, a rather less successful idea than his greenhouse thoughts.

Venus' greenhouse conditions were also contemplated in a 1940 popular book by the English Astronomer Royal, Harold Spencer Jones [97],[8] who interpreted the Adams/Dunham results (and exploited measurements by Adel and Slipher at Lowell Observatory, who measured the spectral absorption of light passed through a 45-m long absorption cell holding some 47 atmospheres of CO_2 – a formidable laboratory experiment! [98][9]) as suggesting CO_2 might be the dominant constituent of Venus' atmosphere, providing a strong greenhouse, and that Venus' atmosphere today may resemble a stage that the Earth's atmosphere once passed through, before life removed much of the carbon.

A recently discovered essay from this period by none other than Winston Churchill articulates much the same outlook as Spencer Jones[10] – while recognizing that the process of planetary formation was not understood, and so the abundance of planets in the universe was not clear, notes "*I am not sufficiently conceited to think that my Sun is the only one with a family of planets*" and that some of these worlds will be "*at the proper distance from their parent Sun to maintain a suitable temperature*" [100].

World War II intervened at this point, bringing a new world order, as well as a vast demand for meteorological products to support military operations. The needs for aviation were obvious, and meteorology even attained a planetary scale in its application.[11] Although the discovery of Earth's jetstreams is often credited to the high level of Allied bomber activity, which yielded consistent observations, in fact the northern hemisphere jet was discovered by Japanese meteorologist Wasaburo Ooishi in the 1920s [102]. Launching small, passive balloons (pilot balloons or pibals) from Tateno, Japan, and tracking them by theodolite (Mt. Fuji was 160 km away, and served as a useful target to assess atmospheric visibility before launch), he measured winter westerly winds at up to 70 m/s at 10 km altitude. In fact, these winds were exploited during the war to launch some nine thousand incendiary-carrying balloons against the USA in 1944–5 – the first

---

[8] This book also discusses the observation that ultraviolet images of Mars are larger in diameter than those in the near-infrared, suggesting the atmosphere is discernable up to a height of 50 miles or so, although the surface pressure is no more than a small percentage of Earth's. The book discusses the challenge of spectroscopic detection of gases on other planets, suggesting the Moon can be used to estimate the telluric absorption.

[9] The paper concludes: "*In the upper strata alone, Venus possesses 10,000 times as much CO_2 as is present in the entire atmosphere of the earth.*" Wexler visited Adel at Lowell in 1939 with a view to supporting his work for terrestrial applications. In related work, they used a long gas absorption cell to study methane, and thereby show that this gas was by far the dominant hydrocarbon in the atmospheres of Jupiter, Saturn, Uranus and Neptune – acetylene, ethane and other gases had to be much less abundant.

[10] Churchill was a talented historian and journalist, as well as a politician. He was also commendably receptive to introducing scientific ideas into policy-making and warfare.

[11] The development of meteorology in the USA as a profession and as an academic discipline in the first half of the twentieth century, and the subsequent initial developments in numerical weather forecasting, are described in Harper [101].

intercontinental weapon [103]. About 300 are known to have reached the Americas, but caused little damage.[12]

Another wartime development of note was the methods by Harald Sverdrup and Walter Munk to predict wave heights at sea, to determine whether amphibious landings would be feasible [105]. Despite the best efforts of mathematicians for over a century, the fundamental prediction of wave growth in response to sea-surface wind stress is a challenging area, and Sverdrup and Munk's semi-empirical approach struck a useful balance between theoretical rigor and empirical expedience: 70 years later I was to follow largely their methodology in predicting waves on Titan's sea Ligeia Mare for a proposed capsule to float there [106].

In 1944 the Dutch-born astronomer Gerard Kuiper made spectroscopic observations of the planets and Titan, and found remarkably that Titan's spectrum showed dark bands that matched laboratory measurements of absorption bands of methane. Titan had a significant atmosphere, making it unlike any other satellite (and promoting it to being a subject of this book). While Comas Sola's visual observation had been correct, it was not reproducible, whereas a photographic spectrum was much more solid evidence.[13] Kuiper also showed with spectra he acquired in 1947 that carbon dioxide was present in the Martian atmosphere (nitrogen and argon were suspected by analogy with Earth, as noted by Stoney, but these simple gases do not have easily-detectable spectroscopic signatures and so were not actually detected until spacecraft measurements were made much later) [108, 109].

In 1943 a couple of reports were made of possible planets around other stars, detected by the astrometric method (measuring the star's path across the sky, which would wobble slightly due to the mass of the purported planet), but these observations proved to be spurious. However, in 1952, in a remarkable two-page paper [110], the Russian–American astronomer Otto Struve pointed out that a different technique, spectroscopy to measure the wobble of the star along the line-of-sight by the Doppler effect, could detect Jupiter-mass planets if they were close enough to their star. He also noted that eclipses (transits) might cause dips in the light from the star, which would be measurable with photoelectric means. Half a century later both of these techniques would become central in exoplanet studies.

World War II had stimulated many technological developments that directly or indirectly influenced the study of planetary climate. Rocketry, notably the

---

[12] A little-known British campaign was less meteorologically ambitious, but more militarily successful. Operation Outward launched small balloons trailing wires from Britain towards Germany, where the wires shorted out power cables [104].

[13] And in fact with a relatively modest modern amateur astronomy setup, a small CCD camera with a diffraction grating, and an 8-inch telescope, you can repeat the discovery for yourself in about a minute of observing time. See Ref. [107].

Figure 4.5 The 50-foot spun aluminum dish at the US Naval Research Laboratory, used by Mayer in 1956 to measure the high surface temperature of Venus. (Image provided courtesy of the Naval Research Laboratery.)

liquid-fueled and gyro-stabilized German V2, would pave the way for the vehicles that would make planetary exploration possible. The war, and the Cold War that followed, saw improvements in electronics of all kinds, which eventually filtered into astronomy. Detectors of infrared radiation intended for heat-seeking missiles, together with improved models of the propagation of such radiation through the atmosphere made major progress, especially in the 1950s and 1960s, both for improving telescopic measurements of planetary temperature, and in modeling the effect of gases on climate. Similarly, the development of radar provided a platform for major advances in radio astronomy, which opened a new window into the study of planetary temperatures.[14]

When turned to our nearest neighbor planet, radio methods indicated Venusian temperatures of many hundreds of degrees. The first reliable estimate was made using observations in 1956 with a 50-foot radio dish at the Naval Research Laboratory in Washington, D.C. [111] (Figure 4.5) (in fact the radio telescope was calibrated with a radio transmitter mounted at the top of the Washington Monument!).

[14] Among WWII scientists working on radar were those working on systems for aircraft landing in fog, including Arthur C. Clarke (who in 1945 proposed that satellites in geostationary orbit would be useful telecommunications relays) and Luis Alvarez (who, with his son Walter, would interpret an iridium-rich sediment layer at the Cretaceous–Tertiary geological boundary as evidence that an asteroid impact was responsible for the extermination of the dinosaurs).

This result was a puzzle, as infrared measurements by Sinton and Strong in 1960 showed that the day and night temperatures in the atmosphere were about 240 K [112], in agreement with earlier work by Pettit and Nicholson [113] (who actually took their measurements in the 1920s[15]). The infrared measurements were probing the cloud tops, but what was the radio seeing? Jupiter had been discovered in 1954 to be a strong radio source [114], not because it was hot but rather because plasma interactions in that planet's powerful magnetic field produce sporadic radio emissions via a rather exotic cyclotron mechanism. Could Venus' radio-brightness be explained this way? C. Mayer and colleagues [111] estimated the radio brightness of Venus at their 3-cm wavelength to be about 600 K, not too far from what would turn out to be the right value for the surface temperature. This very high brightness temperature was a great puzzle, and would not be resolved without traveling to Venus itself.

Meanwhile wartime advances in nuclear physics paved the way for radiocarbon methods and the isotopic analyses that would ultimately be the key to reconstructing Earth's climate. But it was the development of electronic computers, stimulated in part by the need to compute ballistic trajectories for gunnery, that finally permitted the first practical experiments in weather prediction by numerical expressions of physical laws, as envisioned by Abbe, Bjerknes and Richardson, rather than the educated guesswork by graphical methods of human forecasters.[16]

Influential individuals in these steps included the Swede Carl-Gustav Rossby[17] and the computer scientist John von Neumann, but the person most identified with the "first" electronic computation for weather prediction is Jule Charney, hired by Neumann at Princeton, who broke down the atmospheric dynamics

---

[15] It is not obvious why it took them a quarter of a century to publish their results in full. Planetary astronomer Dale Cruikshank noted to me that Nicholson was primarily interested in stellar astronomy, which might explain the slow emergence of the work. Nicholson has the distinction, with Galileo, of having discovered four Jovian satellites.

[16] Of course, advanced computer developments were also underway in Britain for code-breaking; these advances, however, remained under strict secrecy long after the war, but have become the stuff of recent movies, noting in particular the role of mathematician Alan Turing. Actually the famous "Enigma" naval cipher was decoded with electromechanical machines: it was a much more advanced electronic machine named Colossus that was developed at Bletchley Park to crack the even stronger coding performed by the Lorenz (no relation, as far as I know) machine on messages from German High Command. The Colossus is considered the world's first programmable electronic digital computer, but is much less known, as its existence was only declassified in the 1970s.

[17] Rossby was a prominent figure in both Scandinavian and American meteorological (and oceanographic) circles. He also established the journal *Tellus*, in which many important climatological papers would subsequently be published.

equations into something tractable for computation by the ENIAC computer.[18] These first steps in weather computation were regional, not global, with the initial conditions set by meteorological observations rather than first principles, and the output propagated only 24 or 48 hours ahead.[19] Although numerical climate prediction is in principle the same problem, the practical details of the approach are different. In weather prediction, the assimilation of vast amounts of observational data, and the forward propagation of the system by approximations that are "good enough" but streamlined to work at very small scales (high resolution) is the overall approach, whereas for climate prediction, it is more important to get the physics right, albeit at lower resolution (bigger "boxes"). The first global simulation (i.e. a global circulation model, GCM), by Norman Philips soon thereafter, found just this challenge with the physics – the abstract to his paper concludes *"Truncation errors eventually put an end to the forecast by producing a large fictitious increase in energy."* [117]. Ultimately, and with enough model refinement and computer power, the two modeling philosophies converge, but the emphasis in the 1950s at least was on weather forecasting and dynamics, rather than climate.

In the 1950s some graphic insights into heat transfer in a rotating fluid were developed with laboratory experiments, first by Dave Fultz and Herbert Riehl in Chicago and later by Raymond Hide in Cambridge [118–120]. These so-called "dishpan experiments", with cold centers and heated rims, generated long waves, vortices, and meandering jetstreams. While these analog experiments were not predictively useful in the same way computer models were, they helped thinking about the dependence of the circulation regime on the atmospheric and oceanic boundary conditions such as rotation rate (which would be useful in understanding the differences between different planets) and served to give tractable benchmarks for computers to reproduce.

Some basic aspects of the ocean circulation began to be understood around this time, with Henry Stommel of the Woods Hole Oceanographic Institution developing some basic ideas on how wind stress acting on the ocean surface of the rotating Earth would form Western Boundary Currents (such as the Gulf Stream and the Kurishiro Current at Japan) on the western margins of the midlatitude oceans. He further proposed simple models of the way salinity and temperature drive the deep ocean circulation, with cold salty water at high

---

[18] Electronic Numerical Integrator and Computer – developed with gunnery applications in mind, this was at the US Army's Ballistics Research lab in Aberdeen, Maryland, a couple of hours drive from Princeton.

[19] Reference [115] nicely sets these first experiments in context with Richardson and others. The scientific documentation of the first numerical experiment is found in Ref. [116].

latitudes (salty because, as sea ice forms, the salt stays in the liquid) sinking into the abyssal depths whence it rolls equatorward. He constructed a simple box model of this "thermohaline" circulation [121], and found that it had two steady states, wondering *"whether other, quite different, states of flow are permissable in the ocean or some estuaries, and if such a system might jump into one of them with a sufficient perturbation. If so, the system is inherently fraught with possibilities for speculation about climate change."*

The post-war years saw the first polar ice coring – in 1949–52 British–Norwegian–Swedish, USA and French teams drilled ~100-m-long cores in Antarctica, Alaska and Greenland respectively [122]. The idea that annual cycles might be discerned in the firn (packed snow) and ice layers, forming a nice consistent record of the past climate, was first demonstrated in a 15-m pit hand-dug by Ernst Sorge in 1930, on Wegener's expedition to Greenland. Drilling, and especially sample handling and analysis instrumentation, progressively improved. One stimulus was the International Geophysical Year (IGY, 1957), and US–Danish projects (Greenland being Danish territory) could drill over a kilometer down to bedrock by the early 1970s, claiming *"One thousand centuries of climate record from Camp Century on the Greenland Ice Sheet"* (Figure 4.6). These records provide our best understanding of climate over the last million years or so – not only does the isotopic composition of the water shed light on

Figure 4.6 A rather forbidding International Geophysical Year Ice Drilling Station, Greenland, in 1957. (Courtesy of the US Army Cold Regions Research Laboratories.)

temperatures of the past, but tiny bubbles of gas trapped in the ice allow variations in the greenhouse effect to be quantified, and layers with sulfate acidity attest to volcanic eruptions.

In 1952 in Ireland, the Estonian astronomer Ernst Öpik accounted for heat transport in a simple model of the Earth's climate[20] (which he was attempting to use, in conjunction with the record of glaciation, to demonstrate his rather bizzare theories on fluctuations of solar luminosity[21]). While these papers are – probably justifiably – virtually unknown, among them is a notable attempt at a global climate model (specifically, a one-dimensional zonal energy balance model) that predates the more famous incarnations by Budyko and Sellers (see next chapter) by two decades.[22]

Other scientists also favored solar variations and/or volcanic aerosols as the cause for climate change. Harry Wexler in 1956 advocated these effects, yet barely mentions carbon dioxide [126].[23] But after a decade or two of inattention, the hypotheses advanced by Callendar received renewed examination. First, Gilbert Plass, a US physicist, examined infrared absorptions by carbon dioxide more carefully (using a new-fangled electronic computer to analyze the hundreds of thousands of spectral lines), and largely retired the objection that the CO_2 and water absorption bands overlapped,[24] such that the atmospheric absorption at a

---

[20] Opik's rather wide-ranging career (both geographically as well as intellectually) is summarized briefly in Ref. [123].

[21] The nuclear fusion reactions that power the Sun were elaborated by Hans Bethe in 1938/9. Arthur Eddington in 1920 had suggested that fusion of hydrogen is what produced helium, and energy, in the Sun. George Gamow in 1928 had explained with quantum mechanics how the hydrogen nuclei might furtively tunnel through the barrier to join together. With this understanding of the basic process fueling the stars – physics that was simply unknown to Kelvin and Helmholz when they confronted the ages of the Sun and the Earth at the turn of the century – and models of stellar structure starting with Schwarzschild's, the evolution of stars could be predicted.

[22] e.g. Ref. [124]. Opik published even more obscure papers than this in the Communications of the University of Tartu – but his more easily obtained Icarus paper twelve years later summarizes them [125].

[23] The paper doesn't mention carbon dioxide at all. A chapter in Ref. [127] does mention CO_2 in passing, but largely to dismiss it.

Wexler is often claimed to be the first scientist to deliberately fly in an aeroplane into a hurricane to collect scientific data in 1946; he was the chief scientist of the US Antarctic expedition in the IGY, and a major player in introducing meteorological satellites such as TIROS.

[24] The bands are made up of lots of narrow lines. When viewed at low spectral resolution, the bands of CO_2 and H_2O overlap; but the overlap becomes less when looked at more closely as the individual lines that make the band do not overlap. The picture rapidly gets even more complicated, as the lines get broadened by the presence of even transparent gases such as nitrogen, and start to overlap after all.

given wavelength was not already "saturated"; e.g. [128, 129].[25] Thus, adding more CO_2 beyond what was already there would cause warming after all. Second, radiocarbon dating had been devised by 1949, and could be used to assess the amount of carbon being added to the atmosphere. (Basically, a tiny but near-steady fraction of radioactive carbon-14 atoms results from cosmic rays interacting with nitrogen; when this carbon is locked up in a plant, the fraction of carbon-14 slowly declines, with a half-life of about 7000 years. But fossil fuels are very old, all their carbon-14 has decayed, so adding CO_2 from the combustion of fossil fuels dilutes the radiocarbon in the atmosphere.) Radiocarbon analyses enabled Hans Suess[26] to deduce that carbon was indeed accumulating in the atmosphere, although with Roger Revelle, the initial estimates suggested that the oceans might be absorbing much of it [131]. More refined later studies showed this was not the case, however, but those studies were in part stimulated by Suess and Revelle's work – the evolving terrestrial greenhouse was becoming a respectable topic of study.

That, and the healthy scientific support stimulated by the International Geophysical Year in 1957, enabled Charles Keeling to begin a systematic series of carbon dioxide measurements in locations that would be unperturbed by local industrial or agricultural effects, notably on the summit of Mauna Loa in Hawaii. Within just a couple of years, his data showed a nice seasonal cycle of a few parts per million (as vegetation, of which most is in the northern hemisphere, took up and released CO_2 over the course of a year) but also that the cycle didn't quite repeat – the peak value in one year was one or two ppm higher than the previous one [132]. Although sometimes challenged to maintain funding for the program, Keeling managed to sustain it, yielding what has become an icon of climate change, the Keeling Curve (Figure 4.7). When his measurements began in 1958, the concentration was 315 ppm; when Keeling died in 2005 it was 380, and recently it passed the 400 mark.

So much for the Earth. It was still speculated at this time that Mars might yet be habitable and that the seasonal changes in Mars' appearance might still be caused by vegetation – indeed, the title of the book *The Green and Red Planet* nods to these waves of changes seen by Lowell and others, whose influence was impressively persistent. This book [133], by aviation medicine specialist Hubertus Strughold,[27] notes the similarity of likely Martian conditions (he assumes a

---

[25] The contributions of Plass, Revelle, Suess and Keeling as well as Callendar, Tyndall, Arrhenius and others, are surveyed in Spencer Weart's book [130] and many articles by that author.

[26] Hans was the grandson of Eduard Suess, who coined the term "biosphere".

[27] Strughold was brought to the USA as part of "Operation Paperclip", to exploit for the US Air Force his expertise in human response to high-altitude conditions. This expertise

Monthly mean CO$_2$ concentration

Mauna Loa 1958 - 2017

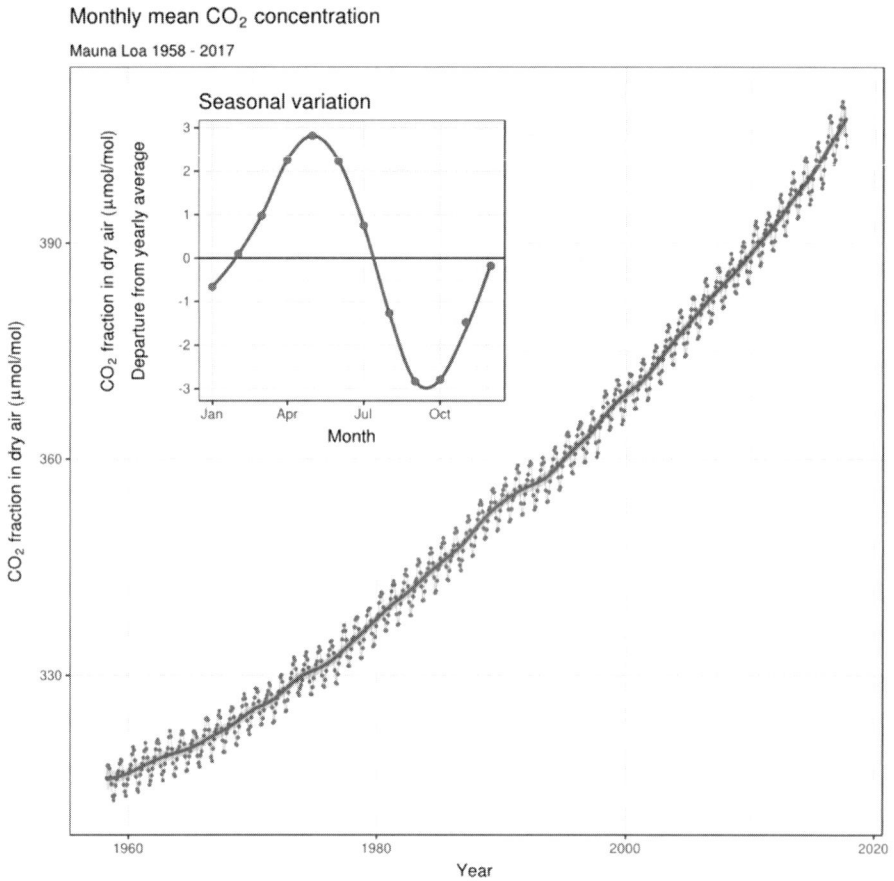

Figure 4.7 The Keeling Curve of carbon dioxide abundance in the Earth's atmosphere – an achievement of sustaining an unglamorous program over decades against bureaucratic pressure as much as an achievement in measurement precision. An annual cycle is visible as the boreal forests inhale CO$_2$ from the atmosphere in northern spring (rather less land surface covered by trees is present in the southern hemisphere, so the net effect is an annual, rather than biannual cycle), but year by year a steady creep upwards in CO$_2$ levels occurs. (Source: Wikimedia/Delorme. Data from P. Tans, NOAA/ESRL and R. Keeling, Scripps Institution of Oceanography.)

Lowell-like surface pressure of about 100 mbar) to those at high altitude (56,000 feet) on Earth, and considers the physiological implications for how simple plants might function there, recognizing that oxygen amounts would be too low ($<$1 mbar) for higher forms of life. Strughold (who also invoked the term

may, unfortunately, have been obtained in questionable circumstances – Strughold has been implicated in possible involvement in involuntary experiments on the effects of cold and low pressure on concentration camp prisoners. Nonetheless, his book is an interesting read.

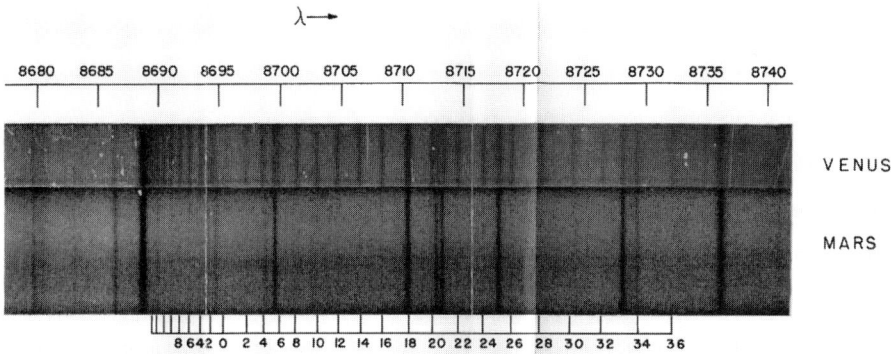

Figure 4.8 Near-infrared photographic spectra by Kaplan et al. in 1963. This shows a large number of spectral lines of carbon dioxide. Note that many more lines are visible in the (much thicker) atmosphere of Venus. (From Kaplan et al. [136] © AAS, used with permission.)

"astrobiology"[28]) even suggested building a "Mars Chamber" to study how lichens or other lifeforms might tolerate Martian conditions.

It was known that there was carbon dioxide on Mars and Venus, and the final discovery of their true amounts by spacecraft will be discussed in the next chapter. But how much water vapor? Water after all is a strong greenhouse gas and is ultimately the determinant of habitability. In principle, spectroscopy would be able to give the answer, but the challenge is to measure the absorption of spectral lines or bands in the remote atmosphere through the Earth's atmosphere, which is itself drenched in the same vapor. Astronomers on both sides of the Atlantic tackled this challenge in two different ways. Hyron Spinrad eventually had the most robust success, using a very high resolution spectrograph on Mt. Wilson in 1963, in a 4-hour exposure with special ammoniated film, taken not when Mars was closest, but when it was at quadrature – i.e. at its widest angle from the Sun [135] (Figure 4.8). This meant that Mars was moving towards Earth at 15 km/s, enough to Doppler shift the narrow Mars water lines such that they did not overlap with the terrestrial ones (a technique suggested originally by Lowell, but difficult to implement with the instruments available at the time). The measurement was robust because several different lines were measured – and they indicated that only a few precipitable micrometers of water vapor existed in the Mars atmosphere, a tiny amount.

In Europe, Adouin Dollfus of the Paris Observatory went to great lengths, or at least dizzying heights, to answer the same question, by getting above as much of the terrestrial atmosphere as he could. After some initial experiments in 1954–7 to

---

[28] The term "astrobiology" was introduced by the Soviet astronomer Gavril Tikhov in a book of that title in 1953. See Ref. [134].

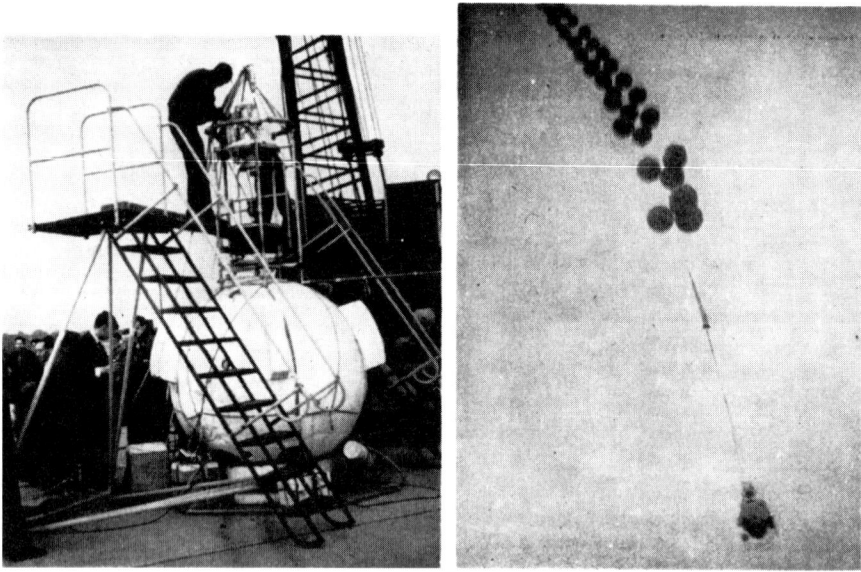

Figure 4.9 Dollfus' efforts to measure Mars' and Venus' moisture included high-altitude observations using a photometer operated from a small pressurized gondola (left) equipped with a telescope (above). The gondola, in a flight seemingly meritorious of a "Darwin Award", was lofted by 104 small weather balloons. (From [138].)

7 km in a conventional balloon with an open wicker basket [137] (to observe the convective granulation of the Sun, and with a spectrometer to study the sky brightness), Dollfus and his aeronaut father developed a remarkable balloon-mounted telescope with a special photometer using a birefringent crystal to select light in and out of a water band, with the telescope operated from inside a pressurized aluminum gondola [138].[29] Dollfus took the system to an altitude of 13.5 km (also in 1959, breaking French ballooning records) suspended, almost comically, beneath a string of 104 weather balloons, which were easier and cheaper to obtain than a single large purpose-built balloon (Figure 4.9). It is rumored that controlled deflation to descend was effected with a small rifle.

In the end, these quixotic attempts were too much at the mercy of the weather, both from the balloon dynamics point of view, and from variations in the water vapor content of the Earth's atmosphere. Dollfus ended up getting better results from his photometer when it was helicoptered to the top of the Jungfraujoch in the Swiss Alps in 1963 to observe Venus, as it was in the same

---

[29] Charles Dollfus was the first Frenchman to cross the Atlantic in both directions by dirigible, and was instrumental in founding an aviation museum in Paris. He gave Adouin his first balloon ride at the age of 8.

I met Dollfus at a conference dinner in 2004 in the Netherlands – even well into his 80s he was bright and engaging.

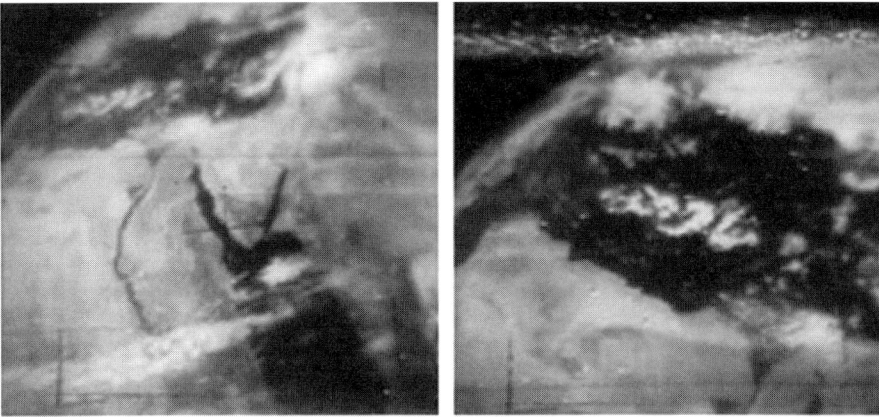

Figure 4.10 A pair of frames from TIROS-1, the first weather-imaging satellite. The spin-stabilized satellite is moving north over the Red Sea, seen in the left frame with the Nile river and a light streak of cloud. The Mediterranean, with a prominent cloud system in its center, is visible at upper left, and is seen (with the Nile delta) in more detail in the frame on the right. Compare the appearance of these images with weather images of Titan in Chapter 8. (Source: NASA.)

part of the sky as (indeed, was occulted by) the Moon, so that the Moon could act as a calibration reference, peering through the same amount of the Earth's atmosphere as Venus, but without its own water absorption.

In November 1957, with the launch of Sputnik-1, the space age began. Within two years satellite measurements of the Earth's climate from space were attempted. The February 18, 1959 headline in the *New York Times* was "Vanguard fires Satellite into Orbit to Scan Weather" – this battery-powered Vanguard 2 satellite was followed up by the solar-powered Explorer 7 in October, which yielded better measurements, allowing Verner Suomi to quantify the outgoing thermal emission and cloud cover [139]. The potential for orbital observations to improve weather forecasting was demonstrated by the satellite TIROS-1, which beamed back the first TV images of Earth in 1960, showing the evolution of storms from above (Figure 4.10). (Wexler had been a significant advocate of the use of satellites for meteorology [140].[30]). But spacecraft would not remain earth-bound for long – within only five years, we would begin to study the other planets up-close, with the first measurements at Venus in 1962, by a spacecraft whose name evokes many of the early explorations described in this book – Mariner 2.

---

[30] Wexler published in the *Journal of the British Interplanetary Society*. This journal is typically a forum for more outlandish ideas in space exploration, suggesting that the use of spacecraft for meteorological monitoring, so routine to us today, was not as widely obvious at the time.

5

First Contact: The Dawn of Planetary Exploration

1960 to 1979

1962 – Mariner 2 visits Venus, obtains crucial data on the state of the atmosphere. Sagan and others assess greenhouse effect

1963 – Lorenz discovers chaotic behavior of climate equations

1964 – Manabe and Strickler introduce convective adjustment model; later estimate 2.3 K response to atmospheric CO_2 doubling

1965 – Mariner 5 measures surface pressure and temperature on Mars

1965 – Lovelock explores ideas of atmospheric chemical equilibrium and life detection

1966 – Leighton and Murray devise model that captures key aspects of the Martian climate

1967 – Komabayasi investigates runaway greenhouse, soon thereafter applied to Venus by Ingersoll and others

1969 – Leovy and Minz build global circulation model for Mars

1969 – Budyko and Sellers use energy balance models to assess climate stability on Earth

1969 – Pollack develops nongrey model of Venus greenhouse effect

1970 – Venera 7 reaches the Venus surface, measures temperature *in situ*

1972 – Mariner 9 reaches Mars, discovers polar layers and traces of past river flow

1972 – Sagan and Mullen pose modern version of the "faint young Sun" problem

1972 – First measurements of infrared flux from Titan pose climate conundrum

1973 – Astronomical climate change on Mars explored as an explanation for polar layered deposits

1974 – Hansen and Hovenier demonstrate that Venus clouds are sulfuric acid

1974 – Mariner 10 photographs Venus cloud structure in the ultraviolet

1975 – Veneras 9 and 10 return first pictures from the surface of another planet; measure winds

1978 – Pioneer Venus measures thermal and radiance structure of Venus' atmosphere

1979 – Voyager spacecraft fly past Jupiter

While the exploration of the airless Moon quickly became the immediate focus of attention of the superpowers' space programs, the scientific question of conditions on our neighboring planets was of great general interest. Venus was the easier to reach, and while there was radio evidence that its surface was hot, it wasn't yet known for sure how inhospitably hot it might be, or why.

A young Carl Sagan argued that a thick atmosphere with a strong greenhouse effect was the most probable explanation, calculating that 4 atmospheres of water vapor and carbon dioxide might do the job [141].[1] The first successful planetary mission, NASA's Mariner 2 in 1962 (Figure 5.1), made microwave measurements that supported that interpretation.[2] Specifically, the two main ideas for Venus' radio brightness were a hot surface or some kind of ionospheric emission (Jupiter was known as a source of radio noise due to energetic particles trapped in its magnetosphere), and these two ideas could be neatly discriminated by measuring the radio brightness with a much higher spatial resolution than could be made from the Earth. Ionospheric emission would be strongest at the limb of Venus with a darker center, whereas a warm surface would be brightest looking straight down at Venus' center and would dim at the edges (limb-darkening). (In fact, similar ambiguities about warm tropospheres versus hot upper atmospheric emission would confront scientists studying Titan a decade later.)

Histories of planetary exploration usually suggest Mariner 2 neatly solved this question at a stroke. That wasn't quite true – Mariner 2 observed at two radio wavelengths and while indeed the longer wavelength was limb-darkened, suggesting a hot surface, this trend was not so obvious at the shorter wavelength, so some small doubt remained.

In fact, ingenious groundbased radio observations in 1965, combining signals from two 30-m radio telescopes at the Owens Valley observatory in California (and combining the efforts of an unlikely, for this cold-war period, pair of American and Soviet astronomers, P. Clark and Arkady Kuzmin [142]) to attain high resolution and to measure the polarization of the radio emission. This work not only gave a very good estimate of the surface temperature at 630 ± 70 K, but also measured the diameter of the hot radio-emitting region to be 6057 ± 55 km (the true average values are 735 K and 6052 km). This radius for the solid surface was about 63 km smaller than the "ephemeris" diameter, measured optically: it followed that the cloud tops, known from infrared measurements to be at 240 K,

---

[1] It is odd that this now-famous study was only published as an internal NASA document. Sagan's ample writings thereafter perhaps made up for this report's obscurity.

[2] Mariner 1 was lost in a launch accident 5 weeks earlier.

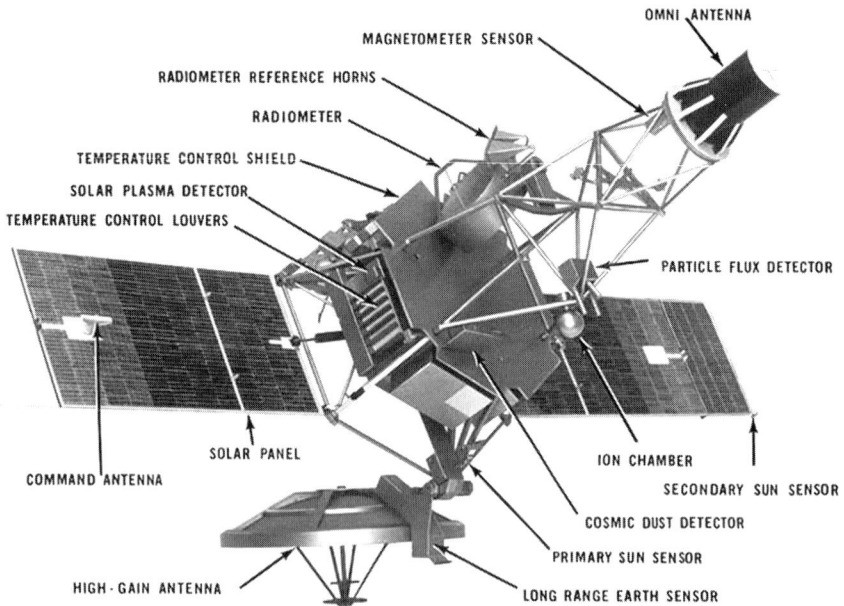

Figure 5.1 Mariner 2 spacecraft with components labeled. The radiometer is the infrared instrument used to measure the cloud-top temperatures: like those on Magellan and Cassini decades later, the microwave radiometer used the high gain antenna whose primary purpose was to transmit the data to Earth. (Source: NASA/JPL.)

were 63 km above this and so the lapse rate should be about 6.5 K/km – rather similar to the Earth.

Even ignoring the ionospheric emission idea (which lingered for several more years), the explanation of the high surface temperature took some time to determine robustly. Öpik had the rather odd idea that wind friction near the ground could deposit heat, yielding warm temperatures caused by a greenhouse effect due to lofted dust [143].[3] This theory, while wrong, had a certain self-consistency. Öpik argued that there wasn't enough water vapor to provide hundreds of degrees of greenhouse warming, and that CO_2 didn't do the job either (the best guess, due in part to imperfect estimation of the pressure at the cloud-top level, was that the surface pressure on Venus might be 5–10 bar). But perhaps dust clouds could. However, if dust could block the heat coming up from Venus' surface, it would be even better at

---

[3] Not only was this wrong, it doesn't even make a lot of sense to me. But science – especially planetary science, where our intuition and terrestrial experience sometimes fail us – often benefits from having a few ideas out on the "lunatic" fringe, just to test the boundaries of the possible and make sure that more fashionable thinking hasn't missed something.

blocking sunlight heading down there. So there had to be a process to deposit heat at the surface. Öpik pointed out that about 2% of Earth's sunlight is deposited at the surface as wind friction – in fact, an effect ignored in some climate models even today – and that perhaps a similar amount could be deposited at Venus. Wind friction in a thick atmosphere might be effective at kicking up dust, so while the idea lacked some basic details (such as how heat deposited at the cloud tops was supposed to stir winds at the surface) it was difficult to refute.

The dust idea was pursued a few years later by James Hansen [144], a Ph.D. student at the University of Iowa. While work by Seymour Hess and others had dismissed the ability of winds to provide enough heating, Hansen ignored that and wondered if internal heat flow might be high enough to inject heat under the thick dust blanket: his interest was in the scattering and absorbing properties of particles in the atmosphere. At the time, Sagan and Pollack, making a more detailed attack on the gas greenhouse effect, were still struggling to find a strong enough greenhouse from gas and water clouds. This competition of ideas is ironic, since Hansen was to go on to become a senior scientist at NASA and prominent – even strident – in his warnings about the effects of rising greenhouse gas concentrations on the Earth.

The idea that cloud scattering alone, without greenhouse gases, could provide enough greenhouse warming was tackled at great analytic length by Bob Samuelson, a scientist at Goddard Space Flight Center (this work, like Hansen's, also in 1967) [145]. He found that an arbitrarily thick layer of cloud with the right scattering properties could let in enough sunlight down to the surface to avoid the need for Öpik's wind friction, but warned that water clouds might not work because of their volatility, so he continued to entertain the idea of dust clouds.

The presence of dust or haze was also suspected on Saturn's moon Titan in 1972 by Danielson and Caldwell [146].[4] Titan was known to have an atmosphere with some methane, but how much, and were other gases like hydrogen or nitrogen present? If there were a lot of gas, Titan should look blue, like Earth or Neptune, but it was an orange-brown. However, absorption by suspended dust could suppress the Rayleigh scattering by gas molecules (which makes our sky blue): such dust would also locally warm the upper atmosphere, creating a warm stratosphere just as ozone does on Earth. The polarization of light reflected by Titan also hinted at an opaque cloud and Sagan argued that the "dust" clouds were made of organic chemicals formed from the destruction of methane by

---

[4] Sometimes this aerosol was referred to as "Danielson dust", or "Axel dust" after Leon Axel who had proposed a similar elevated dust layer at Jupiter shortly before this.

ultraviolet light or lightning discharge [147]. Such a process would also liberate hydrogen, which had been detected tentatively by Lawrence Trafton.[5]

Meanwhile, at Mars, flyby measurements beginning with Mariner 4 in July 1965 indicated that the atmosphere was terribly thin, with a surface pressure only about one percent of Earth's, and that surface temperatures were accordingly quite low. Further, the first pictures showed a cratered, Moon-like surface. These findings finally laid the Lowellian dream to rest.

The technique used to measure the surface pressure was rather elegant – so-called radio occultation. Just as light bends when crossing from air into water, so a light beam or radio beam is deflected slightly (refracted) as it passes into denser air from a vacuum. When a radio transmitter with a precise frequency passes behind a planet with an atmosphere, the frequency history measured on the ground is a transformation of the refractivity of the gas that the signal has passed through, and so, after taking the elaborate geometry into account, a profile of the density of the gas can be obtained. With some assumptions about composition, the pressure and temperature from the surface up to some altitude (limited by the precision of the frequency measurement and other effects) can be constructed.

Unlike the seemingly never-ending arguments in planetary science about the interpretation of near-infrared spectra, the elegant radio-occultation technique yields rather robust results. Even a brisk analysis of the Mariner 4 data, the first time this measurement was ever made (Mariner 3 was supposed to fly to Mars too, but failed en route), published less than two months after the encounter, could confidently declare that the surface pressure at Electris (about $55° S$) was 4.1–5.7 mbar, and it had a temperature of $180 \pm 20 \, K$ – in other words, the density was almost 100 times thinner than that of air on Earth (Figure 5.2) [149]. (These values were for a pure CO_2 atmosphere; allowing 20% of nitrogen or 50% of argon broadened the range to 4.1–7 mbar, and allowed a little more slop in the temperature, but the overall conclusion didn't change.

It was soon realized that the measurement had good and bad news for Mars exploration. The surface pressure of ~5 mbar – the same as that at about 34 km altitude on Earth – was rather lower than the 10–20 mbar that NASA had been planning on, and a thinner atmosphere would make landing more challenging (parachutes would need to be bigger, for example). On the other hand, the profile showed that the upper atmosphere was quite cold – CO_2 being a good absorber of infrared energy means it is also a good emitter, and so (as on Earth) CO_2 cooled

---

[5] The papers by Trafton and Sagan, as well as many others, together with an illuminating record of the discussions that followed, are compiled in a comprehensive proceedings edited by Don Hunten [148].

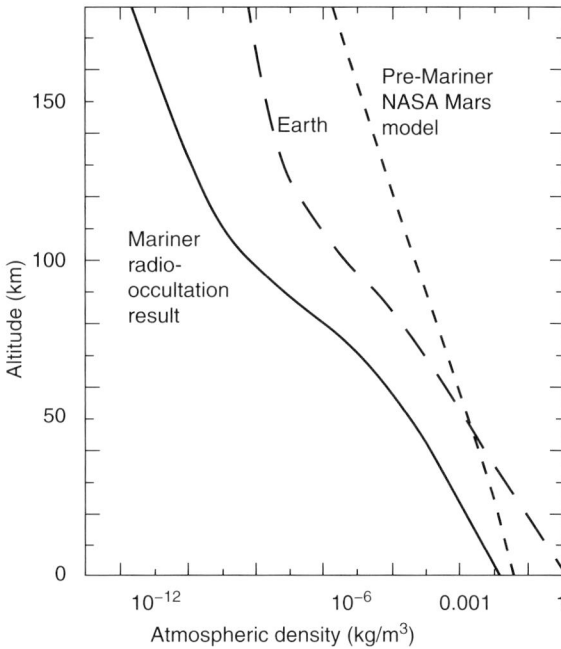

Figure 5.2 The density profile of the Mars atmosphere determined from the Mariner 4 radio-occultation experiment. The NASA pre-Mariner Mars model (left) crossed the profile of the Earth because it was expected that in Mars' low gravity, the scale height would be larger and so the slope on this graph would be steeper, the density falling off more slowly with altitude than on Earth. In reality, the CO_2 cooling of Mars' upper atmosphere keeps the scale height small and so the density is everywhere lower than that of Earth at the same altitude. The slope of the density graph indicates the scale height, which in turn can be mapped (with assumptions about composition) into temperature. (Author, from data in Kliore et al. [149].)

the upper atmosphere more efficiently, making the scale height smaller.[6] Thus, even though Mars' gravity is less than Earth's, which would cause the atmosphere to tail off more gradually into the vacuum of space, the cold temperatures counteract this. The density of the upper atmosphere is a limiting factor on satellite orbits – if the density is too high, the drag caused by the air will sap energy from the orbit, bringing the satellite down to a fiery entry before it can complete its mission. Whereas previous models suggested that satellites would need to orbit at several

---

[6] The scale height is the height increment over which the atmospheric density decreases by a factor of $e = 2.718$. At the Earth's surface it is about 8 km, so the density at the top of Everest (~10 km) is a bit less than $1/e$ times that at sea level. The Mariner 4 data showed the scale height near the Martian surface to be 8–10 km.

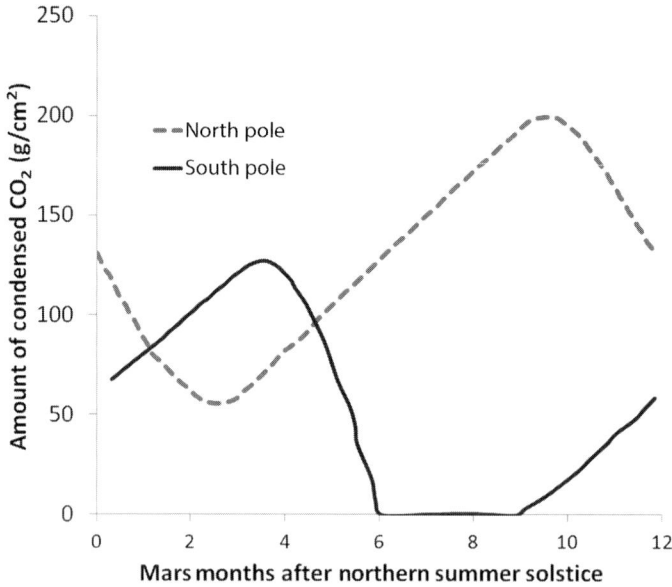

Figure 5.3 By balancing the rate of CO_2 condensation or sublimation against the balance of sunlight and heat radiated away from the Martian surface, Leighton and Murray were able to predict the key features of the Martian seasonal cycle, and in particular the seasonal frost caps. Their model also predicted the atmospheric pressure would go up and down twice each year, as Viking observed later. (Figure by author, using data from [151].)

thousand kilometers altitude to have orbits with lifetimes of 50 years or more, the Mariner 4 data showed that only a few hundred kilometers would be enough [150].

Bob Leighton and Bruce Murray at Caltech in the USA in 1966 soon proposed an elegantly simple model of the Martian climate wherein the thin CO_2 atmosphere was in equilibrium with polar deposits of CO_2 frost (Figure 5.3) [151]. This model (an energy balance model without explicit heat transport) correctly predicted that the Martian atmospheric pressure would have a pronounced seasonal variation. It also determined that there must be very little water vapor in the system, but that deposits of water ice as permafrost might be extensive.

Thus, while debates about habitability continued (indeed, go on to the present day) the key features of the climates of the two closest planets were determined within a few years of the start of planetary exploration. Meanwhile, our understanding of our own climate continued to improve.

While operational forecasting models were generally limited in extent, research into fundamental properties of numerical simulations of the terrestrial global circulation was underway. Syukuro Manabe and Robert Strickler at the US Weather Bureau introduced in 1964 a trick that rapidly became commonplace in climate modeling of all worlds [152]. This related to solving the vertical profile of temperature in an atmosphere as a function of the absorption and rejection of heat by gases, clouds and the surface. As Simpson had warned Callendar, the problem is not just of radiation, but also of convection. At ever-increasing depths in a greenhouse atmosphere, the temperature gradient due to radiation balance becomes steeper and steeper. In a real atmosphere, this temperature gradient becomes unstable, such that the warm lower air buoyantly rises to transport heat by convection, like a pot of soup heated from below. And so Manabe and Strickler introduced the scheme of "convective adjustment" – when the radiation balance indicated a gradient (lapse rate) steeper than some critical value (in the case of the Earth they chose 6.5 K/km) then the profile was clamped to that gradient and the radiation balance recomputed. The profile (Figure 5.4) would then be adjusted again until the fluxes balanced. This procedure has some physical basis, and the obvious virtue that it yields profiles that are similar to those observed, and quickly became the standard approach in atmospheric modeling.[7]

Several scientists were now examining the CO_2 greenhouse warming on Earth, and in 1965 a report by an august panel to US President Lyndon B. Johnson noted the likely effects of CO_2 rise [153]. This report overall was on environmental pollution (Rachel Carson's *Silent Spring*, raising awareness of the ecological impact of pesticides and stimulating the environmental movement more generally, was published in 1962) but called attention to consequences of CO_2 rise specifically (Keeling and Revelle – see Chapter 4 – were among the authors). The report noted some positive effects (e.g. increase in photosynthetic production by plants) as well as the negative ones such as sea-level rise. The report also conceded that the warming observed by Callendar had not continued during 1940–60, such that other factors – climate "noise" – must be masking the effect of CO_2.

Further, while the warming effect of CO_2 alone was somewhat straightforward, the effect of the CO_2 warming on the water vapor greenhouse and thus the overall climate sensitivity was not so obvious. F. Möller in Germany in 1963 suggested from surface energy balance calculations that doubling CO_2 from 300 ppm to 600 ppm might increase temperatures by 1.5 K (rather less than the ~3.8 K predicted by Plass in 1956), but if the relative humidity were

---

[7] There is of course a dangerous circularity here, not discussed in polite society, in that the choice of critical lapse rate in convective adjustment is somewhat arbitrary – it will depend on humidity for example, and so should not be the same everywhere. In the modeling world, these representations of processes such as convection that are not simulated explicitly are called "schemes", a word that has, perhaps appropriately, a slightly dishonest connotation. But these are often the best we can do.

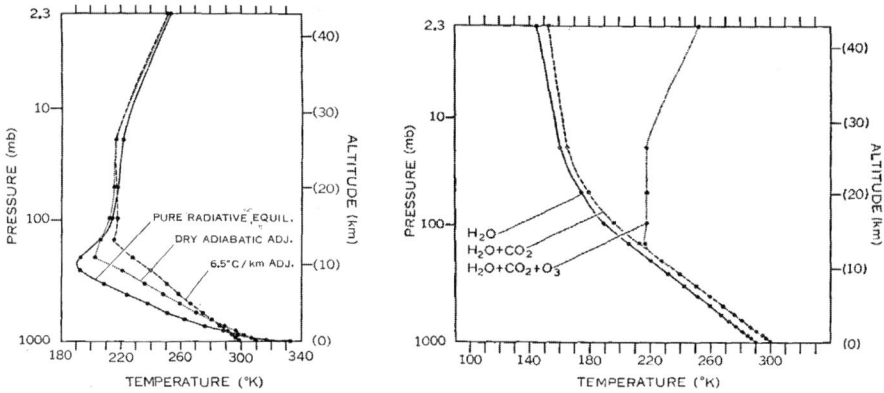

Figure 5.4 Results from Manabe and Strickler's radiative–convective model showing (left) the steep gradient predicted by purely radiative equilibrium, and the more realistic profile generated by clamping it to a convective lapse rate in the lower part, with two possible lapse rates shown – the empirical 6.5 K/km and the dry adiabatic rate of 10 K/km. Their model experiments (right) show the warming contribution of CO_2 throughout the atmosphere, and the effect of ozone in heating the stratosphere. (From Manabe, S. and Strickler, R.F., 1964. Thermal equilibrium of the atmosphere with a convective adjustment. *Journal of the Atmospheric Sciences*, **21**(4), pp. 361–385 © American Meteorological Society. Used with permission.)

fixed such that this temperature rise added more water vapor greenhouse effect, the rise could be some 10 K [154]. His tone was not one of alarm, however, but rather that climate change calculations couldn't be trusted because small changes in cloud or moisture could have a much bigger effect.[8] Manabe and Wetherald found in 1967, with their convective adjustment model (which accounted for the full profile of the atmosphere, not just the surface), that Möller's extremes were not reproduced, but that indeed cloud and water vapor assumptions had a huge effect [155]. Their headline result, though, was that a fixed relative humidity gave a higher response (~2.3 K) to CO_2 doubling than did a fixed water vapor amount (~1.3 K).

The intent of Manabe's work was that these vertical calculations would be incorporated into general circulation models (the presidential report noted that the relative effects of water vapor and CO_2 would be different at high latitude, for example, and so clear predictions needed to combine the approaches). But the dynamical models had uncertainties too – a surprise discovery was that the equations used to describe fluid motion had fundamental limitations on the predictability of their outcomes. In 1961 Edward Lorenz[9] at MIT found that even

---

[8] To quote his paper: "the entire theory of climate change by CO_2 variations is becoming questionable."

[9] No relation to myself, as far as I know.

a simple set of nonlinear equations gave very different results after marching forward when the initial conditions were changed only slightly [156].[10] Since there are limits on the accuracy with which the starting condition of a simulated atmosphere can be measured, and limits too on the numerical precision with which the atmosphere can be represented on a computer, it followed that chaotic fluctuations in results would eventually drown out any predictive ability of a model.

Nonetheless, there would be plenty of room for numerical models to improve before they hit such limits, and much practical progress was made on modeling weather and climate: from minutiae such as the tricky problem of how to divide up a spherical world into the little square boxes on which the equations like to operate,[11] to understanding better the absorption and scattering of light and heat in the atmosphere, to (for Earth) coupling models of the atmosphere to models of the ocean and the biosphere (the first coupled ocean–atmosphere model was by Manabe and Bryan in 1969 [158]).

The first application of a GCM to Mars was in fact reported very soon after the basic properties of the Martian atmosphere were found by Mariner 4: Conway Leovy and Yale Minz in southern California simulated the wind patterns on Mars with a two-layer model (adapting a terrestrial model developed by Mintz and Arakawa), dividing the globe into boxes $9°$ in longitude and $7°$ in latitude, 922 boxes in all [159]. They found some basic features of the present-day circulation, including the overall zonal wind pattern and wave cyclones in the middle and higher latitudes. They also found a large diurnal tidal component of the circulation as the atmosphere warmed up and cooled down over the course of the day – a much more prominent effect in the thin Mars atmosphere than at Earth.

However, for questions of deep climate, rather simpler models were still the most useful tools, on Earth and Mars. Here the important factor to book-keep is not the atmosphere's momentum and arrangement of pressure systems such as cyclones (the basis of short-term weather forecasting) but the budget of heat at different locations, taking into account the reflectivity of the ground, the cloud cover and so on. Progress in this direction was made by the climatologist Mikhail Budyko in St. Petersburg, with his book *Heat Balance of the Earth's Surface* in 1956. Julian Adem of the US Weather Bureau in 1964 showed that this approach could predict temperatures months or seasons in advance.

It was Budyko who applied simple energy balance models in 1969 to study the stability of the Earth's climate. Russia feels its winters keenly, and in the three or four

---

[10] The story of this discovery and its wider impact is well-told in Ref. [157].

[11] Innovations in this and many other areas were made by Akio Arakawa; like Syukuro Manabe, he graduated in the post-war years from the University of Tokyo, and moved to the USA.

decades after 1940, the Earth's average temperature fell slightly (superimposed on the general warming trend of the past century or two) which some took to presage the onset of a new ice age. Budyko and others were keen to explore how the evolving climate might behave – it might affect them directly. Russia was also considering using giant mirrors in space to illuminate its frozen north during winter.

Budyko used measurements and estimates taken from the ground to provide the absorbed sunlight numbers for his model. At each latitude he knew the average sunlight absorbed, and he parameterized how the outgoing heat radiation varied as a linear function of temperature.[12] Budyko made a crude guess at the influence of heat transport by the ocean and atmosphere by adding a term to the heat balance that was proportional to the difference between the local temperature and the global average.

The aspect that caught people's attention in Budyko's 1969 work was his incorporation of a feedback in his model. Specifically, he recognized that when the sea got much below the freezing point it would form sea ice, and that sea ice was much more reflective of sunlight than is open water. This increased reflectivity or albedo would therefore tend to reduce the amount of heat absorbed by the climate system, and via the heat transport incorporated in the model, this would also cool adjacent latitude belts. These might then freeze over too, and the ice cover would expand (as noted by Croll, in Chapter 2). In fact, he found that if the ice line were nudged just too far equatorwards (by reduced sunlight, or a weaker greenhouse effect), the ice would expand catastrophically, jumping all the way to the equator and plunging the planet into a deep-freeze, a runaway glaciation. This ice–albedo feedback introduced hysteresis into the system – once it had been caused by a perturbation such as reduced sunlight, removing that perturbation didn't allow the system to go back to its original state – there were multiple stable states,[13] and knowing which one it should be in relied on knowing the history – the system had a "memory" (Figure 5.5). Quantifying such feedback effects is one of the principal challenges in climate modeling. Similar calculations were performed by William Sellers at the University of Arizona, also in 1969, who determined that a drop in sunlight by 2–5% might precipitate a new ice age. Although others had also used similar models before (and, specifically, Öpik found in his 1965 *Icarus* paper [160] that while the ice line was a continuous function down to $50°$ N, with 88% of present luminosity, "*for lower*

---

[12] In the absence of an atmosphere, as Stefan showed in the 1870s, the outgoing flux is proportional to the fourth power of temperature. But the greenhouse effect and other factors modify the function. Rather than simulate all these (not always well-known) effects explicitly, it is common to make a linear approximation with a proportionality of about $2\,\mathrm{W\,m^{-2}\,K^{-1}}$. A danger in climate modeling, especially at other planets, is to forget that such approximations may only be valid for specific situations.

[13] In 1961 the oceanographer Henry Stommel found that a simple model of the thermohaline convection in the ocean also exhibited this kind of bistability [121].

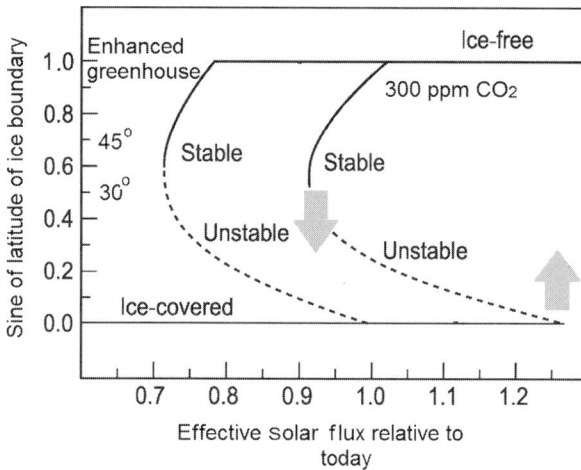

Figure 5.5 A schematic of hysteresis in the climate system from a Budyko–Sellers type model. Starting at upper right, on an ice-free world with a similar atmosphere to today and stronger sunlight, reducing the solar constant leads to a gently cooler planet and the ice cap grows along the solid, stable line down towards about 30 degrees latitude, and increasing the solar constant would cause the ice to retreat back upwards. If, however, the solar constant were to fall to less than about 92%, the ice sheet would grow unstably (jumping as shown by grey arrow) to completely cover the whole planet (i.e. "Snowball Earth"). At this point even increasing the forcing back up to 20% higher than today does not cause the ice to retreat (there is "hysteresis" in the system – the state depends on the system's history as well as the present forcing (what Lorenz called an "intransitive system"). Only by pushing the system to 27% higher does the ice retreat, and does so unstably – jumping to a totally ice-free state. Enhancing the greenhouse effect pushes the curve to the left as shown, but the dynamical behavior is the same. (Figure by author.)

values there is a breakdown . . . resulting in global ice cover") this type of model became known as a Budyko–Sellers energy balance model.[14]

The thought that cooling might run away was topical because, while carbon dioxide and greenhouse warming in the Earth's atmosphere was slowly creeping up, there were indications that the amount of aerosol (dust, sulfates, and carbon) scattering and absorbing sunlight to cause surface cooling was increasing much faster, due in part to coal burning and forest clearing. (The initial evidence was rather indirect: aerosols affect the electrical conductivity of air, and measurements

---

[14] For example Opik [160] uses an energy balance model with parameterized transport and essentially articulates the faint young Sun problem. He also argues that orbital variations cannot explain the glacial history of the Earth correctly, and suggests fluctuations in solar output are responsible for climate changes. Another paper by Eriksson [161] adopts a similar approach and speculates that solar variations might be responsible, and explicitly describes the hysteresis in the model, noting that "*the system thus behaves in the manner of a bistable multivibrator, or 'flip flop' mechanism*". An even earlier rumination on meridional heat transports is by Fritz [162].

by research ships on ocean cruises in the 1960s yielded results that differed from those recorded by the research ships *Galilee* and *Carnegie*[15] in 1908 and 1929) [163]. Stephen Schneider and Icthiaque Rasool at the Goddard Institute for Space Studies (GISS) in New York pointed out in 1971 that, if aerosol pollution increased by a factor of 6–8 over the next 50 years, it would have a much stronger effect than the CO_2, resulting in a net cooling of the Earth's surface by 3.5 K, "*enough to trigger an ice age*" [164].[16] (Although a number of scientific and popular reports warning of a new ice age appeared in the 1970s, in fact the scientific consensus was even then towards greenhouse warming [165].[17])

Around the same time, feedback effects also began to be recognized in the climates of other worlds. Several scientists had recognized the possibility that if condensed reservoirs of greenhouse gases (such as water) existed on a planetary surface, then increasing temperatures would evaporate more of it and lead to stronger greenhouse warming, perhaps to boil off yet more vapor and accelerate the process (e.g. [166]). M. Komabayashi in Japan laid this idea out succinctly in a short theoretical paper in 1967 [167], while Andrew Ingersoll at Caltech independently published a slightly more elaborate treatment, with specific application to Venus in 1969 [168]. In addition to noting that a "wet" atmosphere had an upper limit on the amount of heat it could radiate to space (actually a result discovered in 1929 by George Simpson), Ingersoll observed that this runaway greenhouse mechanism showed that an ocean-covered world at Venus' distance from the Sun would boil dry, and explained the fundamental difference between sister planets purely as a function of early Venus' stronger solar heating. Rasool and de Bergh at GISS also noted that the hot surface temperatures on Venus would prevent liquid water from mediating the silicate weathering of carbon dioxide, so that gas would progressively accumulate, forming the dense greenhouse atmosphere observed today [169].[18]

Peter Gierasch and Owen Toon, then at Cornell University, noted in 1973 that the Leighton–Murray model of Mars' climate ignored heat transport in the

---

[15] In fact the *Carnegie* caught fire and exploded in Samoa on its cruise in 1929, killing her captain.

[16] Around this time the Royal Swedish Academy of Sciences hosted a study of Man's Impact on Climate (SMIC), published by the MIT Press as *Inadvertent Climate Modification*, 1971; this volume is a nice summary of the state of the art in climate science at the time.

[17] Perhaps the deep-sea core data by Hays and others (see later) confirming the ~20-kyr glacial cycle, last ending ~20,000 years ago, was a factor. I remember, myself, following the biggest snowfalls I'd seen in my life (I was 7) in winter of 1976, news media talking about ice ages. There is even some quite good science fiction on the topic: D. Orgill and J. Gribbin *The Sixth Winter*, Macdonald, 1979, describes the sudden onset of an ice age. Scientist/journalist Gribbin also edited a volume on climate change around this time.

[18] Rasool, from Pakistan, had strong connections with France, but his main impact was in NASA, initially working on ozone monitoring on Nimbus, on Tiros-1 and on the Venus greenhouse effect. As Deputy Director for Planetary Programs in 1971 he selected the instruments for the Voyager mission.

atmosphere, which has the effect of limiting the minimum temperatures encountered in polar winter. Since it is these lowest temperatures that control the pressure of the atmosphere, and the heat transport depends on the density (i.e. pressure) of the atmosphere, these coupled effects could lead to an instability, wherein the atmosphere "flips" from a thin "collapsed" state to a thicker one. Like the ice–albedo feedback, removing the perturbation that kicked the system from one state to the other didn't mean the system would flip back – this history-dependent characteristic was termed by Lorenz an "intransitive system", and greatly challenges the ability to attribute climate changes to specific causes; e.g. [170, 171].

The problem of understanding planetary climates was now under attack on several fronts – the worlds were being divided up ever more finely, into belts and boxes each with their own albedo values etc. (much as Halley had recognized would be necessary), and the physical processes such as condensation and freezing were becoming understood and incorporated. But whereas for Mars' thin atmosphere, simple models of the greenhouse effect sufficed (or indeed, it could be ignored altogether), for Venus and Earth, the book-keeping of absorption of infrared radiation by the atmosphere was critical.

The point here is that the blanketing effect of infrared absorbing gases is not at all simple. For a single "color" or wavelength of light, adding a small amount of an absorbing gas will produce a corresponding small reduction in the amount of light that will get through, and the effect is initially linear. But it doesn't stay that way – imagine that some amount of absorption will cut the light in half: adding that amount of gas again will only produce a 25% reduction in the light that gets through, an exponentially decreasing transmission effect that is named Beer's law.[19] So far, so good. But the problem is this applies only to a single wavelength of light, and the variation of absorption with wavelength is very complex for molecules such as water and CO_2, and, moreover, depends on the pressure and temperature conditions (which of course depend on altitude). The problem gets even worse when there are wavelengths at which both gases absorb. So better quantifying the greenhouse effect relies on dividing the spectrum of sunlight and radiated heat into ever-narrower wavelength bands, and the atmosphere into thinner layers, with the attendant computational complexity that parallels the GCM challenge of dividing the world horizontally into boxes. Then the different compositions of the layers must be taken into account – while some gases such as carbon dioxide are generally well mixed, water vapor is not. And then knowledge of the absorbing properties of the gas at relevant conditions – which demands careful laboratory work – must be obtained.

---

[19] Named after August Beer circa 1852, not to be confused with Wilhelm Beer, who made the first globe of Mars a couple of decades before. Beer's law is sometimes called the Beer–Lambert law, or even Beer–Lambert–Bouger law, as Pierre Bouger articulated the idea in 1729.

Figure 5.6 One of the Nimbus satellite series that made fundamental observations of the Earth's atmosphere and climate, including outgoing thermal radiation to understand the greenhouse effect, and observations of ozone abundance. (Source: NASA.)

In this epoch, measurements of the outgoing thermal radiation from the Earth had been compiled by instruments on, among others, the NASA Nimbus series of satellites (Figure 5.6) – and indeed the same measurements were being made at Mars by the Mariners at the same time (Figure 5.7).

While the first models of planetary climate (such as those by Sagan, Komabayasi and Ingersoll) usually consider infrared light averaged over the whole spectrum, such that there is only one "color" of heat (and since a perfect emitter is a color-neutral "black body", such a single-color model of how an absorber reduces the emission is called a "grey" model[20]) accurate models need to explore the full rainbow of heat. Much of Arrhenius and Callendar's work was devoted to quantifying the different absorptions of CO_2 and water vapor on Earth, using the best data to hand (and considering a handful of wavebands – e.g. Callendar used eight). Recognizing the difficulty of accurately modeling Venus' giant greenhouse effect with grey methods, Jim Pollack[21] made a nongrey model [172] to explore the contributions of water vapor, carbon dioxide and nitrogen at different wavelengths. He assumed an atmospheric composition compatible with the best data available in 1969,

---

[20] Strictly, a grey model has the same opacities for heat and light. Usually the values are different, and the model is called "semi-grey", but the "semi" is often omitted. Radiation propagates in all directions, but often for simplicity the atmosphere is considered to be a thin slab, so that only an up and a down direction need to be book-kept, a scheme called the "two-stream approximation".

[21] Pollack was Sagan's first graduate student.

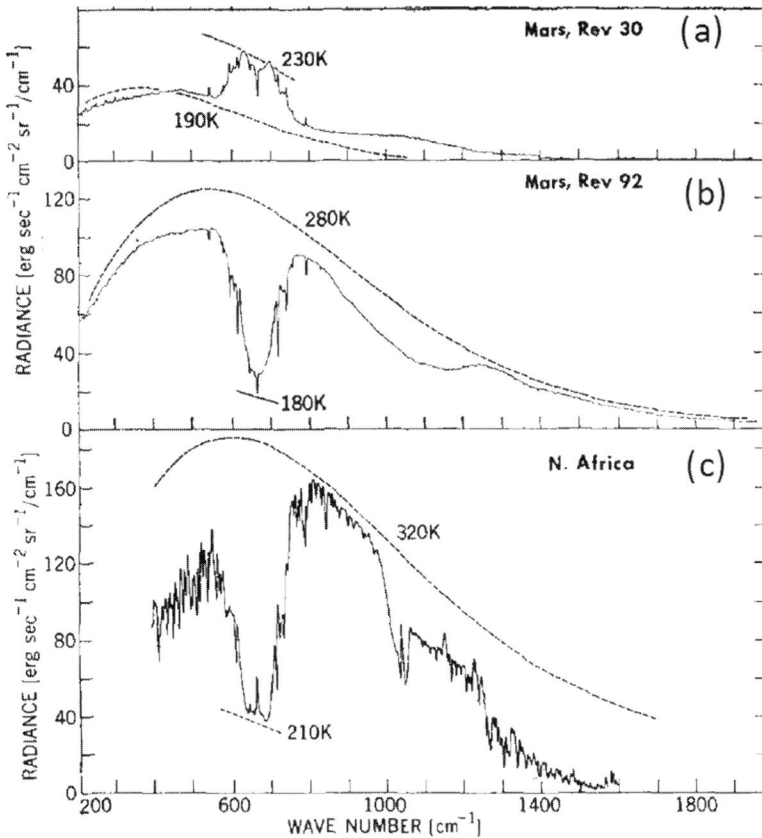

Figure 5.7 Thermal emission of Mars (a,b) recorded by the IRIS infrared spectrometer on Mariner 9, compared with data from a similar instrument at Earth, on the Nimbus 4 satellite (c). Black-body (Planck function) emission curves for different temperatures are shown. In (c) a relatively clear "window" in the atmosphere at 800–1000 wavenumbers indicates the surface temperature of nearly 50 °C (323 K), while the middle of the CO_2 absorption band at 650 cm^{-1} (15 μm[22]) radiates at roughly the tropopause temperature of −60 °C (213 K). On Mars (b) a somewhat similar pattern is seen, although both temperatures are much cooler. In (a), showing the north polar hood (water ice cloud) during polar winter, the CO_2 band radiates at a temperature higher than the underlying emitters, and so the band is seen as a bump rather than a trough. These data can be interpreted to infer the vertical temperature structure of the atmosphere, as well as information about haze and clouds. (Source: NASA.)

---

[22] Infrared spectroscopists sometimes use "wavenumbers" or inverse centimeters to express wavelength. There is some climate justification for showing spectra in this way, as it more properly shows the energy distribution, although the tradition comes in part from the use of such spectroscopy in the chemical laboratory, just because it is easier to talk in integers. One can fit about 650 lengths of 15 μm in one centimeter, hence "650 wavenumbers".

of 90% CO_2, 10% N_2 and 0.5% water vapor. This model broke down the absorption and radiation of infrared radiation not only by layers in the atmosphere (as had Simpson's of Earth, 40 years earlier – i.e. a 1-D radiative transfer model), but also by wavelength. In many respects this exercise was the prototype for most planetary climate models. Pollack found that the contribution of N_2 was unimportant.

But the surface conditions were not finally known for sure until 1970, after Pollack's first model efforts.[23] In 1967, the Soviet probe Venera 4 made the first measurements of temperature and pressure from within Venus' atmosphere, but its results were initially confused – it was thought at first to have reached the surface, but in fact failed 22 km or so up, at a pressure of 18 bar – it had been designed considering the more hopeful interpretations of Earth-based remote sensing suggesting a surface pressure of 5–10 bar. The US Mariner 5 spacecraft made a radio-occultation experiment at Venus, much like that of Mariner 4 at Mars, but the signal could not penetrate all the way to the ground in Venus' thick atmosphere, fading to noise at a pressure of 7 bar. Its temperature profile at least matched that of Venera 4, which made it to 18 bar. These results confirmed that the radio astronomy indications of a hot surface were correct, but quite how hot was still unknown. The Soviets beefed up their probes – already over 400 kg in mass and able to tolerate the crushing 450 g deceleration of entry from space – improved their instrumentation, and made the parachutes smaller so the probes would fall faster so as to reach the ground before their batteries became depleted and their innards too hot. Venera 6 in May 1969 got down to 27 bar, and, beaming its data back to Earth at 1 bit per second, at last revealed the bulk composition of the atmosphere – some 96 percent CO_2, a couple of percent nitrogen, and only a tiny amount of water vapor – less than a part per thousand. The greenhouse effect would have to be re-thought, there was much less water vapor than Pollack had assumed. Finally, toughened up even more than its predecessors, Venera 7 reached the surface in December 1970, smacking into the ground at 17 m/s and confirming the surface temperature to be about 460 °C (733 K).

These were heady times at Venus, with various climate scientists writing several papers a year,[24] and early reports at meetings of spacecraft results were being assimilated into modeling studies before they were confirmed. A similar to-

---

[23] This illustrates one of the principal challenges of climate modeling, which is a fundamentally underdetermined problem. There are often dozens of ways of getting the right answer – sometimes one can be right for the wrong reasons.

[24] A key journal for papers in planetary science, then as today, was *Icarus*, established in 1962. Sagan served as the editor of this journal during the epoch of this chapter, 1968–1979.

and-fro, with many of the same protagonists, was underway at Titan. Gases were clearly doing something at Titan – while observations in 1971 by Dave Morrison and Dale Cruikshank at 20 μm (using a 2.2-m telescope on Mauna Kea in Hawaii) indicated a temperature of 94 K, there were "*significant departures from a black body emission*" [173]. Infrared measurements by F. Gillett at Kitt Peak Observatory near Tucson, Arizona, using 2.1-m and 4-m telescopes showed in the early 1970s that, at some wavelengths in the 8–13 μm range, Titan's atmosphere had a brightness temperature as high as 160 K [174, 175]. That was consistent with the "Danielson dust" model (if dust was fine-grained it would be a poor emitter, and so would need to be warm to reject the heat it absorbed from the Sun). On the other hand, perhaps a deep atmosphere with a strong greenhouse effect could make the surface warm, and this was the temperature being sensed by the 8–13 μm emission. This scenario was much preferred by Sagan [176], who, ever optimistic that a greenhouse due largely to lots of hydrogen might create comfortable conditions at the surface, suggested, "*temperatures as high as 200°K are not excluded*".[25] Don Hunten in Arizona also advocated this greenhouse model. Frank Low and George Rieke (also in Arizona) marshalled the available infrared data in 1974, finding a lack of a hydrogen spectral feature and thereby limiting a greenhouse surface temperature due to that gas to no more than 120 K, whereas if the surface were cooler than 75 K, the methane that was observed would freeze out. They suggested a methane greenhouse with a surface temperature of 80–90 K, with some warm cloud layers above [177]. But Titan is a dim object, and measuring its brightness near the glare of Saturn is challenging; and as with Venus radio observations, the thickness of its atmosphere (and thus the area of the surface responsible for the surface emission) was not and could not be known. Sagan's conclusion "*Because of its high temperatures and pressures and the probable large abundance of organic compounds, Titan is a prime target for spacecraft exploration in the outer solar system*" set the stage for the mystery being resolved much as Venus had been – by a close encounter.

Meanwhile, Mariner 9, the first spacecraft to orbit Mars in 1972, transformed our perspective of that world with new instruments and higher-resolution pictures. In fact it arrived during a global dust storm (Figure 5.8), but after the atmosphere cleared, Mariner observed sand dunes, showing that the atmosphere could sculpt the surface. It also observed the polar caps, finding layered structures (Figure 5.9), suggesting alternating periods of deposition of ice- and dust-rich materials: Mars climate must have had variations not unlike the Croll–Milankovitch cycles on Earth. And indeed studies of Mars' orbit

---

[25] This report was before the 1972 convention that determined the degree symbol should not be used with kelvin.

Figure 5.8 Mariner 9 image of Mars – almost completely obscured by a global dust storm – the three dark patches emerging from the murk are the three Tharsis volcanoes. A thin haze layer is visible on the limb. The regularly spaced dots are reseau marks for geometric correction of images (characteristic of the vidicon cameras used early in the planetary program before modern charge-coupled devices (CCDs) were developed). (Source: NASA.)

suggested that insolation variations with periods of ~95,000 and ~2 million years might occur.[26]

But the most striking discovery from Mariner 9 was the observation of river channels,[27] seen better by the Viking missions a few years later. These demanded an explanation of how, perhaps, Mars could have been warmer in the past. But a past equable climate for liquid water was even more challenging than one in the present, as it was now generally accepted that the Sun evolved in luminosity, by a

[26] Murray et al. [178] is an early report making the connection. More detail on the insolation history is by Ward [179]. A good review of the polar caps is by Byrne [180].

[27] A vast canyon system, roughly the size of the entire United States, was one of Mariner 9's early discoveries, and was named Vallis Marineris as a result.

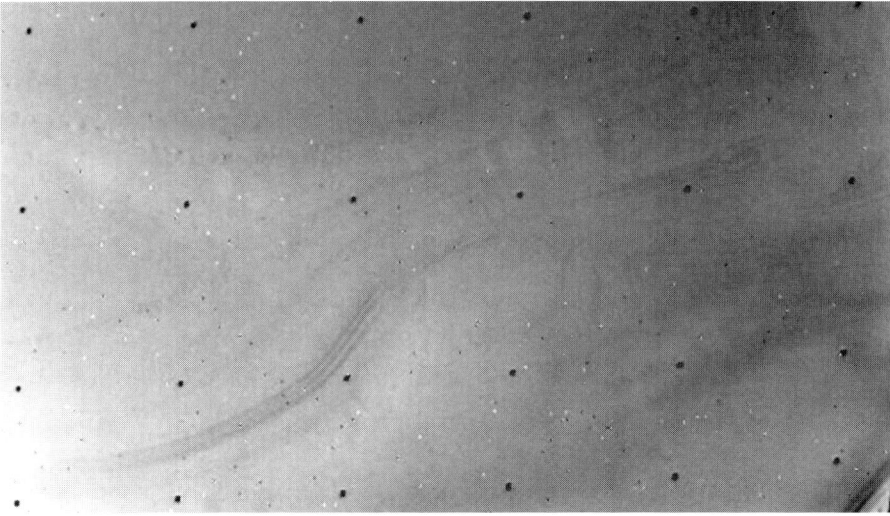

Figure 5.9 Parallel streaks showing layers in the south polar terrain on Mars. Note that the original image quality is rather poor, with only a few shades of grey varying across the scene (the contrast has been stretched in this rendering) and quite a few speckles of noise. Nonetheless, these layers kicked off study of orbit-driven paleoclimate on Mars in 1972. (Source: NASA.)

considerable amount, when geologic time spans of billions of years were considered. Specifically, for the early Earth and Mars 3–4 billion years ago, the Sun may have provided about 25% less light than today – the "faint young Sun" problem, as Carl Sagan popularized it [181].

The sewers of speculation opened up, allowing various evil-smelling gases to offer themselves as prospects for enhancing the early greenhouse on Earth or Mars. Ammonia, sulfur dioxide and hydrocarbons such as methane were all entertained (and still are). Indeed, these compounds were also considered as possibilities for the liquids that carved the valleys on Mars.

Laboratory data on the absorptions of these gases were not as good as for carbon dioxide and water vapor, so somewhat simplified radiative transfer approaches were used. One, by Ann Henderson-Sellers in the UK, used a sort of step-function model of atmospheric transmissivity, to explore the possible roles of methane and ammonia on early Mars and Earth [182, 183].[28] In general, introduce enough of anything and you can fix the problem, since most gases have some contribution to the greenhouse effect; the challenge is now a geochemical one – are such amounts plausible?

[28] Henderson-Sellers later went on to be a prominent terrestrial climate scientist in Australia and in the Intergovernmental Panel on Climate Change (IPCC).

On present-day Earth, better quantification of the greenhouse effect demanded that the contribution of ever-rarer gases needed to be considered [184]. Not just water vapor, carbon dioxide and ozone, but nitrous oxide and even the tiny parts-per-million amounts of chlorofluorocarbons. These molecules, used as refrigerants, aerosol propellant and in fire extinguishers, have a disproportionate absorption in the infrared in windows not absorbed by water vapor and carbon dioxide, making them potent greenhouse gases even at low concentration (Figure 5.10).

The inertness (and thus low toxicity) of the chlorofluorocarbon gases makes them useful, but also makes them long lived in the atmosphere, where they began to accumulate. In fact, their ubiquity, albeit in trace amounts, was discovered in the early 1970s by James Lovelock, a British scientist who hitched a ride on a research trip in the South Atlantic to try out a sensitive gas sensor, the Electron Capture Detector, he had invented.[29] It was only some years later that the role of chlorofluorocarbons in ozone destruction was recognized (the gases diffuse into the stratosphere, where ultraviolet light liberates the chlorine which then attacks ozone).

Lovelock had been a consultant to NASA in the 1960s on life detection experiments. He reasoned that living things rely on (and may enhance) chemical disequilibrium in the execution of the functions of life, and therefore that detection of such disequilibrium might be an indicator of life [185, 186]. Specifically, the Earth's atmosphere contains abundant free oxygen, yet there is also methane. These compounds should (i.e. it is thermodynamically favorable for them to) combine to yield free energy and carbon dioxide, such that both would not be observed simultaneously unless life – in the form of photosynthesizing plants for oxygen and bacteria in marshes and the guts of termites and cows for methane[30] – were producing them rapidly. Conversely, the dominance of Venus' and Mars' atmospheres by thermodynamically dead carbon dioxide offered little prospect for abundant life. The corollary, which Lovelock refined into the "Gaia Hypothesis" with the biologist Lynn Margulis,[31] is that life can make large-scale

---

[29] I saw Lovelock's detector on display in the Science Museum, London, in July 2014. It was a matter of some pride to me that my own little instrument, the Huygens penetrometer (see next chapter), was on display a few meters away.

[30] It is fun to think of cow farts as a methane source – indeed in 1992 when I gave a talk on Titan to an astronomy society on the Isle of Wight, UK, an elderly lady noted, "You said Titan has lots of methane, but doesn't methane come from cows?" – I have in fact since learned that most bovine methane is actually emitted as belches.

[31] Margulis, a brilliant thinker who also developed the idea that mitochondria in cells began as endosymbionts, was married to Carl Sagan 1957–64. The endosymbiont idea, now widely accepted, was at the time seen as somewhat heretical, as was Gaia. The name "Gaia" was suggested to Lovelock by his neighbor, the novelist William Golding. Golding's novel *Lord of the Flies* is one of the few works of fiction I have ever encountered where a character named Ralph is anything like a hero.

Radiation Transmitted by the Atmosphere

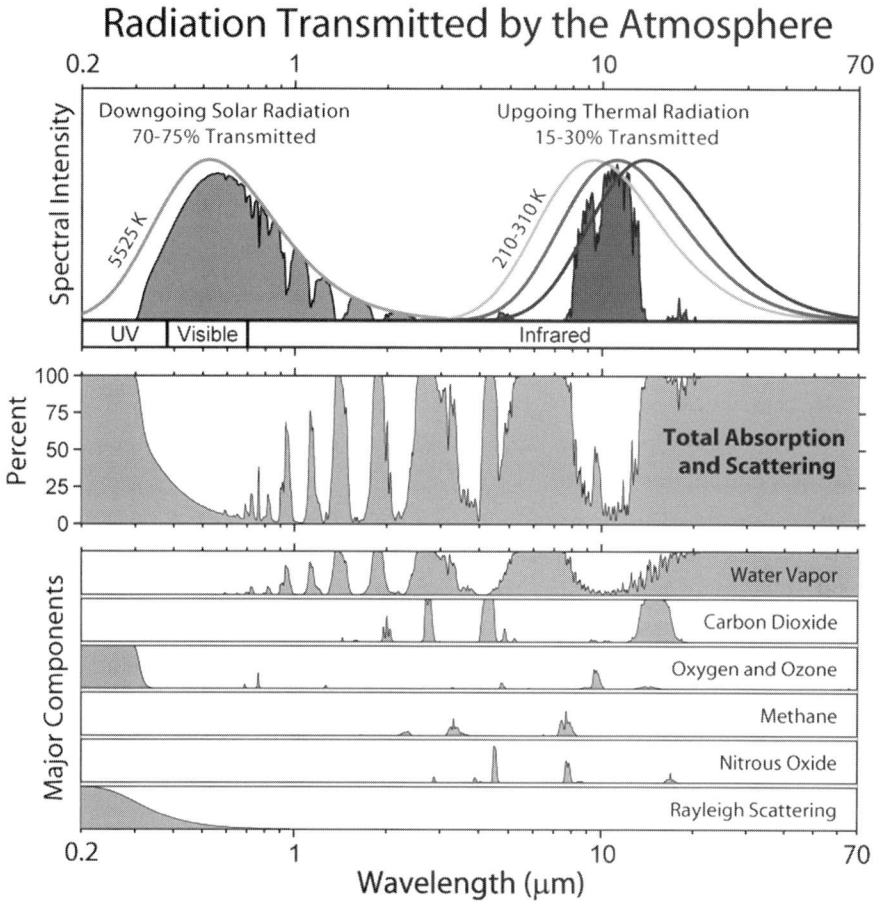

Figure 5.10 The greenhouse opacity of the atmosphere depends on how absorptions by different gases overlap. Water vapor and carbon dioxide are responsible for most of the Earth's greenhouse absorption, but by filling in "windows" not covered by these two, even trace gases can have a significant effect. Ozone is a strong absorber of ultraviolet light (0.3 μm and shorter). Methane has little optical effect at its tiny concentrations in Earth's atmosphere, but its vast amounts on Titan give it much more prominence at short wavelengths there. On cold Mars, there is very little water vapor, making its absorption much less there. (Image by Robert A. Rohde for Global Warming Art project – CC-BY-SA-3.0.)

changes to the planetary environment, a capability that at the time was not a widely held view but today is mainstream bioclimatology/biogeochemistry. The notion – also inherent in Hutton's writings – had been advocated in the 1920s by the Ukrainian scientist Vladimir Vernadsky, who introduced the term "Biosphere". Lovelock also drew attention to the long-term regulation of the salt composition of the sea,[32] and the possibility that biota may have influenced climate via cloudiness mediated by sulfur aerosols produced at the sea surface. The Gaia Hypothesis, as described by Margulis and Lovelock [187–189], is sub-titled "homeostasis of and for the biosphere".[33]

Sulfur proved to be important at Venus too. Whereas its clouds had initially been assumed to be of water, like Earth's, the low abundance of water vapor in the atmosphere ruled that out. The key measurement was a telescopic one, of the polarization of light reflected by Venus' cloud particles. Analysis by James Hansen and others showed that the cloud particles were spherical, had sizes of about 1 μm, and had a refractive index of 1.45, rather higher than water's 1.33 [193]. The answer was that the clouds were made of rather concentrated (~75%) sulfuric acid, which not only had the right optical properties, but would be liquid at the cloud-top temperatures of 230–240 K, and would be consistent with the rather low abundance of water vapor.

Venera 8 in 1972 carried a photometer to record light levels during the descent, and found that the cloud layer didn't stretch all the way to the surface, thinning out below 45 km or so, and that a small percentage of the incident sunlight filtered through the thick clouds to illuminate and warm the surface – this would be enough to drive the greenhouse effect without requiring Öpik's wind friction. Mariner 10, flying past Venus in 1974 on its way to airless Mercury, showed that Venus' bland appearance in visible light belied dramatic cloud structures that could be seen in ultraviolet light (Figure 5.11). The Soviets followed up the success of Veneras 7 and 8 with more elaborate probes (Figure 5.12). The massive Veneras 9 and 10 thudded onto the surface in October 1975, and were able to survive for a torrid hour or so before they were cooked. They carried cameras, showing for the first time the rocky surface. They also carried cup anemometers (the challenge in ensuring moving parts in a 750 K acidic atmosphere should not be underestimated) showing the surface winds were about 0.6–1 m/s.

An obvious aspect of the clouds was that they swirled along rapidly (Figure 5.13), moving east–west some 50 times more rapidly than the surface rotates (the clouds circle the globe in about 4 days, in contrast to the 243-day rotation of the solid planet). This "superrotation" seems a common feature of optically thick atmospheres on slowly rotating worlds – Titan has it too. The question confronting dynamicists is by what process does the angular momentum get distributed in

---

[32] Of course, if exchange or feedback processes (whether biological or not) regulate the saltiness of the sea, then Halley's idea to measure secular changes in this quantity to estimate the age of the Earth was doomed to fail.

[33] Many have written much on the idea and its history; one recent related perspective is Schwartzman [190], but Lovelock's own books [191, 192] are probably the best.

Figure 5.11 The swirling cloud belts of Venus, revealed in the ultraviolet by Mariner 10. (Source: NASA.)

Figure 5.12 Venera 9, with its electronics inside a heavy metal pressure sphere, more like a diving bell than a spacecraft. With some uncertainty about the light levels, a set of lamps was carried just above the ring-shaped landing shock absorber to ensure that pictures could be taken. The spiral antenna on top sent data back at 256 bits/second (about as fast as someone texting) and is mounted above a drag brake – the upper disk – which slowed and stabilized the descent (the atmosphere was dense enough not to demand a parachute). Two small cup anemometers are just visible mounted on the disk. (Source: NASA.)

Figure 5.13 Jupiter with its belts and zones, and swirling wakes behind long-lived vortices, provided new challenges to atmospheric dynamicists. Image from Voyager 1 on February 25, 1979, at a distance of 9.2 million kilometers. (Source: NASA/JPL image PIA00014.)

the atmosphere to cause this. Large numerical models would explore the question a couple of decades later, but some conceptual ideas were advanced in the mid-1970s. Peter Gierasch suggested that the meridional ("Hadley") circulation was responsible [194], while Stephen Fels and Richard Lindzen favored the interaction of gravity waves and thermal tides [195]. One flavor of the thermal tide idea had been advanced several years earlier by Gerald Schubert in a neat tabletop experiment where a bunsen burner flame was moved around an annular bath of mercury, which caused it to develop a circular current [196].[34]

---

[34] This is just the sort of physically simple but intellectually challenging experiment that I love. It would likely be prohibited by health and safety rules today.

Much of the theoretical work and observations were summarized at a work-shop at GISS in New York in 1974.[35] Although general circulation modeling of planetary atmospheric circulation was in its infancy, Eugenia Kalnay de Rivas, an Argentine scientist working at the Massachusetts Institute of Technology, pre-sented two-dimensional and three-dimensional numerical models that showed wind speeds comparable with those observed at the cloud tops and near the surface [197]. Kalnay later went on to become prominent in numerical weather prediction on Earth, and in particular on "assimilation", the continuous refine-ment and correction of models by observations.

Understanding the profile of deposition and transmission of sunlight and infrared radiation in Venus' cloudy and sulfurous atmosphere was an evident priority, and this question was tackled with the Pioneer Venus mission in 1978, which featured an orbiter and no less than four descent probes. Probe measure-ments not only profiled the clouds and atmosphere structure, but also hinted that Venus had lost much of the water with which it had presumably been endowed. The addition of three small probes to the large "Sounder" probe allowed measurement of the differences in light absorption and temperature profiles between day and night, and low and high (north) latitude. The large probe carried more elaborate instrumentation to measure the cloud properties and the gas composition. Armed with the Pioneer Venus probe data, Pollack and colleagues refined their greenhouse model of Venus, isolating the contributions of different gases and finding that CO_2, the clouds and SO_2 mattered most, with CO and HCl being minor factors [198]; H_2O was important not so much directly, but in influencing the clouds.

The headlong race to the planets of the 1960s and early 1970s saw the major characteristics of our two neighboring planets discovered. The following decade and a half would introduce another climate altogether – that of Titan – and reveal the atmospheric characteristics of the outer planets, would bring a more in-depth examination of Mars, and expose better the recent history of our own world.

---

[35] The proceedings of this workshop "The Atmosphere of Venus" edited by Jim Hansen and published as NASA SP-382 can be found online. This workshop (and indeed others in the same series, like that on Titan the same year, NASA SP-340, edited by Don Hunten) make particularly fun reading because the discussion after the talks is recorded. Reading the banter between luminaries such as Sagan and Ingersoll gives a sense of how much different models were believed, hard to extract from the literature itself. A remark by Hess, who led the Viking meteorology investigation, reminds us of audio-visual chal-lenges before Powerpoint: "There are eight ways to put in a slide, only one of which is right."

6

Return of the Ice

1975 to 1989

1975 – Paltridge proposes thermodynamic principle controlling heat transport in Earth climate system

1976 – Deep-sea mud cores provide evidence of Croll–Milankovitch forcing of the ice ages

1976 – Viking landers make first meteorological measurements on the surface of Mars, but find no convincing evidence of present-day life

1980 – Voyager 1 measures Titan surface pressure and temperature by radio occultation; spectrometers detect many organic compounds, but surface obscured from observation by thick haze

1980s – Speculations about nature of Mars paleoclimate and paleohydrology, and the faint young Sun on Earth

1981 – Walker and colleagues propose temperature-dependent chemical removal of CO_2 provides feedback to regulate climate over geologic time

1983 – Watson and Lovelock propose Daisyworld model of biological feedback control in climate

1983 – Nuclear winter scenario explored by Turco and colleagues

1984 – Kasting and others explore evolution of Venus climate and loss of water from atmosphere

1985 – Farman in Antarctica discovers strong ozone depletion, soon confirmed in earlier satellite data

1987 – Montreal Protocol brings 43 countries together to regulate ozone-depleting chemicals

1988 – Intergovernmental Panel on Climate Change founded

1989 – McKay and others develop foundational one-dimensional climate model of Titan

1989 – Voyager 2 encounters Neptune and Triton, images wind streaks, geyser-like plumes

Even as late as 1975, rather exotic ideas were proposed to explain the ice ages – British astronomer William McCrae suggested that the Sun passing through the lanes of dust in the galaxy might lead to variation in the solar constant and thus ice ages [199] (resurrecting an idea by Hoyle and Littleton from 1938, and indeed ideas by Shapley from 1921).[1]

While Croll–Milankovitch cycles – the astronomical forcing of ice ages on Earth – had faded somewhat into the background for some decades because of weak evidence, new analytic techniques (measuring the isotopes of oxygen in the chalky shells of plankton in seafloor sediments; the amount of "heavy" oxygen O-18 being related to the volume of the ice sheets) and the collection of large numbers of abyssal mud cores finally allowed the recovery of a well-dated high-resolution timeline of terrestrial climate. Hays, Imbrie and Shackleton in their 1976 paper "Variations in the Earth's orbit, pacemaker of the ice ages" [201] were able to construct a record spanning 450,000 years, which was fine enough to detect variations of only a few thousand years. By analyzing this record, they found distinct cycles with periods of 23,000, 42,000 and 100,000 years, with about half of the variation occurring at the 100-kyr period, the same period at which the Earth's orbit varies in eccentricity. Moreover, the ~23-kyr variation showed two sub-peaks (Figure 6.1), at 19,000 and 23,000 years, just as the precession of the Earth's poles was predicted to have. This fingerprint at last demonstrated convincingly that astronomical changes played a role in Earth's climate. The challenges that greenhouse, volcanic and other variations happened too, and that some kind of feedback was necessary to amplify the astronomical signal still remained, however.

The "faint young Sun" problem also remained. Given the obvious effect that cloud cover can have on our local weather, even subtle climatological changes on the extent, altitude or optical properties of water clouds on Earth could have profound effects on the average temperature. Henderson-Sellers in the UK [202], and later William Rossow and colleagues at GISS [203], suggested that minor changes in cloud feedbacks could compensate for changes in the solar constant, an idea that continues to recur to this day.

---

[1] Oddly, the idea is that material from the cloud accreting onto the Sun would actually increase the luminosity, thereby increasing the rate of precipitation, which Simpson had argued would cause more snow accumulation and thus growth of ice cover. Hoyle's science fiction story "The Black Cloud", about an interaction of the solar system with an (intelligent) interstellar dust cloud, was more successful than his climate suggestion. H. Shapley's short 1921 paper [200] (interestingly an astronomer publishing in a geology journal) notes the existence of dust clouds in the Orion Nebula and that "A change of 20 per cent in the solar radiation, if maintained for a considerable period of years, would sufficiently alter terrestrial temperature to bring on or remove an ice sheet; an 80 per cent change, unless counteracted by concurrent changes in the terrestrial atmosphere, would completely desiccate or congeal the surface of the earth."

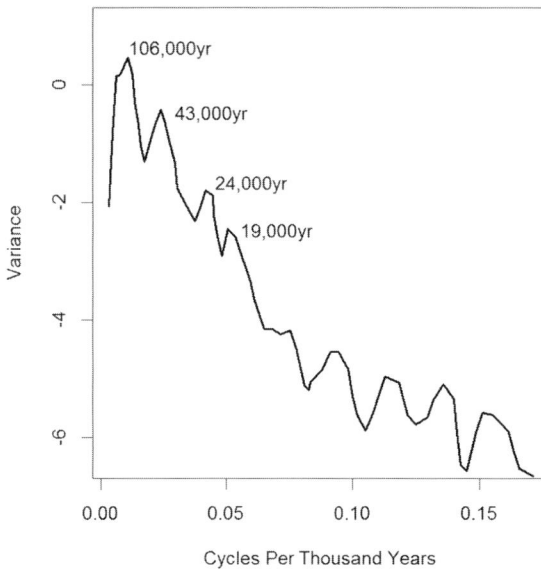

Figure 6.1 The power spectrum of variations in oxygen isotopes in deep-sea cores, recorded by Shackleton and team, with a clear fingerprint at multiple frequencies demonstrating that the Pleistocene glaciations were paced by astronomical factors. (Author, from data in Hays et al. [201].)

Meanwhile, at Mars a bonanza of data was obtained by the Viking mission, comprising two formidable landers (Figure 6.2) and two orbiters. The landers' primary objective was to search for evidence of life. Although experiments designed to detect "respiration" in the soil did measure gas exchanges, the landers' gas chromatographs found no evidence of organic compounds in the soil. The best interpretation was that the regolith had some oxidizing component, perhaps a peroxide, which spoofed the gas-exchange experiment – such an oxidant would also destroy organic material. As Lovelock had expected, Mars seemed dead.

However, the mission generated massive fodder for other science. The orbiters mapped the surface, and measured temperatures, dust and water vapor in the atmosphere. The landers obtained what remains – 40 years later – the best meteorology record from the Martian surface [204], over 1000 sols (Mars days) long in the case of Viking 2, recording in detail the dramatic seasonal pressure cycle (Figure 6.3) predicted by Leighton and Murray, and the thermal tides anticipated by Goody and indicated in Leovy and Mintz's model [159]. In general the wind and pressure patterns were rather repetitive day to day, and winds were evident not only to the wind sensors, but also to Viking 2's seismometer as gusts shook the lander gently [205]. Wind speeds reached about 25 m/s [206] (actually rather weaker than a GCM prediction for the Viking lander sites made by Jim Pollack working with Leovy and Mintz [207]).

Figure 6.2 Carl Sagan posing (apparently in Death Valley) with a test model of the Viking lander. The vertical cylinder with the dark line is the scanner used to make images. The wind and temperature sensors are on a short arm at the upper right, not quite long enough to eliminate perturbations to their measurements induced by the lander's warmth. The high-gain antenna used to beam data back to Earth is to the right of Sagan's head. (Source: NASA.)

The landers saw dust storms come and go, the sky darkening and reddening. The Viking 2 lander was further from the equator than its sibling, and in fact saw frost (Figure 6.4) deposited during winter [208]. Much of the Viking observations – especially the deluge of data from the orbiters – took many years to be fully assimilated and understood.

While not yet a political issue, scientifically the accuracy of climate models was a subject of intense debate. The choice of convective parameterization in models was one area of contention. In a widely cited 1981 evaluation of the impact of CO_2 rise, Jim Hansen noted that using a lower lapse rate in convective regions reduced the sensitivity of the climate [209], while a study by Richard Lindzen at Harvard noted that replacing the convective adjustment procedure with a more physical (but admittedly more elaborate) scheme of cumulus convection reduced the global climate sensitivity by a third, but interestingly made the low-latitude sensitivity even weaker [210]. Essentially, he argued, tropical convection acted somewhat as a local thermostat.

Stephen Schneider and Starley Thompson warned that grand conclusions about "cosmic questions", such as Snowball Earth, runaway greenhouse or

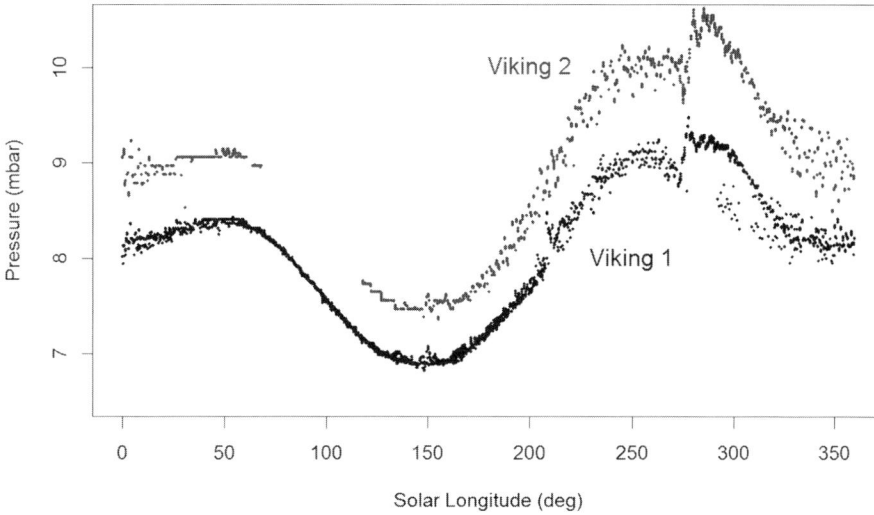

Figure 6.3 The atmospheric pressure recorded by the two Viking landers showing the annual cycle as the formation of the seasonal polar frost caps remove gas from the atmosphere. The record from Lander 1 actually lasted for three Mars years, but the data nearly all collapse onto a single curve when plotted against solar longitude, a measure of Martian season. There is an offset between the two curves – Viking Lander 2 was at higher elevation. Note the jump at longitude ~270 degrees, the southern summer solstice when dust storms tend to occur. (Created by author from NASA data.)

Figure 6.4 An image of the surface of Mars taken by Viking Lander 2 at Utopia Planitia on May 18, 1979, showing part of the lander in the foreground, and a thin deposit of water ice frost on the rocks and sand. The view is to the southeast, and the large boulder at right is about 1 m wide. The frost remained on the surface for about 100 days. (Source: NASA.)

habitability of planets around other stars, depended rather critically on such parameterization details [211]. One such parameterization is that of horizontal heat transport applied in zonal energy balance models such as that by Budyko and Sellers, which were now being applied to Mars [212, 213]. The success of the Leighton–Murray model for the present Mars climate relied on the atmosphere being very thin and so being unable to transport much sensible heat (which was simply not included). But that assumption broke down in the investigations now being applied to explore Mars' past, when the atmospheric pressure might have been much higher. In the Budyko–Sellers models for the Earth, this horizontal heat transport (which is what controls how cold the poles get, and thus the prospect for runaway glaciation on Earth, or atmospheric collapse as explored by Gierasch and Toon on Mars) is treated as a diffusion process, with heat flow from equator to pole simply proportional to the temperature gradient. But the constant of proportionality, the effective horizontal "conductivity" of the climate system, was just chosen to be a number that worked – as it happens, a value of the order of 1–$2\,\mathrm{W\,m^{-2}\,K^{-1}}$. That choice – like the 6.5 K/km critical lapse rate in convective adjustment – is purely empirical, it's the value that gives the best fit to terrestrial conditions today. But the 1–$2\,\mathrm{W\,m^{-2}\,K^{-1}}$ had no fundamental basis – after all it is a combination of many processes, ocean circulation as well as atmospheric heat transport.

The philosophy of evading such empirical parameterizations by ever-increasing model resolution (to explicitly resolve eddies, for example) appeared to Garth Paltridge, an Australian climatologist, as ultimately absurd, like trying to estimate the pressure in a room by tracking every molecule of a gas. He thought that perhaps some thermodynamic principle might be useful in estimating the heat transports, and in 1975 found some success with a simple energy balance model for the Earth by selecting heat transports such that they were somehow "optimum", maximizing the production of entropy [214, 215]. We will return to this idea later – it received much attention, both favorable and otherwise.[2] In fact, Edward Lorenz had recognized in 1960 that the terrestrial climate system generates available potential energy at about the maximum possible rate, given the amount of sunlight absorbed by the surface [216]: not only was Carnot's

---

[2] And in full disclosure, it is an idea I have explored myself, as discussed in Chapter 7. The idea, often referred to as a "Principle of Maximum Entropy Production" (MaxEP, or MEP), arouses such strong feeling among dynamical meteorologists that a reviewer of the proposal of this book warned the publisher that the book might be a "Trojan Horse" for the principle. Readers are therefore warned to maintain a healthy skepticism about the MEP concept, as for that matter they should about everything they read anyway. Paltridge's original paper referred to the concept as minimizing entropy exchange, but it is more easily thought of as MEP, or perhaps a maximum in mechanical work output.

vision of the atmosphere as an engine a good analogy, the engine appeared to be running at maximum power.

For the moment, Mars energy balance modelers assumed (reasonably) that the heat transport would be somehow proportional to the atmospheric pressure, or maybe fudged upwards by a factor to account for the smaller size of Mars [217–219]. And because the sensible transport was small anyway, there was nothing to show that this approach was wrong. And they found, happily, that with enough CO_2 they could get Mars warm enough to have liquid water. (Liquid water lasting long enough to carve river channels and streamlined islands – Figure 6.5 – not only needs temperatures above the freezing point, but also a pressure high enough to stop the liquid from boiling.)

Then, in 1980, a new world was added to the pantheon of climate worlds to study – Titan. The presence of an atmosphere (which turned out to be dominated by nitrogen) made Titan a priority target in the Voyager mission to the outer solar system, and Voyager 1 (launched in 1977, rocketing past Jupiter in 1979) was targeted in 1980 to measure the thickness and composition of Titan's atmosphere.

As at Venus in the 1960s, while the spacecraft encounter is considered the landmark event that solved the mystery, radio-astronomical observations pretty much got the right answer shortly beforehand at Titan too. W. Jaffe, John

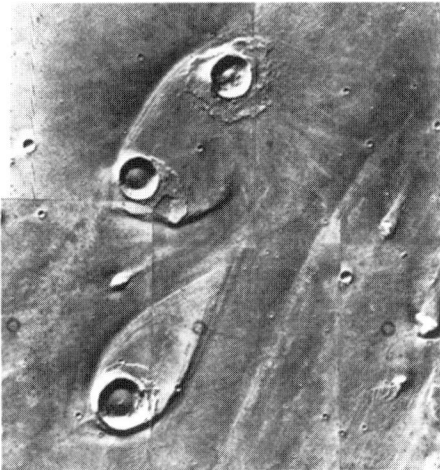

Figure 6.5 Viking images showed even more prominently than Mariner that surface flow – imagined to be water – had shaped the Martian surface. The streamlined island formed in the wake of a crater is about 45 km long, and rises about 600 m above the plains. This area in Ares Vallis is near where Mars Pathfinder landed 20 years later. Notice the seams between parts of the scene – although the images were transmitted digitally, image processing at the time generally relied on cutting-and-pasting photographic prints! (Source: NASA.)

Caldwell and Toby Owen, using the newly constructed Very Large Array interferometer in February 1979[3] (the observatory was officially inaugurated in 1980), measured the brightness temperature and radius of Titan, using 14 dishes, each 25 m in diameter, separated by up to 20 km [220]. Their measurement gave a radius of 2400 ± 250 km and a brightness temperature of 87 ± 9 K. Obviously, Titan is a much smaller target than Venus, much further away, and colder – all of which make it a more challenging measurement. Nonetheless, this observation measured the conditions quite correctly.

The radio-occultation experiment, the precise geometry of which drove the Voyager 1 encounter (and precluded it from making further planetary encounters as Voyager 2 was later able to do), yielded precise answers at Titan, as it had done at Mars 15 years earlier: radius 2575 ± 0.5 km, with surface pressure and temperature (assuming a pure nitrogen atmosphere) of 1496 ± 20 mbar and 94 ± 0.7 K [221]. This was not as warm as Sagan and others had perhaps imagined – another hoped-for habitable world failed to materialize. The atmospheric profile (measured near the equator), enabled at last the choice between the dusty-stratosphere thin-atmosphere idea of Danielson and Caldwell, and Hunten and Sagan's deep warm greenhouse. But in fact it had been a false choice – both features were present (somewhat as Low and Rieke had suggested). This would not be the first time that Titan would laugh in the face of Occam's razor, the fourteenth-century heuristic principle that the simplest explanations are best.[4]

In fact, Titan's atmospheric profile (Figure 6.6) looked much like Earth's, with the height and temperature axes stretched and shifted. Of course, Titan being so much further from the Sun, it was much cooler. And Titan's low gravity, about seven times weaker than Earth's, meant that its scale height was several times larger, so Titan's atmosphere extends far into space.

Titan's troposphere had a temperature rising towards the ground, just like Earth's, due to greenhouse warming from methane, nitrogen and the trace (about a tenth of a percent) of hydrogen; in Titan's low gravity the lapse rate is about 1 K/km. While perhaps disappointing in not being as thick and warm as Sagan had hoped, the atmosphere nonetheless was four times denser than our own, and thus had a significant greenhouse effect. Moreover, the surface temperature of 94 K was rather close to the triple point of methane, raising the possibility that methane might participate in a hydrological cycle of clouds, rain and seas, like water on Earth (which is also close to its triple point).

---

[3] Although the observations were taken in February 1979, the paper was submitted in February 1980, accepted in May and finally published on December 1, 1980, which was just after the Voyager 1 encounter with Titan on November 12.

[4] Or rather the most parsimonious theories, i.e. those that entail the fewest assumptions, are most likely to be true. Named after William of Ockham, although Aristotle and other philosophers advocated the same principle.

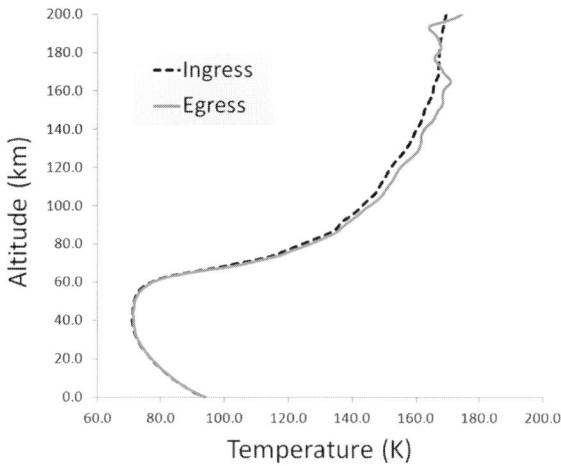

Figure 6.6 Voyager 1 radio occultation profiles of Titan's atmosphere, showing the cold (~70 K) tropopause at about 40 km and a warm stratosphere above. The surface temperature was estimated at 94 K. The two profiles, about 180 degrees apart in longitude (one near dawn, one near dusk) were almost identical. (Author, from data in Lindal et al., 1983 [221].)

Voyager used an infrared spectrometer to identify around a dozen different organic compounds in Titan's atmosphere, and ultraviolet observations indicated that molecular nitrogen was the dominant constituent.[5] The infrared instrument (Figure 6.7) indicated how the temperatures at a few levels in the atmosphere varied between the equator and pole: in particular, the polar surface temperature appeared to be a couple of degrees cooler than the tropics.

Perhaps the most profound observation was what was not seen by Voyager – Titan's surface. The surface and lower atmosphere were completely obscured by a thick layer of red organic smog (Figure 6.8). Material that appeared similar was made in the laboratory by Sagan and a colleague, Bishun Khare, by sparking or irradiating mixtures of methane and nitrogen gas, as one might expect on Titan, and a brown material formed, sometimes powdery or sticky, which Sagan dubbed "tholin", after the Greek word for mud.[6] This material, produced high in Titan's atmosphere, formed a thick deck of haze which absorbed much of the

[5] The best post-Voyager summary of Titan observations is the chapter by Hunten in Ref. [222].

[6] Similar experiments were first done by Stanley Miller in 1953, with a slightly different mix of gases (water, methane, ammonia and hydrogen), thought to resemble that on the early Earth. Working with Harold Urey, Miller found that amino acids and other important prebiotic molecules were produced by sparking. The Titan flavor of these experiments leaves the water and ammonia (solids at Titan conditions) out and substitutes nitrogen for hydrogen. The tholin material produced seems spectroscopically to resemble Titan's haze, and in fact when liquid water is added to it, amino acids are produced.

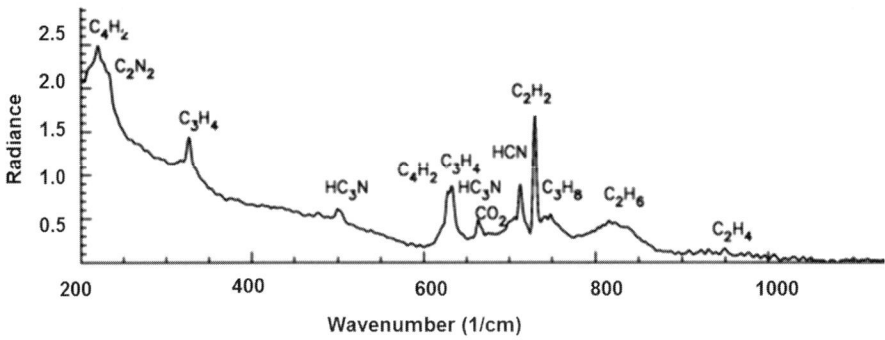

Figure 6.7 Titan's thermal emission spectrum, recorded by the IRIS instrument on Voyager 1, is rich in information on the abundance of organic molecules in the atmosphere, including not only hydrocarbons such as ethane (C_2H_6) and acetylene (C_2H_2, whose prominent emission spikes and bands account for the "nonthermal" character noted by Morrison and Cruikshank in 1971 [173]) but also nitriles such as hydrogen cyanide (HCN) and cyanogen (C_2N_2). (Figure by author.)

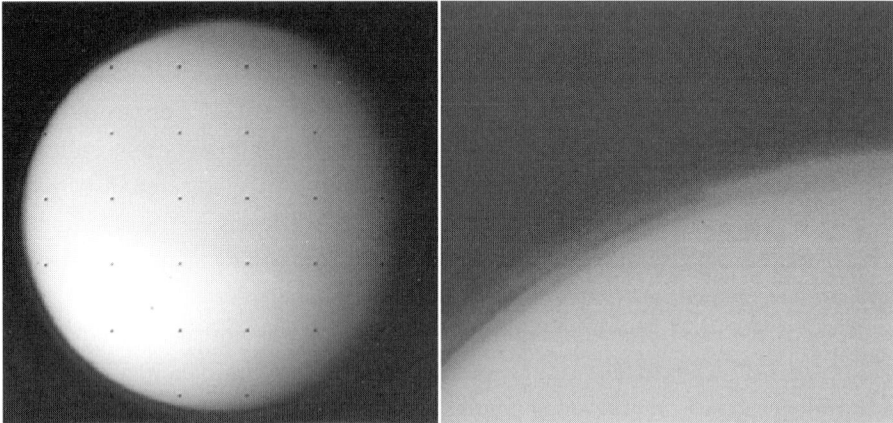

Figure 6.8 While Voyager's cameras showed neither the surface nor discrete clouds, there was a distinct difference in albedo of the north and south hemispheres as well as a dark "hood" or collar above the north pole, analogous in some ways to the polar stratospheric clouds on Earth that are involved in ozone depletion. A close-up at right shows some structure in the haze, with a "detached" haze layer above the main deck. (Source: NASA.)

incident sunlight (which is only 1% of what Earth receives, since Titan is ten times further from the Sun).

This solar absorption leads to a (relatively) warm stratosphere, in a somewhat analogous way to the absorption of ultraviolet light by ozone in Earth's stratosphere. But the attenuation of sunlight reduces the surface temperature below

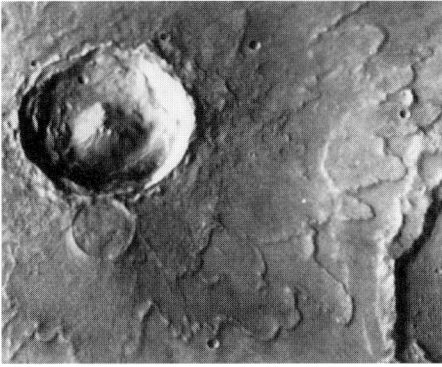

Figure 6.9 A Viking image of a "rampart" crater, where the ejecta thrown out from the impact resembles a "splat", suggesting ice or water in the surface material. (Source: NASA.)

what it could have been – an "anti-greenhouse effect". Bob Samuelson at NASA's Goddard Spaceflight Center constructed an analytic model in 1983 showing how Titan's main atmospheric structure could be explained by absorption of blue sunlight by the haze, while red light filtered down and maintained surface temperatures with the help of the greenhouse effect [223].

The mid-1980s saw something of a hiatus in the pace of planetary missions in the USA, with Cold War budgets for space shrinking, and much of NASA's attention occupied by the space shuttle. But the respite allowed the digestion and synthesis of the Mars, Venus and Titan findings of 1976–81.

Pollack and Kasting and others speculated about warmer, wetter conditions on Mars [224]. Evidence emerged on sifting through Viking data that Mars might have had quite extensive deposits of surface water and that much water may still be very near to the surface in the form of permafrost. An early indicator of this, reported by Squyres and Carr, was that the texture of the rims and ejecta from impact craters in the north polar regions was suggestive of mudflow (Figure 6.9), while craters closer to the equator merely had blankets of rocks sprayed out around them [225].[7]

Gary Clow at the US Geological Survey explored the idea that the insulating effect of a Martian snowpack might allow melting at its base – that snowmelt rather than rain formed the channels [226]. With the right amount of dust in the snow, melting could occur at least at noon in summer, although permanent liquid could not be supported. Michael Carr, also at USGS, advocated a water-rich Mars, suggesting that a groundwater system beneath the permafrost might burst out into the northern lowlands, carving the channels and streamlined

---

[7] Squyres would later become the lead scientist for the two Mars Exploration Rovers, Spirit and Opportunity.

islands seen in the Viking data [227]. Steve Clifford at the Lunar and Planetary Institute in Houston later extended the idea to consider melting at the base of the polar cap [228].

Tim Parker and Steve Saunders at the Jet Propulsion Laboratory (JPL) even found hints of what could be fossil shoreline features of an ancient northern sea, but cloaked this outrageous suggestion in a paper with the obfuscating title "Transitional morphology in West Deuteronilus Mensae, Mars: implications for modification of the lowland/upland boundary" [229]. However, slowly the idea of a wet Mars became more and more respectable, allowing Vic Baker at the University of Arizona (a hydrologist, with a fondness for Harlen Bretz's icono-clasm) to venture a rather more brash paper in 1991, with the title "Ancient oceans, ice sheets and the hydrological cycle on Mars" [230]. But the evidence for widespread subsurface ice, and its at least occasional appearance as water on the surface, was being advanced from a number of directions, amid the overall digestion of Viking and other data – these were summarized in a series of workshops and dozens of journal papers in 1991–3 with the theme "Mars Surface and Atmosphere Through Time" (MSATT), setting the stage for future spacecraft exploration. Carr's book *Water on Mars* summed things up.[8]

At Venus, while the Pioneer Venus probes had lasted only an hour or so after their arrival in 1978, the Pioneer Venus Orbiter kept operating until it burned up in 1992. Its ultraviolet spectrometer provided data that could estimate the sulfur dioxide (SO_2) amount above the cloud tops, and Larry Esposito at the University of Colorado found that the amount had systematically declined during 1978–84, while the 1978 value was much higher than some upper limits derived in the 1960s [233]. The SO_2 amount was evidently changing dramatically, and it was tempting to suggest that volcanism was responsible.

While US planetary missions in the 1980s were sparse, the Soviet Union continued to explore Venus.[9] Further Venera landings showed similar, presum-ably volcanic, landscapes with better cameras than before. Veneras 11 and 12 recorded the spectrum of sunlight as it filtered down to different depths in the atmosphere, finding that indeed very little water vapor exists in the lower atmosphere. They also found that much absorption of ultraviolet light occurred above 60 km altitude (a mystery "UV absorber" that has not yet been identified). Venera 13 in 1981 even took a pair of images on the surface about an hour

---

[8] A review article was published in 1987 [231]. Carr published the book of the same title in 1996 [232].

[9] There are many accounts of Soviet Venus exploration. One of the best from the scientific standpoint is Marov and Grinspoon's book [234]. A nice retrospective paper, again by one of the key protagonists, is by Vasily Moroz [235]. The web pages of Don Mitchell, http://mentallandscape.com/V_Venus.htm are a tremendous resource.

apart – the second showed that some dirt thrown onto the lander ring at impact had blown away, suggesting that winds strong enough to move surface particles must be rather frequent. Then, in 1985, two Venera-like VEGA probes made measurements during descent that suggested the Venus clouds were even more exotic than indicated before – the sulfuric acid was spiked by about one percent of iron chloride salt, and there was evidence of other sulfur compounds as well as phosphorus; e.g. [236]. The VEGA[10] probes [237, 238] each also released a small helium balloon, floating near the cloud tops to make rudimentary measurements as they were swept along in the zonal winds. They operated for some 40 hours, tracked by a network of radio telescopes on Earth as they bounced up and down in the turbulent wind.

Meanwhile, Jim Kasting developed in 1984 the most detailed Venus radiative–convective model yet, working at Ames with Pollack and Thomas Ackerman, and focussed on the problem of the loss of water from Venus [239]. This work in particular noted that the escape of water on Earth was limited by a "cold trap" – the lowest temperature part of the atmosphere at the tropopause. Because water vapor comes from below, the upward flow is throttled by the low saturation vapor pressure at this temperature minimum, and as a result the Earth's stratosphere is very dry. However, with strong sunlight and a strong greenhouse, the Venus tropopause temperature would be higher, opening this sphincter and allowing more water vapor to leak up into the stratosphere, where ultraviolet light would destroy it. Andrew Watson and colleagues at the University of Michigan followed similar reasoning with a simpler model that same year [240]. The hydrogen would escape, but would leave some evidence that it had done so. Because hydrogen escapes more easily than its chemically similar isotope deuterium or "heavy hydrogen", an overall hydrogen loss leads to a relative increase in the deuterium:hydrogen ratio. Pioneer Venus probe data indicated that this ratio is 100 times larger at Venus than at Earth, suggesting Venus lost more than 99% of its original water endowment.[11] Thus, if Venus had oceans, their hydrogen was lost to space and the oxygen was somehow mopped up by chemical absorption into the rocks. The Venus runaway greenhouse by this time had become a climate bogeyman, a warning about how mismanaged planets can end up as unpleasant environments.

[10] The name VEGA is a transliteration from the Cyrillic for VEnus-HAlley: the mother craft that dropped off the probes went on to encounter Halley's Comet. In Cyrillic "G" and "H" are not distinct.

[11] See Donahue et al. [241], Kasting and Pollack [242] and Grinspoon [243]. David Grinspoon has written many papers since on the D:H question at Venus, as well as a couple of excellent books on the planet.

The question of why the Earth had apparently not been fatally inclement in the past had not gone away. The Gaia Hypothesis, which strove to answer the challenge with biologically mediated feedbacks, had suffered the criticism that it was too "teleological", implying foresight on the part of biological evolution. Lovelock and Watson[12] came up with a clever "toy" model called "Daisyworld", which merely posited that a grey planet confronting a brightening Sun might be populated by black and white daisies [244]. If the black daisies predominated, the planetary albedo would be low and the planet warmed, with white daisies having the opposite effect. If it so happened (an easy-to-imagine evolutionary accident) that black daisies grew better than white ones at low temperatures, and vice-versa, then a stabilizing feedback naturally emerged that regulated the planetary temperature at a comfortable value over a much wider range of solar illumination than if the daisies were absent.[13] The transparency of this simple model (the model was enabled by the emerging availability of personal computers) gave it substantial appeal, and the model even exhibited hysteresis (e.g. Figure 6.10) – if an already-strong Sun flared up to kill all the (white) daisies, restoring the solar constant to its earlier value didn't give habitable conditions again because the now-grey planet stayed too warm. Later modelers played with one- and two-dimensional Daisyworlds, with various shades of grey, and then put in rabbits to eat the daisies, foxes to eat the rabbits and so on [245]. All these elaborations were rather fun, but did not add much to Watson and Lovelock's principal conclusion beyond the general result that more complex ecosystems were actually more stable to solar changes than simple ones (a possibly disturbing thought, given the tendency to more efficient, but less robust, monoculture in modern farming).[14]

While the Daisyworld model was a pleasing demonstration, a more fundamental feedback had been worked out by James Walker and J.D. Hays in Michigan and James Kasting (then in Colorado). This noted that the rate of weathering – the removal of carbon dioxide from the atmosphere by reacting with silicate rocks – was temperature dependent [247]. Thus, since CO_2 is

---

[12] Yes, the same Andrew Watson mentioned in the previous paragraph! He moved from the University of Michigan, the home of the Pioneer Venus mass spectrometer instrument, to the Marine Biological Association in Plymouth, UK, a couple of hours from Lovelock's laboratory-home in Wiltshire.

[13] A beautifully oxymoronic NASA conference proceedings has the title "Variations of the Solar Constant".

[14] In fact some interesting analytical parallels have been drawn with elements of control theory, specifically the regulation of blood sugar in humans and the implications for diabetes [246].

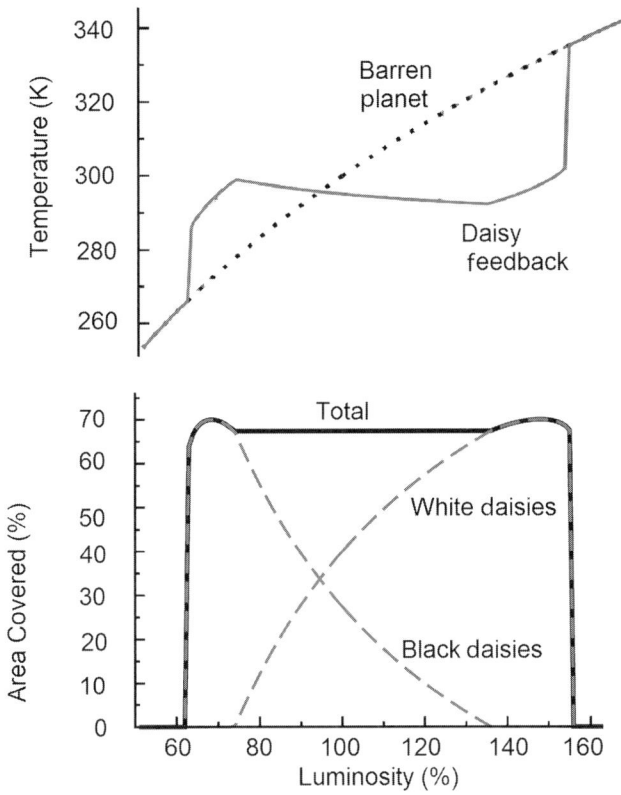

Figure 6.10 Features of the Daisyworld model. Top shows the temperature variation on a barren (grey) planet as a dotted line: introducing the Daisy feedback stabilizes the temperature over much of the luminosity range. The lower panel shows the relative amounts of black and white daisies (dashed lines, with total cover as solid line) as a function of luminosity (solar constant). Below 60% and above 155% in this version, the planet is uninhabitable, but for intermediate values the relative areas of black and white can compensate for the external forcing. The system exhibits hysteresis (not shown), and it is noticeable that the curve is actually negative in slope in the stabilized region. (Figure by author.)

continuously supplied by volcanoes, if conditions were cold (as, for example, with a faint early Sun), then the lower removal rate would allow CO_2 to build up and the greenhouse effect would strengthen, warming the surface until the removal could keep pace with the volcanic supply.

Some planetary climate modelers in the USA strayed a little into politics in this era, applying the same tools and models developed for planets to explore the climate perturbation to the Earth in a scenario where a nuclear conflagration caused widespread city and forest fires, injecting soot high into the atmosphere, which would cause a Titan-like antigreenhouse effect. The effects of nuclear conflict had been considered before, most notably the possible destruction of the ozone layer by nitrogen oxides produced in fireballs, but a combination of Cold War paranoia and a high-profile study in a prominent journal by well-known scientists gave the topic some new impetus [248].[15]

Titan's antigreenhouse came under close scrutiny now too. Chris McKay, at the NASA Ames Research Center, with Pollack and Regis Courtin, constructed in 1989 a detailed radiative model of Titan's atmosphere, book-keeping the absorption and scattering of sunlight (which had a complex wavelength dependence, due to the red color of Titan's haze particles, their varying size, and the absorption bands of methane – see Figure 6.11) and the broad absorption of thermal radiation by methane, nitrogen and hydrogen [249]. This model more accurately quantified the greenhouse and antigreenhouse, but was also self-consistently calculated with a model of how the haze particles, formed by methane destruction high in the atmosphere, descended and coagulated, and perhaps were washed out by methane clouds and rain. It also became clear – literally – in this and similar studies, that despite the opacity of Titan's haze to the visible light sensed by Voyager's cameras, an appreciable amount of sunlight penetrated the atmosphere in red and near-infrared wavelengths. It might therefore be possible to observe the surface at these wavelengths – and indeed in 1991 and 1992 evidence of a near-infrared "lightcurve", a variation in brightness as Titan's rotation exposed different longitudes to Earth, was detected in telescope observations [250].

McKay's model showed that the greenhouse effect on Titan raised the surface temperature by about 21 K, but the antigreenhouse due to the haze dropped it by 9 K, for a net warming of 12 K above the effective temperature set by Titan's albedo and distance from the Sun [251]. The greenhouse effect was due mostly to methane and nitrogen, but with an important contribution from hydrogen (even though that gas comprises only 0.1% of the Titan atmosphere, its opacity plugs an important window in the infrared welling up from the surface).

---

[15] The principal paper, in *Science* magazine, was sometimes referred to as the TTAPS paper, after its authors Turco, Toon, Ackerman, Pollack and Sagan.

[16] The full text at http://www.margaretthatcher.org/document/107817 is well worth reading – appealing to Charles Darwin, and pragmatically recognizing that industry and economic growth would be the solutions to the problem. Thatcher had scientific training as a chemist. Although her views on climate change and the right policies to address it were to change in her later years, her early warnings were likely instrumental in raising awareness

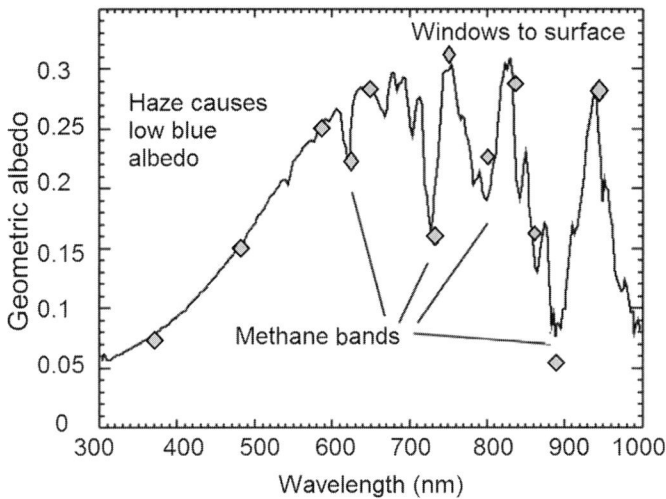

Figure 6.11 The visible and near-infrared spectrum of Titan from blue at left (0.35 μm), to red at 0.65 μm, to 1 μm in the near-infrared (a TV remote control operates at 0.94 μm). The blue albedo is low, due to haze absorbing the sunlight. The brown haze is bright in the red and near-infrared, but methane causes strong absorptions at some wavelengths (the bands Kuiper used to detect the atmosphere). The diamond symbols indicate albedos predicted by McKay's model with certain assumptions about the haze. By tuning the model to match the observed curve, as well as the radio occultation profile, McKay could recover key properties of Titan's atmosphere. (Figure by author.)

Around this time, the implications of climate change on Earth began to be recognized politically. Britain's Prime Minister, Margaret Thatcher, was notably outspoken – in an address to the United Nations in 1989 she warned [252]:[16]

> *Mr President, the evidence is there. The damage is being done. What do we, the International Community, do about it?. . . Before we act, we need the best possible scientific assessment: otherwise we risk making matters worse. We must use science to cast a light ahead, so that we can move step by step in the right direction . . . no issue will be more contentious than the need to control emissions of carbon dioxide, the major contributor — apart from water vapour — to the greenhouse effect . . .*

Thatcher's remarks supported the work of the recently instituted Intergovernmental Panel on Climate Change (IPCC), raised awareness of climate change

of climate change issues on the global political stage. An interesting perspective on the early days of the IPCC, and on Thatcher's interest, can be read in the autobiography of one of the key players: J. Houghton, *In the Eye of the Storm*, Lion Books, 2013.

[17] The Vienna Convention in 1985 and the Montreal Protocol in 1987. By the mid-1990s, the frightening drop in the minimum Antarctic ozone levels had leveled out.

concerns in the international political community, and pointed to the success of international agreements in reducing the emission of ozone-destroying chloro-fluorocarbons (CFCs).[17]

Although no new US planetary launches took place for almost the entire decade of the 1980s, the Voyager 2 spacecraft continued to explore. Relieved by its sibling of the responsibility of making a radio-occultation measurement at Titan, it was targeted to exploit the remarkable opportunity in the 1980s (not to be repeated for centuries) of using Saturn's gravity to swing it towards Uranus, and then Neptune. These two planets, smaller and denser than Jupiter and Saturn (earning the outer pair the name "ice giants" rather than the warmer "gas giants") and their systems of satellites, rings and ringlets, gave the only new NASA planetary encounters for almost a full decade after Voyager 2 at Titan in 1981.

Uranus has the distinction of an exceptionally high obliquity, its spin axis being tilted almost at 90° to the ecliptic, giving a very strong seasonal cycle, and, as at the other planets visited by Voyager, the radio-occultation experiment during its flyby in 1986 profiled Uranus' atmosphere. Neptune received similar treatment in 1989. While the gas giants and ice giants have interesting photochemistry and dynamics, it is not obvious that they have climates. (Or rather, you could say a given giant planet either has no climate, or it has an infinite number – if you like it warmer, just go deeper!) Since they don't have accessible surfaces I think they are less interesting than Titan, Mars and Venus, although of course some of the processes such as haze formation, convection and dynamics have relevance to planetary climate. Tyler Robinson and David Catling [253] neatly demonstrated that the temperature structure (Figure 6.12) of the giant planets is relatively easy to model: the minimum (tropopause) temperatures are almost invariably at a pressure of about 0.1 bar. The temperatures here, lacking much greenhouse warming, are simply set by albedo and distance from the Sun (and so have values more or less exactly as Christiansen had calculated in 1885). Below this level, the atmosphere is increasingly opaque to thermal radiation and so the temperature increases with depth according to the convective lapse rate which depends mostly on the planet's gravity. Above the tropopause, the complex relationship of solar and thermal absorption can vary a lot depending on the details of hazes and gas composition, but this region is only of practical significance for the design of heat shields and parachutes on entry probes.

While its giant parent had an atmosphere but lacked an observable surface, Neptune's largest satellite Triton (diameter 2700 km, about half that of Titan) proved to have a thin atmosphere, perhaps just enough to qualify as having a climate. As had also been the case with Mars and Titan (perhaps

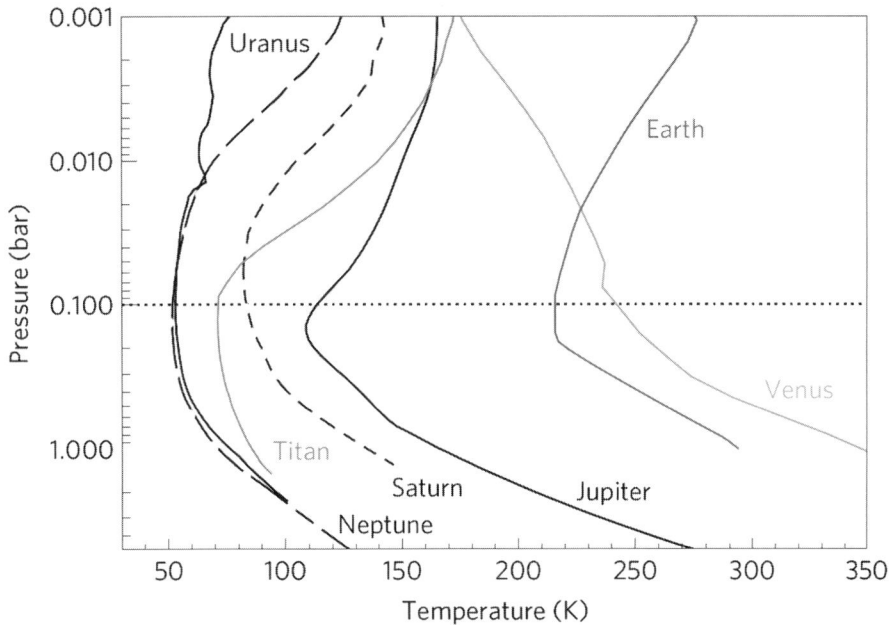

Figure 6.12 The temperature–pressure relationships of the major atmospheres (most revealed by radio occultation). All except Venus have a tropopause – a minimum temperature – of roughly 0.1 bar, as most atmospheres become somewhat opaque to infrared radiation at deeper levels (higher pressure). The slope below the tropopause is generally set by convection and depends mostly on gravity (Titan's gradient is shallow, Jupiter's is large). Venus lacks strong solar absorption like Titan's haze or Earth's ozone, and so doesn't have a warm stratosphere. (Figure courtesy of T. Robinson.)

indicating that wishful thinking is a more powerful factor in planetary science than we might like to admit), the atmosphere was thinner and colder than had been expected. Triton is so bright [254], being partly covered in nitrogen ice, that it essentially represents a "collapsed" state in a Gierasch/ Toon sense, with the atmosphere frozen out on the surface. Indeed, Triton's reflectivity or albedo[18] of 0.82 is one of the largest in the solar system – larger even than Lowell's misguided guess at Earth's. Voyager's radio occultation experiment found that the surface pressure was only about 17 microbar – 70,000 times lower than Earth's – and the surface temperature a bracing 38 K [255].

---

[18] The important quantity is called the Bond albedo, which accounts for reflection in all directions. This is not always the same as the geometric albedo, which is much easier to measure, being the reflectivity in a single direction.

Figure 6.13 A mosaic of Voyager images of Triton. The surface has terrain textures that may indicate sublimation of volatile ices (nitrogen, carbon monoxide, methane) in the seasonal cycle, and the dark streaks are material blown by wind from plumes. (Source: NASA/JPL.)

One can enter an interesting debate as to what constitutes a "climate" and thus whether Triton has one. On one hand, the atmosphere is too transparent to meaningfully affect the transmission of sunlight or heat between the surface and space, and is too thin to carry any significant heat by virtue of its temperature (i.e. sensible heat). On the other hand, perhaps more persuasive is the fact that the atmosphere is able to shunt around a significant fraction of the heat absorbed by the Sun by subliming in one place and freezing in another, conveying latent heat.[19] Another aspect is that there is some haze suspended in the atmosphere, and that streaks of dark deposits on the surface (Figure 6.13) attest to the downwind advection of particles lofted in plumes (sometimes incorrectly referred to as "geysers"): thus, the influence of the atmosphere is manifest in the appearance of the surface. On balance, then, I'm inclined to say Triton does have a climate.

Triton does see significant seasonal changes as the evolving latitude of the Sun chases frost from one hemisphere of the body to the other, and has been the subject of study with Leighton–Murray type energy balance models [256] (and even, more recently, GCMs [257]). Stellar occultation measurements in the

---

[19] Sublimation/migration occurs on Callisto, where antisunward crater walls were seen by Voyager to be paradoxically brighter – it is thought water ice migrated to those cold shadowed areas. Permanently shadowed craters near the poles of the Moon and Mercury, seeing only the bitter cold of deep space as Fourier imagined, similarly act as traps for water and significant deposits of ice exist there, of potential importance (in the Moon's case) for future human exploration. However, in these cases, and that of Iapetus (Chapter 9) the sublimation happens very slowly and does not itself transport any significant amount of heat.

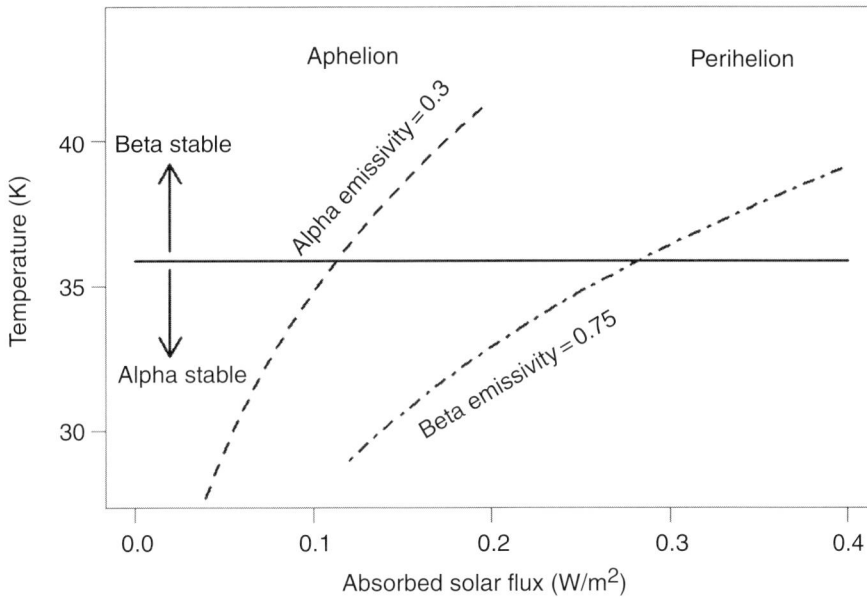

Figure 6.14 As described by Stansberry and Yelle [258], the surface temperature of a nitrogen-ice-covered world, in this case a theoretical Pluto, may be controlled by the alpha–beta phase transition at 35.6 K. Near perihelion, temperatures are high and the emissive beta phase is stable, but when the solar flux drops towards aphelion, the temperatures are cold enough to switch to the less emissive alpha phase. Thus, a Daisyworld-like feedback can pin the temperature at the transition temperature for much of a Pluto year. (Figure by author.)

mid-1990s would show that the atmospheric pressure at the surface could be changing (not unlike the seasonal variation of the Mars surface pressure).

Of course, while the familiar tools of GCMs and energy balance models can be applied to Triton, this world involves such extreme conditions that some unexpected physical effects can arise. A rather cute feature of nitrogen ice noted by John Stansberry and Roger Yelle, which may or may not be important for Triton or Pluto, is that nitrogen ice exists in two different solid phases, which have different emissivities.[20] Below 35.6 K, the ice has a cubic structure (alpha), while above 35.6 K the beta phase has a hexagonal lattice and a much higher emissivity [258, 259]. Thus it behaves almost like a thermochromic paint, changing its "thermal color" depending on temperature. Unlike bright snow melting on Earth to accelerate warming by becoming darker liquid, however, the nitrogen phase transition is stabilizing (Figure 6.14) – the high emissivity cools the planet, but if

---

[20] The heat radiation from a surface is proportional to the fourth power of absolute temperature, and to the emissivity, which is a surface property between 1 and zero in much the same sense as albedo. A "black body" or perfect radiator has an emissivity of 1.

the phase transition is reached, the emissivity drops. Thus the temperature is buffered at 35.6 K – rather like a crystallographic Daisyworld. Hence, over a range of absorbed sunlight flux, Pluto or Triton conditions may be regulated at this transition temperature.

And so, a mere three decades after it began, the initial reconnaissance of the planets (except Pluto) was complete, giving climate scientists a pantheon of worlds to contemplate.

7

Mars Attracts

1990 to 2000

1990s – Various one-dimensional model studies of Titan and Venus climate stability

1990 – Magellan maps Venus surface by radar

1991 – Pinatubo eruption injects massive amount of sulfur aerosol into Earth's atmosphere, perturbs climate

1991 – Quantitative assessment of terraforming on Mars

1992 – Kirschvink proposes Snowball Earth

1992 – Greenland ice core yields high-quality climate record over last 400,000 years

1995 – First Titan global circulation model constructed

1995 – First exoplanets discovered around "normal" stars

1996 – Antarctic meteorite proposed to offer evidence of life on Mars, galvanizes US planetary program

1997 – Mars Global Surveyor, Mars Pathfinder arrive; MGS discovers magnetic stripes and measures global topography, a vital input to circulation models

1998 – Strong evidence for Snowball Earth laid out by Hoffman, Schrag and others

1998 – Climate effects of dinosaur-killing asteroid – soot and sulfur aerosols – laid out

1999 – Climate-focussed missions Mars Climate Orbiter and Mars Polar Lander both lost

1999 – Links between tectonics and climate on Venus recognized

1999 – Mars Global Surveyor altimeter suggests extent of Mars paleoseas, later measures clouds, snow depth

2000 – Mars Global Surveyor images suggest present-day groundwater seepage

The nineties opened new windows into the Earth's past and present climate. New deeper ice cores from Greenland [260] and Antarctica [261] provided high-quality climate records (Figure 7.1) extending back 400,000 years; not only indicating temperatures via oxygen isotopes, but also providing a history of carbon dioxide and methane concentration as tiny bubbles of air were trapped in the accumulating snow and ice. Layers of slightly acidic ice attested to volcanic aerosols – even as the ice record supported the astronomical forcing of climate, it gave insight into the other factors at work. Exquisitely sensitive chemical analyses of the cores even measured tiny traces of lead pollution, not only from petrol additives in the twentieth century, but from the rise and decline of Roman lead and tin mining centuries ago [262].

It is evident in the record that the carbon dioxide abundance changed appreciably with time, and correlates with the temperature changes. But what that means is still a subtle question – presumably astronomically driven sunlight

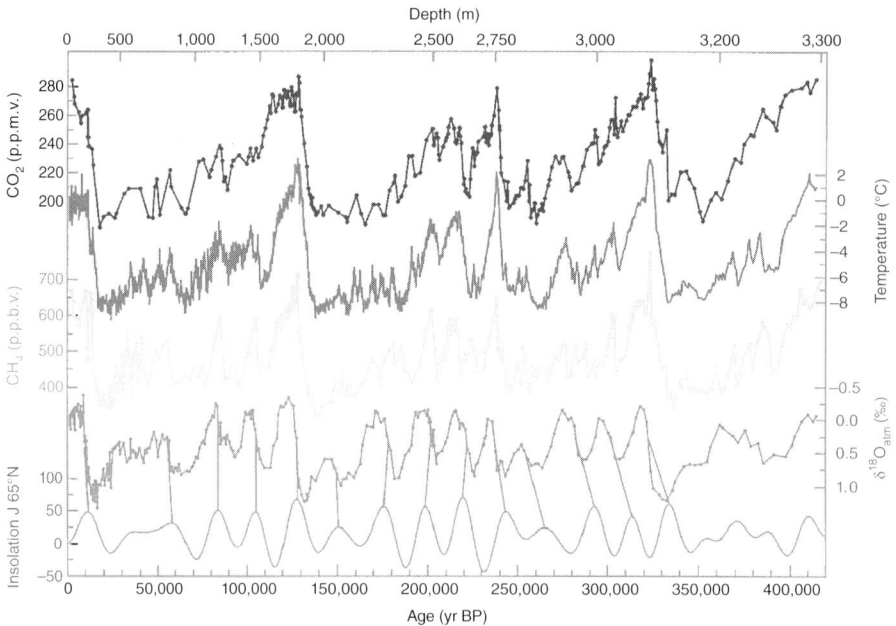

Figure 7.1 The record of temperature, methane and carbon dioxide concentrations (and other parameters) in the Vostok ice core from Antarctica, yielding a high-quality climate record over the last 400,000 years. The 100-kyr eccentricity forcing, as well as the ~20-kyr obliquity cycle, is evident in the methane and CO_2 record as well as in the oxygen isotope (ice volume) signal. Note that the depth scale of the ice core is not linear in time – the snow accumulation rate varies, snow is compressed into ice, and the ice flows, all factors that have to be modeled to reconstruct the time history. (Source: US Global Change Research Program.)

changes have both a direct effect on climate and an influence on the carbon cycle and thus the carbon dioxide abundance, and thus climate indirectly. But the fact that methane tracks many of the same variations shows that it is not merely chemical weathering that is being mediated (volcanic resupply and silicate weathering wouldn't affect methane) – biological activity is presumably involved as a cause or effect or both.

The Greenland core in particular (specifically, the 3028-m-long Summit ice core of the European-funded Greenland Ice-core Project, GRIP) yielded a record dominated in the last 400,000 years by the 100,000-year eccentricity cycle (as in Shackleton and Hays' deep sea cores in the 1970s). But the much higher time resolution showed that there were very rapid and violent swings in the conditions in the North Atlantic [263].[1] The climate had switches as well as knobs – and many of these switches turned out to be in the ocean.

Croll and others recognized the significance to European climate of the heat transported in the ocean, but in climate and weather models up to this point these transports were generally considered fixed, usually by the simple expedient of fixing sea surface temperatures to their present observed values. But to consider climate at the Last Glacial Maximum, the ocean might have been quite different. Not only might the sea ice extent have been different, but with so much water locked up in ice sheets on land, the sea level was about 100 m lower than today. This meant that more land was exposed, and also that parts of the ocean where warm currents like the Gulf Stream can flow more or less unimpeded today (notably a shallow region called the Greenland–Iceland–Scotland sill) became blocked.[2] Furthermore, continental ice sheets could block or divert rivers from their present course, altering the input of freshwater into the sea. While the temperature and/or buoyancy effects of these river fluxes were by themselves modest, it was shown with quite simple models that different circulation modes in the ocean existed, and only small nudges might flip the circulation from one mode to another; e.g. [121].

The ice sheets themselves have dynamics – when the climate changes to favor accumulation of snow on an ice cap, it takes some time for the ice to adjust – at first it simply thickens, but the thicker ice will flow faster in a possibly nonlinear way (depending on how well the ice sticks to the rock underneath – meltwater

---

[1] Dansgaard–Oeschger events – rapid warmings by ~5–8 °C over a few decades, followed by a gradual cooling. Around 25 such events are evident in the Greenland temperature record for the last glaciation. These events are specific to the northern hemisphere: there is rather less variation of this sort in the Antarctic ice core records.

[2] and many landmasses isolated by the sea today were linked: notably, the British Isles were linked to continental Europe, and Siberia and Alaska were not separated by the Bering Strait, allowing humans and animals to migrate into North America.

can lubricate the flow in unpredictable ways). Doug MacAyeal in Chicago devised a "binge–purge" model of the Laurentide (North American) ice sheet, showing how it could grow for about a millenium and a half, then surge outwards as its base gave way, launching an armada of icebergs across the Atlantic each time, before starting to accumulate again [264]. This model sought to explain periodic deposits ("Heinrich events") of ice-rafted Canadian gravel found in Atlantic seafloor mud cores, 3000 km away [265].

The extent to which the Heinrich events and the less frequent Dansgaard–Oeschger events might be linked, and to what degree the respective proposed ice-sheet and ocean circulation mechanisms are responsible for each, continues to be debated. The problem in modeling all this is that the timescales are dramatically different. While the atmosphere has dynamics with characteristic timescales of a few days, the oceans take years to centuries to respond to changes in their boundary conditions, and ice sheets take centuries and longer to grow. With these different timescales for different parts of the Earth system, it is very difficult to perform a simulation that couples them together (these elements of a model have very different tempos, to continue Richardson's paradigm of these calculations as an orchestra). This challenge was addressed in earnest in the late 1990s, stimulated in part by the overall growth in climate science, but by the GRIP ice core results in particular. Stefan Rahmstorf in Kiel (and later, Potsdam), Germany, led a prominent effort, showing that indeed an ocean circulation model could produce the rapid changes suggested by the ice cores [266]; later matching an ocean model to a simplified climate model (i.e. a coupled model of "intermediate complexity") yielded results that could be compared directly with the paleoclimate record [267, 268].

Seas of another kind altogether were the subject of much speculation at this time. Not long after the Voyager encounters (Voyager 2 flew past Titan 9 months after Voyager 1, finding some small changes in the polar haze), speculations about Titan's surface had centered on methane, and whether seas of it or even a hydrological cycle (Figure 7.2) might exist.[3] A photochemical model by Yuk Yung and colleagues at Caltech [269] (who had improbably suggested hydrocarbons for carving the channels on Mars a decade before [270]) showed in 1984 that the amount of methane in the atmosphere would be destroyed in about 10 million years, so either the methane was being resupplied from the interior (perhaps via cryovolcanoes, in a water-as-magma-with-methane-bubbles analog to volcanic resupply of CO_2 on Earth), or it was buffered by a reservoir on the surface. But the radio-occultation data were not compatible with thermodynamic equilibrium with pure methane on the surface (in the way that Mars'

---

[3] Occasional debate ensues as to whether "hydrological" is the right word for the methane cycle. It is: the relevant science is hydrology. The fact that water is not involved does not prevent stellar astrophysicists from referring to "magnetohydrodynamics" for example.

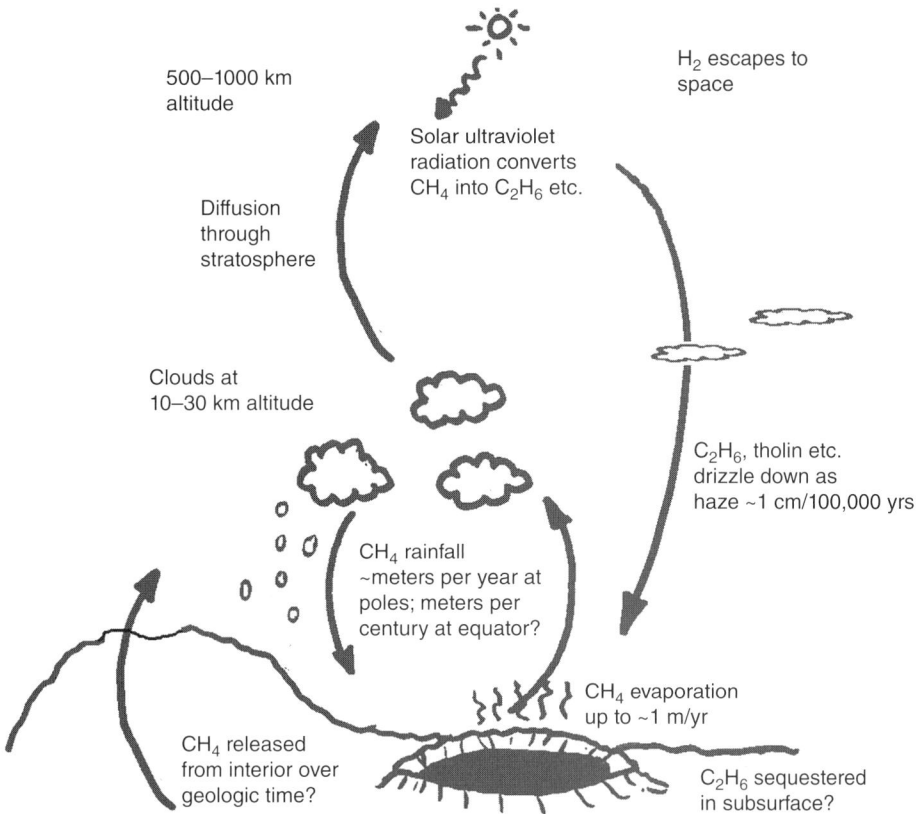

Figure 7.2 Schematic of the methane cycle on Titan. An outer, open cycle saw the irreversible chemical loss of methane by conversion into heavier compounds such as ethane, and the loss of hydrogen to space. An inner, purely physical cycle was once entirely speculative, mirroring the hydrological cycle of clouds, rain and surface liquids that operates with water on Earth, but has now been observed to be active today. The long-term supply of methane and removal of ethane are somewhat uncertain. (Figure by author.)

polar caps are close to balance with the atmosphere) [271]. Straying beyond thermodynamic equilibrium was anathema to planetary physicists, so this was something of a quandry. One way to break the deadlock and yet retain a self-consistent (albeit zero-dimensional) situation was noted by Jonathan Lunine, then a student at Caltech: he pointed out that ethane, which is the main product of methane photolysis, is also a liquid at Titan temperatures, but has a much lower vapor pressure [272]. Thus, if the surface were a mixed ocean of methane and ethane, the ethane as an involatile solute would lower the methane equilibrium pressure (much as spilled syrup takes longer to dry than water, because the sugar as an involatile solute suppresses the water vapor) and thus allow a photochemically plausible ocean composition that could buffer the atmosphere and remain in equilibrium with the observed atmosphere. The models suggested that, if photochemistry had acted on methane for the entire age of the solar system, many hundreds of meters of organics (presumed to be dominated by ethane) would have accumulated on the surface (an idea pointed out by Don Hunten in the 1970s) and so the speculation arose that perhaps Titan might have a deep hydrocarbon ocean. Carl Sagan and dynamicist Stanley Dermott pointed out that Saturn's gravity would cause tides in such an ocean, perhaps wearing down any topography by erosion such that the ocean might cover Titan's surface completely! In these early days of understanding Titan's environment, it was generally considered as a point, with a single temperature defining its climate. A rare exception was Dave Stevenson, also at Caltech, who contemplated what a difference the ~2 K equator-to-pole temperature difference might make, setting up a Leighton–Murray-like seasonal model which indicated that perhaps a methane-rich "liquid polar cap" of a few meters thickness might condense out over the winter pole, floating on Lunine's ethane ocean before boiling off in spring [273].

These speculations heightened interest in exploring Titan in more detail – with a probe like Galileo's and a radar-mapper like those being considered for Venus. The Cassini–Huygens mission, after years of study in the 1980s as a joint endeavor of Europe and the USA, began detailed implementation in 1990, to be launched in 1997 and to arrive in 2004.[4]

The evolution of Titan's conditions became an interesting topic in climatology, because if the Titan atmosphere's methane was buffered by surface seas, then the amount of methane in the atmosphere (and thus the greenhouse effect) would be controlled by the surface temperature, and the possibility of a strong positive feedback might exist (somewhat analogous to the Gierasch/Toon CO_2 cap for Mars). This feedback would operate against a background of steadily increasing solar luminosity, and the possible progressive conversion of a methane-rich sea into a less volatile one as photochemical products accumulated in it (in essence, the paradigm of Titan's ocean composition evolution was like the saltier-with-time paradigm imagined for the Earth's seas by Halley – with the seas steadily becoming progressively enriched in involatile solutes, with ethane substituting for salt). McKay found by coupling his radiative–convective model to a thermo-dynamic model of the ocean–atmosphere equilibrium that the positive feedback might have led to Titan's atmospheric pressure being perhaps a factor of 5 less than today – still a respectable 300 mbar, but with much of the thick nitrogen atmosphere condensed as liquid or even as a partly frozen methane–nitrogen

---

[4] My first job was as a junior payload engineer on the Huygens probe.

slush [274].[5] Similar explorations were made with a grey model by Jonathan Lunine and Bashar Rizk [275].

One claimed motivation for Titan study is that it might resemble in some respects the early Earth. If the early terrestrial atmosphere had large amounts of methane (which it might well have done, since oxygen was not abundant prior to 2 billion years before the present), then perhaps the greenhouse effect of that gas could have offset the faint young Sun. The production of organics by the action of sunlight on methane and nitrogen might have helped produce the prebiotic compounds that led to the origin of life, perhaps in some warm little pond. The similarities between Titan and the early Earth can sometimes be overblown, but are intriguing nonetheless (Earth was likely never as reducing [effectively, hydrogen-rich] nor as cold as Titan today; although in fact early Titan and early Earth might have been interestingly similar).

Radiative transfer models are fun toys. McKay, with Kasting and Toon, ventured in 1991 to explore a Mars climate, deliberately modified by humans – terraforming [276].[6] They quantified what the limits on required gas abundances would be for humans, and for the less exacting demands of plants alone. The temperature-dependent CO_2 greenhouse is stuck in a collapsed state (as described by Gierasch and Toon), and water vapor is even more condensation limited at Mars' present temperatures. The most obvious and effective approach would be to introduce tens of billions of tons of chlorofluorocarbons, which have a powerful greenhouse effect (because their absorptions do not overlap with the already-saturated CO_2 absorptions – in fact, the significance of the greenhouse contribution of growing amounts of trace gases such as methane, nitrous oxide, chlorofluorocarbons and ozone on Earth was recognized in the 1980s [278]). Of course, this amount of gas is far too large to carry from Earth, one would need to set up chemical processing factories to process salt deposits into these gases.[7] Warmed by about 20 K, perhaps eventually release of CO_2 from the polar caps would raise the pressure to a point where plants could flourish. But their conversion of CO_2 into oxygen would take 100,000 years, and, even then, massive amounts of nitrogen as buffer gas would be needed to make Mars habitable. Indeed, nitrogen might be the bottleneck for even allowing plants to survive, and

---

[5] This paper came out mid-way through my Ph.D. research. I thought it was terribly exciting, and was probably instrumental in drawing my interest in planetary climate. I explored doing a postdoc with McKay, but eventually went to Arizona to work with Lunine instead, although we collaborated with McKay extensively in any case.

[6] The high-profile journal *Nature* gave the topic some respectability. McKay had begun thinking about these issues almost a decade before, and was the first person to use the word "terraforming" in the title of a scientific paper – see [277].

[7] As had been suggested by Lovelock and Allaby in their book *The Greening of Mars* in 1984.

a quarter century after this work, we still don't know much about how much nitrogen might be available in Mars' soil.

With the dawn of a new decade, attention of planetary scientists (or geologists, at least) turned to Venus, where the Magellan mapping mission arrived in 1990. Originally a mission wider in scope had been hoped for, but the budget strictures of the 1980s limited Magellan to be a one-instrument spacecraft, a spare antenna from the Voyager program being used for radar mapping and radio occultations. The Magellan and Galileo missions were intended to have been launched in the 1980s, but were delayed by the hiatus in space shuttle launches after the Challenger disaster in 1986. Magellan was designed to map the surface of Venus in about as much detail as Mariner 9 mapped Mars, 20 years before: Venus' thick clouds meant that the only practical way to do that was with radar. The Pioneer Venus Orbiter had carried a crude radar experiment, and a pair of Soviet missions, Veneras 15 and 16, also performed partial radar mapping in 1983,[8] but these results were quickly eclipsed by the superior resolution and quality of the Magellan data, and its near-global coverage. Volcanic features (Figure 7.3) were widespread, and even a few intriguing channels were seen, some hundreds of kilometers long, but these were thought to be due to an exotic kind of lava rather than meteorologically emplaced liquid. Magellan's mapping, which revealed far fewer large impact craters than should have accumulated over the age of the solar system, hinted that Venus may have suffered a paroxysm of volcanism 500 million years ago, which could have "wiped the slate clean".[9]

Indeed, it became recognized around this time that Earth may have been wiped clean of life a number of times. The process of planetary accretion, assembly from smaller planetesimals, was a process that probably didn't end in a smooth, gradual manner, but rather tailed off episodically as the dregs of the planetesimals were finally swept up by the planets. A few final large impacts (of the size that formed the major basins on the Moon) in a Late Heavy Bombardment would have been energetic enough, as noted by Kevin Zahnle, Norm Sleep and others [281], to boil the oceans on Earth, sterilizing the planet in a Venus-like (but wet) pressure-cooker atmosphere of hot steam. Thus, the fact that most of the oldest surviving organisms on Earth, such as archaea found today in hot springs, seem to be "hyperthermophiles" able to tolerate extreme heat, may not be because they were the first life that formed on Earth, but rather they are just the survivors of these violent epochs.

---

[8] These missions also carried the first high-resolution thermal infrared spectrometers to Venus, originally developed in (East) Germany, revealing high-altitude distributions of sulfur dioxide, water and temperature (although the tools to fully extract these results did not emerge for a decade and a half). See [235].

[9] The most comprehensive description of Venus post-Magellan is the 1300-page "Arizona Venus 2" book [279]. A more recent volume that covers some Venus Express results is [280].

Figure 7.3 A synthetic perspective view of Maat Mons, a major volcano on Venus, generated using topographic data from Magellan (the volcano is some 8 km high – heights are exaggerated by 22 times in this view) for the shape, and radar reflectivity for brightness – note the bright lava flow in the foreground. (Source: NASA.)

These transient hot steam atmospheres, studied at length by Yukata Abe and colleagues (e.g. [282–284]), may have been a feature of Venus and Mars (and even Titan) in the first billion years of solar system history as well, and due to the hysteresis of the climate system, would have affected the subsequent habitability of the planets. The calculations of the "habitable zone" (see Chapter 11) by Jim Kasting therefore had to consider "hot start" conditions as well as the "cold start" that the faint young Sun might otherwise have suggested [285].

Smaller, more recent impacts have affected life and climate on Earth. Cratering attracted popular attention around this time, as radar pictures of the Yucatan peninsula in Mexico provided graphic indication in 1995 of an impact structure, named Chicxulub. The existence of the structure (revealed in 1991 from gravity data and oil prospecting drill-core samples) had been suspected from the regional concentration of impact-related rocks and materials bearing the rare element iridium at the Cretaceous–Paleogene geological boundary,[10] 65 million years ago. In the 1980s the father-and-son team of Luis and Walter Alvarez suggested that, because iridium is relatively abundant in meteorites, the worldwide existence of an iridium-rich layer at this boundary implied an extraterrestrial cause for the mass extinctions in the fossil record at this time. In short, that a giant asteroid impact killed the dinosaurs.

This crater, perhaps the third-largest known on Earth, is only about 180 km in diameter – while the devastation by shock, tsunami (the area was a shallow sea at

---

[10] Perhaps still better known outside geology classrooms as the Cretaceous–Tertiary boundary, abbreviated as K-T, via German (Kreide – chalk).

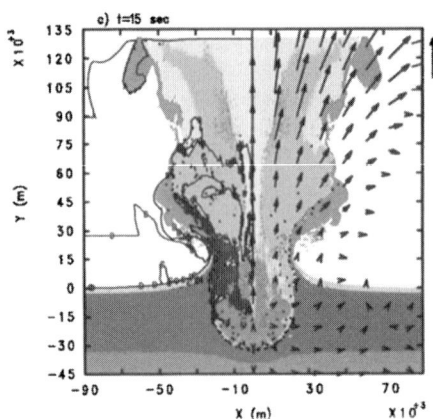

Figure 7.4 Snapshot from a hydrocode simulation of the formation of the Chicxulub crater by the K-T impact: a vast plume of rock vapor (including sulfates) is blasting up through the atmosphere to be dispersed worldwide.[11] The instantaneous cavity ("transient crater") is near-hemispherical – this cavity later rebounds, slumps and widens to become much more shallow (From Pierazzo, E., Kring, D.A. and Melosh, H.J., 1998. Hydrocode simulation of the Chicxulub impact event and the production of climatically active gases. *Journal of Geophysical Research: Planets*, **103**(E12), pp. 28607–28625. Used with permission from Wiley[12].)

the time) and other effects of a ~10-km asteroid hitting the Earth would have been locally unsurvivable, the effects would have been minimal over most of the planet, except for fiercely glowing skies as ejecta thrown into space arced back down in a global meteor shower, broiling the surface for several hours (Figure 7.4). The resultant global wildfires would have been enormously disruptive to ecosystems via soot and pyrotoxins as well as the fire itself, but do not explain the magnitude and breadth of the extinctions (including marine biota such as calcareous plankton).

In this respect, the dinosaurs were especially unlucky. The bolide hit seafloor sediments that were rich in sulfates, such as anhydrite or gypsum: the vaporization of these rocks by the impact and injection of the sulfur material into the upper atmosphere (much more energetically than is possible from wildfires or volcanic eruptions) caused an instant "impact winter", as well as acid rains [286]. Thus, it

---

[11] A hydrocode is a fluid mechanics model, embodying some of the equations also used in GCMs, but concentrating on the thermodynamics of material at high pressure and temperature.

[12] Betty Pierazzo was a dear friend – I remember many great Italian meals at her house in Tucson before she passed away in 2011, aged only 47.

Figure 7.5 A picture of the Earth's limb, taken by a handheld camera through the window of the space shuttle in 1991. The irregular dark horizon shows the towering peaks of convective rainclouds, but well above those are two layers of haze, the sulfate aerosols resulting from the Pinatubo eruption. (Source: US Geological Survey.)

was the climate effects of the impact, rather than the impact itself, that cleared the dinosaur's ecological niche and allowed mammals to take over the land.[13]

In the meantime, conflict on the other side of the world, in Kuwait and Iraq, led to a climate experiment. Oil-well fires set by Saddam Hussein in the first Gulf War in 1991 were feared by Sagan and others to lead to a small-scale regional "nuclear winter". However, very little climate perturbation occurred: the smoke plumes did not rise into the stratosphere where the injected soot would be long-lived, but climbed only a few thousand meters, remaining in the troposphere where they were quickly scavenged away by rainfall. In contrast, later in 1991 Mt. Pinatubo in the Philippines saw a violent eruption, injecting more sulfate aerosols (estimated at 20 million tons) into the stratosphere (Figures 7.5 and 7.6) than in any event since Krakatoa. The aerosols led to a ~0.5 °C temperature drop worldwide for 2–3 years, and enhanced ozone depletion.

---

[13] There was also significant volcanic activity at the time, forming the Deccan Traps in India. A few scientists argue the gas emissions from these eruptions was a significant or even decisive stressor on the dinosaurs and/or the other species that became extinct. A few even argue that the impact caused the eruptions, which are almost antipodal to the Chicxulub site. My view is that the impact-generated volcanism is implausible (the seismic energies communicated by the impact are not large), and that basaltic eruptions do not inject SO_2 high in the atmosphere, so their climate impact would be modest and localized. Conceivably volcanic emissions could cause ocean acidification, in which case they could have contributed to the marine foraminifera extinctions, but not those of the dinosaurs. The impact seems likely to have been the principal cause of their demise.

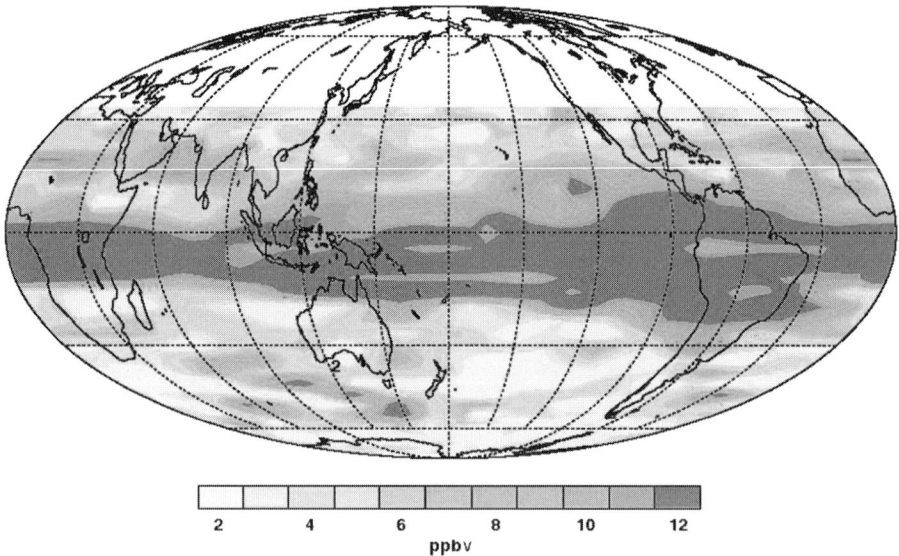

Figure 7.6 Stratospheric sulfur dioxide concentration measured by the Microwave Limb Sounder on the UARS satellite, about three months after the Pinatubo eruption. The amount of SO_2 had declined by a factor of a few from its peak already, and spread into a belt around the tropics. The peak amount even in this exceptional layer is ten times less than the "background" level in Venus' atmosphere. (Source: NASA/GSFC.)

As the Magellan data were digested, laboratory measurements and modeling struggled to make sense of how Venus' climate works. There was nothing to say that the Rasool and Debergh paradigm of "the surface is hot, Venus' entire inventory of CO_2 is sweated out of the rocks and into the atmosphere" – i.e. a CO_2 runaway – was wrong. But it had not escaped scientists' notice that the surface pressure was not far from a possible mineral equilibrium – perhaps the atmosphere was buffered after all.[14] However, while a match of this chemical equilibrium to present conditions could be entertained (although such equilibrium would require gas and rock to interact without the enhancing effect of liquid water or plant roots that assist similar exchange on Earth), Mark Bullock and David Grinspoon at the University of Colorado showed in 1996, with a radiative–convective (R-C) model (improved over Pollack's 1980 version), that the temperature in the R-C model increased faster with pressure than the temperature needed in chemical equilibrium with the rocks had to increase to create that pressure [287, 288]. In other words, the equilibrium was unstable – nudge the system a little and it would run away. George Hashimoto and Yukata

---

[14] The buffering reaction is calcite–quartz–wollastonite $CaCO_3 + SiO_2 \leftrightarrow CaSiO_3 + CO_2$, originally proposed by Harold Urey in 1952.

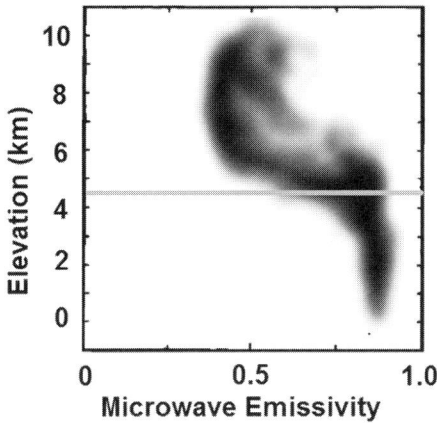

Figure 7.7 Pyrite frost on Venus? Above a threshold altitude on Venus of around 5 km, the microwave properties of the surface show a dramatic change (here, the emissivity drops), consistent with the presence of some high-dielectric-constant material that is stable only in a specific pressure–temperature range. (Figure by author.)

Abe in Japan also examined the same question, and noted that the equilibrium condition defines an elevation, somewhat like a "snow-line", above which carbonates are stable [289].

These workers, drawing on laboratory work by Bruce Fegley and others, also considered another chemical equilibrium – whether anhydrite might equilibrate with calcite to mediate the amount of sulfur dioxide in Venus' atmosphere.[15] However, this equilibrium yielded SO_2 amounts a hundred times less than the amount observed. Another idea was that iron sulfide – pyrite (fool's gold) – might control the SO_2 abundance instead, in a buffer reaction with magnetite.[16] Pyrite is of particular interest on Venus because Pioneer Venus and Magellan radar data showed evidence (Figure 7.7) of some kind of metallic "frost" deposit. At a certain range of altitudes (and, thus, temperatures and pressures) on Venusian mountaintops, the electrical properties of the ground changed dramatically, with the reflectivity increasing substantially and the emissivity dropping [290]. Just as pyrite glitters to the eye, so its dielectric properties fit the radar results. But the agreement is far from unique, many other sulfide minerals such as lead or bismuth, or some exotic metal frosts such as tellurium, might explain the surface properties. Unfortunately, the uncertainties in the near-surface atmospheric composition did not allow the choices to be confidently narrowed, and this remains an important question today.

[15] $SO_2 + CaCO_3 \leftrightarrow CaSO_4 + CO$ [16] $3FeS_2 + 16CO_2 \leftrightarrow Fe_3O_4 + 6SO_2 + 16CO$

The other big-picture possibility, of course, is that the sulfur dioxide is not buffered at all, but merely represents (as the CO_2 amount on Earth) the dynamic balance between supply (presumably volcanic) and destruction. Such a situation would also allow for the excursions in its abundance that had been seen by Esposito.

Taking the modeling further, Bullock and Grinspoon examined what would happen if they kicked their model with a massive outpouring of sulfur dioxide and water vapor that would plausibly have accompanied the putative global resurfacing by volcanism hundreds of millions of years ago [291]. They coupled their radiative–convective code to a chemical/microphysical model (much like McKay's model for Titan) to examine what might happen to the cloud structure given such a large perturbation. Interestingly, they found (Figure 7.8) that the resultant sulfuric acid clouds gave a net cooling for 100–300 million years, until surface rocks (they assumed the ground could exchange gas with the atmosphere) mopped up the SO_2, at which point the atmosphere would clear somewhat allowing surface temperatures to jump to some 900 K as the greenhouse took over, until eventually photolysis and the escape of hydrogen brought the water vapor abundance back down. This scenario intrigued the Magellan geologists and geophysicists – such a prolonged and intense cooling and heating would cause

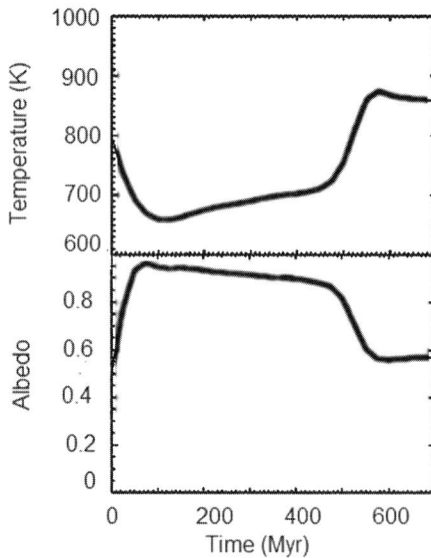

Figure 7.8 Post-volcanic Venus climate scenario explored by Bullock and Grinspoon [291], with rapid outgassing (100 Myr) followed by slower escape (160 Myr). Initially the elevated SO_2 abundance thickens the cloud layers and brightens the planet, leading to cooling. Then, as the sulfur is eaten up by the surface regolith, the clouds thin and temperatures rise.

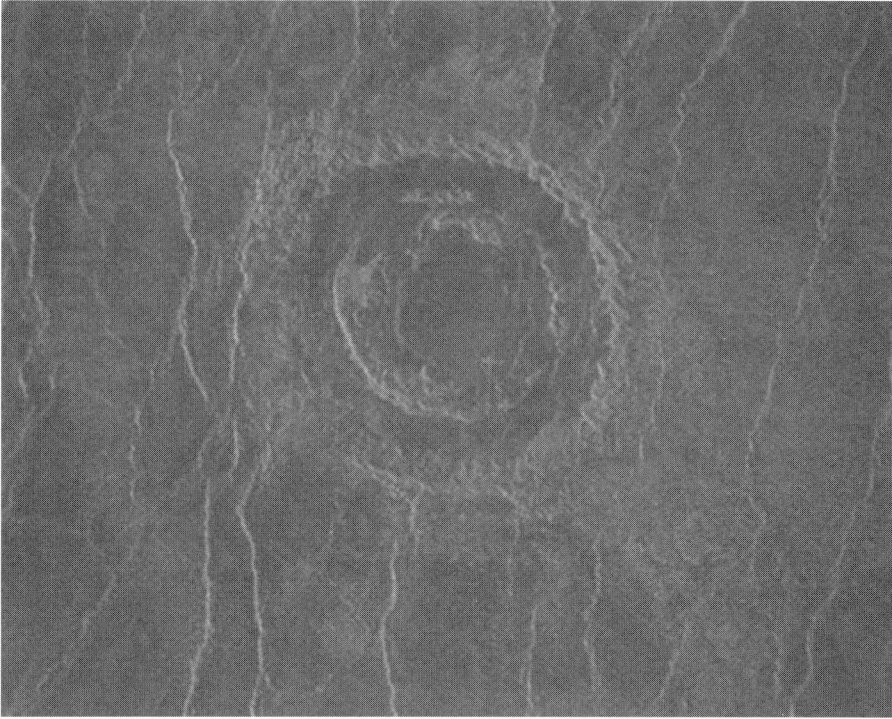

Figure 7.9 The crater Barrymore on Venus, imaged by the Magellan radar, is about 50 km across, and is superposed on somewhat regularly spaced wrinkle ridges that are consistent with what might be expected from compressional stress caused by climate change. (Source: NASA/JPL.)

contraction and expansion of the crust to a depth of several kilometers. Sean Solomon and others noted that the resultant stresses could cause a characteristic pattern of fractures to form (Figure 7.9), which had been seen in Magellan data [292, 293]. Thus, despite lacking a hydrological cycle, Venus' climate could modify the planet's landscape. More generally, surface temperatures might also influence how rapidly heat was lost from the planetary interior – in short, that Venus' climate evolution and the tectonics and internal evolution might be coupled in both directions [294].

While Venus enjoyed sustained attention for a decade after Pioneer Venus, Mars exploration after the post-Viking decade was stillborn. A Soviet plan to send two spacecraft to the Martian moon Phobos-2 in 1988 met bitter disappointment, with one spacecraft lost a few months after launch, and the other failing while tantalizingly close to its target (after at least returning some data). These missions carried significant instrumentation, not only from the Soviet Union, but also from Western European countries, themselves buoyed by the success of comet exploration with the European Space Agency's Giotto probe in 1986. But this investment was cruelly lost in the double failure, and indeed not long thereafter the Soviet Union itself unraveled, bringing a change to the Cold War world order. The upheaval in the Russian economy delayed a follow-up mission, Mars-94 to become Mars-96, but then this too was lost in a launch failure.

The American scientific wish-list post-Viking was to be addressed largely by a large spacecraft, Mars Observer, based around a proven terrestrial satellite design to monitor seasonal changes in dust and water vapor in the atmosphere, image the surface in much more detail and map its topography (compared to the thickness of the atmosphere, the height of Martian terrain varies much more than the Earth; circulation models would need this information to understand Martian winds). But contact was lost just as the spacecraft fired its engines to brake into Mars orbit in 1993. Another item on the wish-list, a global network of small surface weather and seismic stations, was seen as risky until a low-cost landing system using airbags could be proven.[17] So a small, single-station mission called Mars Pathfinder, with a very limited payload of a camera and meteorology package, was developed to demonstrate the approach, and was launched in 1996.

The region around the Pathfinder landing site was strewn with rocks (Figure 7.10), some leaning on each other ("imbricated"), having been deposited by a flash flood like that believed to have occurred in the channeled scablands of Washington State. This was not even the first time that water was "discovered" on Mars, but was at least a vindication of the interpretation of Viking data that had guided the selection of the landing site as an outwash plain, the idea being that lots of different rocks would have been brought there from different places. The instrument to study the compositions of these rocks was positioned by a small rover, Sojourner, which stole the show. In pictures that suddenly everyone in the world could easily view shortly after they were taken via the explosively proliferating World Wide Web, the progress of our robotic presence on Mars could be followed, trundling around in the scene from rock to rock. A few of the other pictures showed transient smudges on the horizon – dust devils.

Astronomical techniques relevant to planetary studies had been advancing, with the first experimental adaptive optics systems on large telescopes coming into operation. The Hubble Space Telescope, whose launch had been delayed to 1990 by the Challenger disaster, had a slow start in imaging the planets. First, it took some years for its ability to track solar system targets to be developed, and its myopic primary mirror had to be corrected: in any case, it was a heavily

---

[17] In fact, the (failed) Soviet Mars landers of the early 1970s had used airbags.

Figure 7.10 Part of the panorama seen by Mars Pathfinder, showing the low hills named "Twin Peaks" about 1 km away. The scene is littered with boulders, brought by a catastrophic flood that also carved streamlined islands nearby, seen in Viking images. At left is the Sojourner rover: accumulation of dust on its solar panels, causing their output to drop ~0.3% per day, may have set expectations for the MER rovers a few years later. (Source: NASA/JPL/U.Arizona.)

oversubscribed asset with only about a tenth of its time devoted to a few carefully chosen planetary studies. Among these were the first near-infrared maps of Titan, and some showcase images of Mars (Figure 7.11). An opportunistic target was the dramatic barrage of fragments of comet Shoemaker–Levy-9 onto Jupiter – a striking reminder that the impact process that did for the dinosaurs is not a historical process, but an ongoing one.

But it was in fact with rather modest telescopes that something of a revolution in astronomy began in 1995 with the discovery of a planet around another star, ushering in the era of exoplanet studies.[18] The innovation that made the discoveries possible was the use of very high resolution spectrographs, with accurate wavelength calibration, which permitted the detection of a sinusoidal Doppler shift as the star was tugged back and forth by the planet. The first definitive detection was announced by Michel Mayor and Didier Queloz of the Observatoire de Haute Provence in France, finding a Jupiter-size planet in a surprisingly close orbit around the star 51 Pegasi. Large, close planets with short periods were the

---

[18] Actually, a set of planets were identified several years before. Whether because this was a set of planets around a pulsar, the exotic spinning core of a massive star that had exploded, or because the discovery was by not-very-photogenic radio astronomy, this did not seem to have nearly the impact that the later optical discovery of 51 Peg b did.

Figure 7.11 The Hubble Space Telescope in the 1990s surveyed a number of outer solar system targets (in fact Venus is difficult owing to safety considerations about aiming the telescope and its sensitive instruments too close to the Sun) and Mars: the earlier image on the left shows both polar caps and light streaks of water ice clouds around the limb in June 2001 and especially around the north pole; the image at right two months later lacks these clouds and has muted contrast owing to a global dust storm. (Source: NASA, J. Bell (Cornell), M. Wolff (SSI), and the Hubble Heritage Team (STScI/AURA).)

easiest to detect by this means, but it was clear that while planets may be common around other stars, the architecture of our solar system was not a universal template. Other planet discoveries trickled in in the subsequent years, becoming a torrent by the end of the 2010s. Many are truly extreme worlds, and are discussed in Chapter 11.

But the extremes of our own world's past were still being uncovered. Paul Hoffman and Dan Schraag of Harvard published evidence in 1998 from rocks in Namibia that low-latitude glaciation had occurred in the Earth's past [295, 296]. Others had noticed similar evidence before, but lacked an explanation of how the Earth could have become un-frozen. They furthermore lacked a catchy name. But Hoffman and Schraag used "Snowball Earth", a name devised in an obscure paper a few years earlier by Joe Kirschvink at Caltech, who was interested in magnetic signatures in rocks, and whether glacial deposits had been laid down at low latitudes [297].[19]

---

[19] The story of the discovery and reception of the Snowball Earth idea is engagingly told in the book of that name by Gabrielle Walker (Bloomsbury, 2003).

The possibility for the Earth to fall into a runaway glaciation was known from Budyko's work.[20] It was possible to contrive ad-hoc perturbations to the Earth system that might kick it out of a frozen state, but the new work pointed out that a very simple effect would cause an inevitable end to the Snowball. The removal of carbon dioxide from the atmosphere relies on liquid water and air acting on silicate rocks: if the earth were substantially frozen, this removal process would be suppressed and CO_2 – emitted from volcanoes, as well as by animal life – would accumulate. Thus, the greenhouse effect would progressively strengthen, until the ice began to melt and expose dark sea, and the glaciation would runaway in reverse. Indeed, the abundant CO_2, suddenly warming temperatures, and prodigious deposits of finely ground glacial sediment would set the stage for rapid formation and deposition of carbonate rocks – neatly explaining the limestone cap rock in Namibia, perched above the glacial sediment.

This Snowball Earth was interpreted to have last occurred around 670 million years ago, but probably several times prior to that. An obvious question is how did life on Earth survive if the whole planet were frozen? One environment on Earth that offered some hints is the McMurdo Dry Valleys in Antarctica, where a wall of mountains prevents the ice sheet from flowing in. The combination of low temperature and minimal precipitation makes this one of the most Mars-like environments on Earth. Yet some organisms survive, cyanobacteria living just under the surface of rocks, which eke enough moisture to survive from the few flakes of snow that blow in and settle and melt on the rocks when they are warmed by the Sun. Similarly, some snowmelt trickles down into a series of hypersaline lakes that are near-permanently ice-covered. And yet, despite the low temperature and high salinity (the salt acts as an antifreeze, so these lakes remain liquid beneath the ice), these lakes have rich ecosystems.

But it was from the "blue ice", part of the Antarctic ice sheet that sublimes away in the dry wind off the mountains, leaving behind whatever was entombed in it, that the real impetus in Mars exploration came. In 1997, the announcement of the discovery of possible indicators of past life in ALH84001, a meteorite retrieved from Antarctica (Figure 7.12) in the mid-1980s, grabbed the world's attention.[21] Ultimately most of this work was challenged, but this meteorite, known to be from Mars, served as a focal point for the Mars program.

---

[20] to which the Snowball paper refers, although as this book has noted, Croll qualitatively recognized the ice–albedo feedback long before, and Opik had found runaway glaciation in his 1965 ice-line model [125].

[21] The meteorite's designation indicates it was the first find of the season, in 1984, in the Allen Hills region. The meteorite was determined to have come from Mars because the noble gas composition of bubbles in the rock exactly matched that measured by Viking at that planet. Beginning with the SNC meteorites (shergottites, nakhlites, chassignites) in

Figure 7.12 Mars Sample Return, the easy way. The US ANSMET (Antarctic Search for Meteorites) program has acquired several samples of the Martian crust, blasted off the red planet by asteroid impacts to fall millions of years later on Earth. A few of those rare arrivals – like ALH84001 – have landed on the ice sheet, been buried and transported in the ice to sites where dry winds ablate the ice away, leaving the meteorites exposed on the glazed surface to be found by teams of researchers camped out for weeks. (Photo: Jani Radebaugh.)

With the Snowball Earth idea stimulating thought on Earth's deep past, indications from the Galileo spacecraft that Jupiter's moon Europa might harbor a habitable ocean of liquid water beneath a relatively thin ice crust, but especially with interest in laboratory analyses of Mars samples such as ALH84001, astrobiology became a buzzword in NASA, which established an Astrobiology Institute to act as a focal point for multidisciplinary studies into the origins of life and related topics.[22]

the 1980s, over a hundred Mars meteorites are now known. The Nakhla meteorite, which fell in Egypt in 1911 as dozens of fragments, is notable for the possibly apocryphal story that one fragment killed a dog.

[22] These topics already received some NASA support, but at a much lower level, through a much less prominent exobiology program. Cynically, the astrobiology juggernaut saw much ongoing work simply rebadged as astrobiology. Nonetheless, the enterprise spawned many interdisciplinary investigations that otherwise would not have happened.

Galileo dropped off a probe into Jupiter's atmosphere in 1995, much like the Pioneer Venus probe, to measure its composition and determine how light and heat were absorbed and transmitted through the clouds. A pair of crossed wires that delayed its parachute deployment meant measurements happened at a lower level than planned, but the hardy vehicle continued to transmit down to a level of about 20 bar. The usefulness of some of its measurements, however, was limited because the probe happened (as indicated by telescopic observations from Earth) to descend in a downwelling region of dry air that was relatively free of water vapor and clouds, making it unrepresentative of Jupiter's atmosphere as a whole.

The giant planets like Jupiter differ from Titan in that their most abundant constituent is hydrogen, whereas that on Titan is nitrogen. While hydrogen has an appreciable greenhouse effect, nitrogen's is fairly minimal (indeed, because nitrogen interacts so little with solar or thermal radiation, it is hard to detect and so was typically the major unknown in planetary atmospheres until they were explored by a probe). Theoretical studies of Titan's climate in 1997 by myself, McKay and Lunine [298, 299] asked the question of what happened to Titan's climate if the methane ran out? We knew from the Hubble maps and radar data that Titan didn't have a global ocean of liquid methane, so perhaps the methane seen today was only episodically introduced by volcanism or some other process. But it would be steadily depleted by photolysis, so maybe it sometimes ran out. The loss of the methane and hydrogen greenhouse warming would lead, as in the faint early Sun cases considered before, to collapse to a thinner-atmosphere state. But even invoking a brighter surface albedo would not cause Titan's atmosphere to collapse completely – there would still be a residual atmosphere of 100 mbar, much thicker than Mars' present atmosphere. But we recognized that McKay's model did not properly handle the nitrogen clouds that would form under such cool conditions, and furthermore that the collapse process depends on the coldest (polar) temperatures, not just the average, and so a 1-D model was not an adequate tool.

The team also pushed McKay's radiative–convective model in another direction, to see what happened when the Sun increased in luminosity to become a red giant (Figure 7.13). The warming would be rather more than one might expect from the increased sunshine alone: the reduction in ultraviolet flux from the reddening Sun would produce less haze, weakening the antigreenhouse effect, and red light would be better able to filter through what little haze there might be. Titan might easily become habitable in future: in fact, Stephen Baxter's novel *Titan*, published in the same year as this study and capitalizing on the publicity around the Cassini launch in October 1997, portrayed (independently) this very scenario.

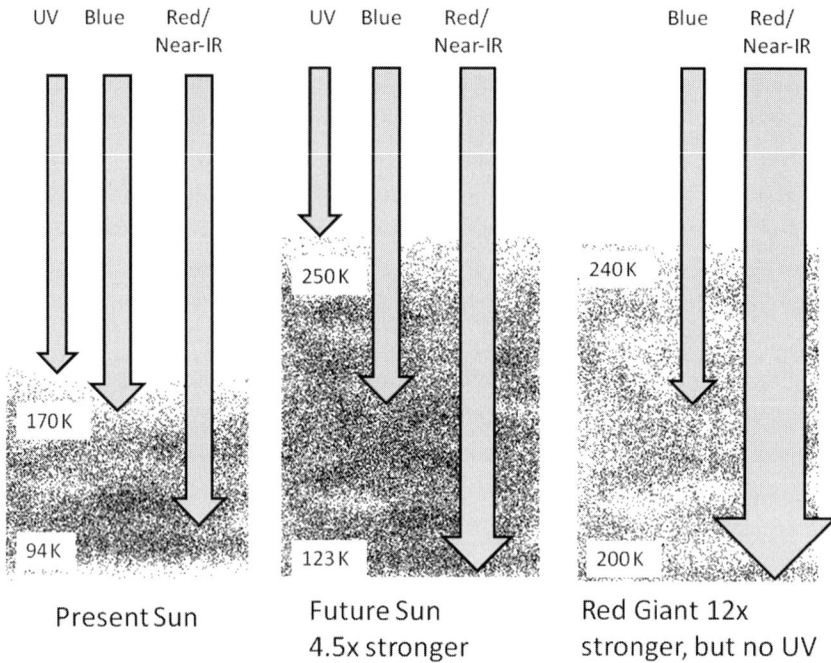

Figure 7.13 Schematic of changing radiative transfer as the Sun warms then reddens. Simply increasing the solar flux increases the temperatures somewhat, but not as much as might be expected because the haze layer thickens, the visible optical depth increasing from about 1 to about 1.6. When the solar spectrum reddens and ultraviolet flux drops, the haze thins, allowing more sunlight to reach the surface, which warms substantially. The thermal opacity is held constant at ~4 in this model: in reality there may be a strong positive greenhouse feedback, which may substantially increase surface temperatures. (Figure by author.)

In 1997, Carl Sagan[23] and Chris Chyba resurrected the old idea that ammonia might have acted as a greenhouse gas on the early Earth to alleviate the faint young Sun. Their new wrinkle was that if there was methane too, then Titan-like haze produced by ultraviolet light would block that light from destroying the ammonia (which had been an earlier objection to the idea). In fact, McKay and colleagues shortly thereafter showed that if the haze was enough to shield the ammonia, it would be thick enough to cause an antigreenhouse effect, defeating the warming.

[23] Sagan himself passed away in December 1996.

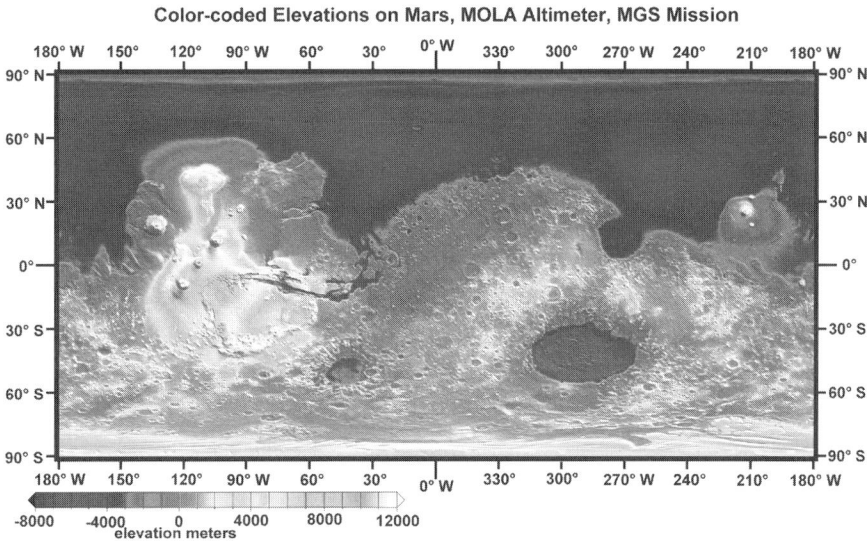

Color-coded Elevations on Mars, MOLA Altimeter, MGS Mission

Figure 7.14 The map that transformed Mars science, showing the tremendous range of elevations on Mars, from the deep Hellas basin to the summits of the Tharsis volcanoes. These topography data were instrumental in building accurate models of the Martian circulation and climate. The smooth northern lowland plains may have been a seabed in a wetter epoch. (Source: NASA Goddard Space Flight Center.)

Picking up the pieces of Mars Observer, NASA's Mars efforts regrouped.[24] The instruments flown on Mars Observer were rebuilt and spread across smaller spacecraft, of which the first was Mars Global Surveyor, launched in 1997. This generated a global topographic map of exquisite detail and precision with a laser altimeter[25] (Figure 7.14) – a map that was important in defining Mars' terrain, which is a significant factor in shaping its winds.[26] The altimeter was in fact accurate enough to measure the thickness of the layer of carbon dioxide frost

---

[24] An excellent review of the Mars program, at least as seen at the Jet Propulsion Laboratory, is Ref. [300]. This well-researched story describes not only the successes of various missions (amply covered elsewhere too) but also the various committees that determined mission priorities, the various false starts of missions that were contemplated but never flown, and the challenges encountered during mission and instrument developments.

[25] A great description of Martian cartography and its culmination in the MOLA map is Ref. [301].

[26] Earth's highest mountains reach ~10 km above sea level, a little over one scale height. Mars' topography, lacking a sea-level reference, is based around a datum level that sees a 6.1 mb annual average pressure. The lowest basin, Hellas, is 8 km below this datum, while the summit of Olympus Mons is 21 km above it. The reference pressure level was chosen because this is the saturation vapor pressure of water at its melting point – i.e. above this altitude, water boils as soon as it melts.

Figure 7.15 Over a dozen gullies nearly 1 km long run down the south-facing slope of the Nirgal Vallis (29° S) wall in a Mars Global Surveyor image from 1999. Each narrow channel starts at about the same position below the top of the valley wall, indicating that there is a layer along which a liquid – most likely, water – percolated until it reached the cliff, then ran downhill to form the channels and the fan-shaped aprons at the bottom of the slope. Some of the apron deposits seem to cover the dunes on the floor of the valley, suggesting that the channels and aprons formed more recently than the dunes. The fact that neither the dunes nor the aprons and channels have impact craters on them suggests that these features are all geologically young, meaning a few million years at most, a few days or weeks at least. Most of the gullies reported in MGS/ MOC data were in fact seen further towards the pole than this example. (Source: NASA/ JPL/Malin Space Science Systems.)

that was deposited in the Martian polar winter. A thermal emission spectrometer began monitoring dust and water vapor in the Martian atmosphere, beginning a nearly continuous record that goes on today.

A camera offering resolution down to a few meters (many times better than Viking) showed close-ups of the enigmatic Martian surface (Figure 7.15). Dust devils were seen in action, sometimes leaving dark wavy tracks on the surface where they had scoured the dust away. But most dramatically, transient streaks and gullies were seen in the walls of midlatitude craters [302]. Perhaps water might sometimes be flowing even today! The public were primed by science fiction (Kim Stanley Robinson's Red Mars trilogy,[27] published 1992–6),

---

[27] Robinson is well engaged with the science community, and even spent a season in Antarctica. I met him at a terraforming workshop held by Chris McKay at NASA Ames.

which told of changing conditions and catastrophic floods. The Arnold Schwarzenegger movie *Total Recall* brought Mars conditions to dramatic, if implausible, life in 1990.[28]

Most of the rest of Mars Observer's planned payload, a radiometer to map the three-dimensional distribution of water vapor and dust in the Mars atmosphere, and a wide-area color camera especially suited to monitor dust storms and the growth and decay of the polar frost caps, were launched on the small Mars Climate Orbiter in 1998. Hot on its heels, the Mars Polar Lander, destined for the layered terrain near the Martian south pole, was sent on its way, carrying an instrument package called MVACS, Mars Volatile and Climate Surveyor. The emphasis of Mars studies was aiming towards climate, but cruel disappointment was soon to come.

The polar regions are particularly interesting for climate history, in that they are more sensitive to changes in sunlight input. Questions that guided design of the lander's instruments mirrored many of those that confronted scientists in the nineteenth century contemplating the Earth – was there water ice in the regolith? Was a once-thicker carbon dioxide atmosphere locked up in carbonate rocks? Meteorological measurements at high latitude would also give insights into the Mars climate that the lower-latitude Vikings and Pathfinder could not give.

But the overriding questions were "how warm?" and "how wet?". Were the channels seen by Mariner and Viking carved by flowing water, or something else? If it was water, was it some occasional anomalous event, like a jökulhlaup (an Icelandic word for a catastrophic flood caused by geothermal melting of water at the base of an icecap – in fact Carr used this term in a Mars context in the 1980s), or was there actual rain on the surface. And if there had been abundant water, where was it now? This latter question has relevance not only for climate history, but for where extant or fossil life might be found, and perhaps one day as a resource for human exploration. As a result of these considerations, and the need for a crisp mantra with which to explain NASA's intent to budget-setting politicians, "Follow the Water" became the succinct rationale for US Mars exploration in the first decades of the twenty-first century.

Although a handful of experiments with numerical models of the Martian general circulation had been made in the 1970s and 80s after the pioneering work of Leovy and Minz, circumstances now opened the way for much more work in this area. First, the new observations at Mars gave new questions – phenomena that modelers could attempt to replicate. Second, a major unknown in Mars meteorology had been the topography – just how tall were the

---

[28] Other Mars movies emerged in the following decade, notably *Mission to Mars* and *Red Planet*. None were especially good, however.

mountains, how deep the valleys? The global topography dataset from the laser altimeter on MGS gave the models a firm place to stand. And, of course, computers had been continuing to grow in capability and fall in price: what had once taken a supercomputer, accessible only to a privileged few, could now be performed on workstations or later clusters of computers (like Richardson's dream, but with an orchestra of microprocessors, rather than human calculators) that were affordable to universities and other research groups. Moreover, the internet made planetary data accessible, and the sharing of computer codes and their results easy.

Several groups became established in developing planetary GCMs, applying them to Mars in particular. Bob Haberle at NASA Ames was among the first, and many postdocs and students fanned out to other institutions as the ability to run models became widespread. The "original" GCM group, where Jule Charney had developed terrestrial models, now the Geophysical Fluid Dynamics Laboratory (GFDL) at Princeton, also got into Mars modeling. In Europe, groups at Oxford University and at the Laboratoire de Meteorologie Dynamique (LMD) developed independent capabilities too.

A few initial experiments began in Titan GCMs too, at LMD, at the Goddard Institute for Space Studies (GISS) in New York, and then at the University of Cologne in Germany, stimulated by anticipation of Cassini. Titan of course lacked the observed details of Mars, but the fundamentals of a thick, hazy atmosphere on a slowly rotating world offered a novel environment to explore. Such a thick atmosphere is much more challenging to model, in that the system has a large "memory" and so it takes a much longer run to "spin-up" than does Mars. A major challenge for these early Titan models was to generate an atmospheric superrotation, hinted at by Voyager's equator-to-pole stratospheric temperature contrast, and by a stellar occultation, which indicated that the stratosphere rotated fast enough to be slightly bulged. An early experiment by Tony Del Genio at GISS suggested that such rotation might be a general feature of optically thick atmospheres on slowly rotating planets – after all, such strong zonal circulation at high altitude was readily observable on Venus [303]. The first Titan GCM experiments at LMD by Frederic Hourdin seemed to generate promising rotation [304], but several subsequent models in Germany and the USA struggled to do so. Was this early result some lucky tuning of the model?

While GCMs grappled mostly with the present-day climate of Mars and Titan, the deeper history and the faint young Sun problem remained unsolved. In 1997, François Forget and Ray Pierrehumbert proposed a new angle on the Mars climate question. The challenge to a simple CO_2 greenhouse as an answer to the faint young Sun problem had been that adding more gas would lead to condensation, as clouds and onto the polar cap.[29] But Forget and Pierrehumbert realized that clouds of CO_2 ice crystals would have quite different optical properties in the infrared than do water clouds on Earth. Specifically, they would scatter infrared radiation, causing a greenhouse effect of their own [305].

On the other hand, Yuk Yung proposed that adding a dash of SO_2 to the atmosphere would not only augment the CO_2 gas greenhouse, but would warm the upper atmosphere to prevent CO_2 cloud formation [306]. Thus it seemed (as had been the case with zero-dimensional models in the 1970s) that one could wave one's hands and come up with any answer one wanted.

The laser altimeter on Mars Global Surveyor showed with stark precision that indeed many features tentatively suggested by Tim Parker in Viking data to be shorelines were indeed at a nearly uniform elevation, consistent with the idea that a sea ("Oceanus Borealis") had covered the northern polar region [307]. The altimeter even was able to measure the thickness of the seasonal CO_2 frost deposits as they grew and shrank [308], and picked up some reflections well above the polar caps as CO_2 clouds formed in the winter darkness [309].

During MGS's early low-dipping aerobraking orbits, it brought a magnetometer closer to the surface than had the Mariner missions. Strikingly, this instrument found stripes of magnetization in the Martian rocks, frozen in from the deep past (Figure 7.16). The discovery of these fields, normally a purely geophysical investigation, was germane to climate and habitability for a couple of reasons. First, the stripes were in some ways reminiscent of those discovered on the Earth's seafloor, which had been primary confirming evidence for continental drift and plate tectonics as originally proposed by Wegener: perhaps the magnetic stripes were (circumstantial) support for an earlier epoch on Mars where the plate subduction is lubricated by water, and carbonates were decomposed to return CO_2 back to the atmosphere. Second, a dynamo field (recorded by the stripes) may have been a profound factor in controlling how the solar wind could have affected the loss of gas (notably water) to space. There was another angle too – the purported biological features in ALH84001 contained crystals of magnetite, which were claimed (by Kirschvink and others) to resemble those in magnetotactic bacteria on Earth [310, 311]. These use the magnetic field essentially to work out which way is up and so to swim to their favored environment – perhaps Mars bacteria had done the same thing. In fact, piece by piece, the various lines of evidence for life in ALH84001 were picked apart by the community (indeed, an early and obvious caution was that the "cells" in the

---

[29] Jim Kasting pointed out another challenge with arbitrarily large CO_2 greenhouse atmospheres: Rayleigh scattering by the thick gas – Mars would become a pale blue dot – would increase the planetary albedo, giving the greenhouse less solar flux to work with.

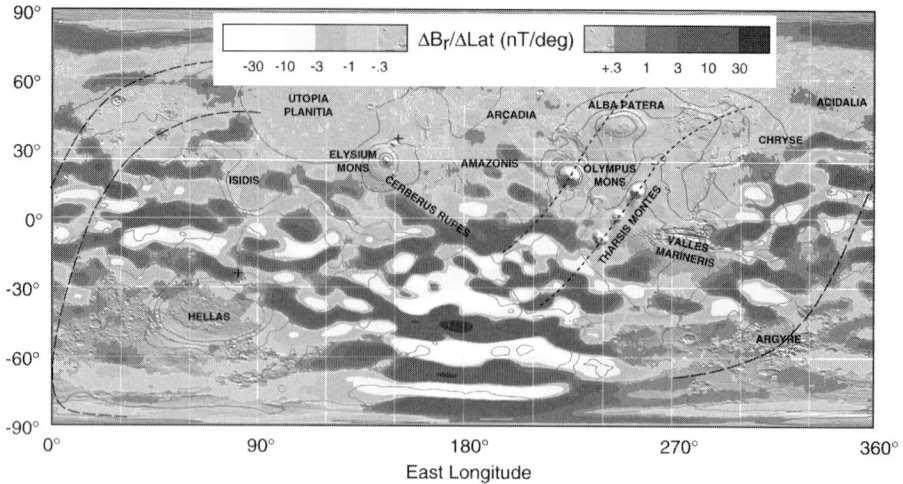

Figure 7.16 A map of Mars' magnetic field, frozen into the rocks. The data are filtered to bring out the stripes, which may be suggestive of some sort of early plate tectonics. A dynamo magnetic field, of which these crustal signatures are just a fossil remnant, may have influenced how Mars lost part of its atmosphere and water to space. (Author, from data in Connerney et al. [312].)

meteorite were much smaller than even the smallest terrestrial bacteria), but this period saw much detailed contemplation of Mars conditions and life.

Meanwhile, Mars was becoming internationalized. With shiny spare copies of powerful instruments left over from the Mars-96 development sitting in European institutions, the European Space Agency elected to start a rapid, streamlined spacecraft development, leading to the name Mars Express. And Japan, which had already dipped its toes into interplanetary waters with two spacecraft to Halley's Comet in 1986, developed a small Mars orbiter called "Planet-B", renamed "Nozomi" after its launch in 1998. This spacecraft was to study Mars' upper atmosphere in particular, and understand how gas was lost to space. However, the Japanese space agency had to ingeniously redesign the mission to cope with underperformance of its rocket engine on Earth departure, leading to Mars arrival in 2003 instead of 1999 as planned.[30] This much longer cruise exposed the spacecraft to more misfortune, and its electronics were damaged by solar flares in 2002, and the mission was unable to get into Mars orbit.

---

[30] Actually, at the time there were two agencies – Nozomi being the purview of the smaller, scientific one, the Institute for Space and Astronautical Science. The agencies were subsequently merged into a single entity, the Japanese Aerospace Agency, JAXA.

8

A New Millennium

1994 to 2004

1994 – Hubble Space Telescope maps Titan, observes changes in haze structure since Voyager

1998 – Transient clouds detected on Titan by groundbased spectroscopy

2001 – Mars Odyssey arrives, measures extensive deposits of ice in near-subsurface

2001 – Observations of secular changes in Mars polar CO_2 deposits

2001 – Suspension of commercial aviation over USA following terrorist attacks provides experiment in cloud effects on climate

2002 – Large groundbased telescopes using adaptive optics observe clouds around Titan's south pole

2003 – Mars Express arrives (Beagle 2 lost); sounding radar probes structure of polar ice caps

2004 – Spirit and Opportunity rovers arrive at Mars, find evidence of hematite and other minerals requiring liquid water conditions to form

2004 – Indications of methane on Mars from groundbased spectroscopy and Mars Express

Spacecraft to the outer solar system were having better luck than those at Mars. Albeit hamstrung with an unfurled antenna that limited its data return, NASA's Galileo mission reached Jupiter in 1995, deploying a probe into Jupiter's atmosphere and entering orbit. The orbiter spacecraft's close encounters with the moon Europa brought that world's ice crust into sharp focus in following years, confirming hints from Voyager that it had a (relatively) thin ice shell overlying a layer of liquid water. Such an "ocean" might in principle be habitable,[1] although even the lower robust estimates of the ice thickness were more than 10 km thick, much thicker than the ice sheet on a Snowball Earth.

Galileo used planetary swingbys to help supply orbital energy to reach Jupiter: during its Venus encounter it used its near-infrared mapping instrument to confirm an Australian telescope observation reported in the 1980s [313, 314] that contrasts could be seen at selected wavelengths on the Venusian nightside (Figure 8.1). In essence, Venus' surface glows nearly red-hot, and at particular window wavelengths around 1 μm, where carbon dioxide and the other gases do not absorb light, this glow filters through the thick clouds to space. Near 2 μm radiation doesn't probe to the surface, but shows cloud structures deeper than the UV cloud tops.

Similar select window regions in the near-infrared also expose Titan's surface to scrutiny. While the methane absorptions that heralded the presence of an atmosphere to Kuiper block much of the infrared spectrum, and the thick tholin haze blocks blue and green light, about 10% of the light filters through the haze in about half a dozen "window" regions. These first showed via lightcurves[2] in the early 1990s that Titan's surface was not uniform, and thus was not covered in a global ocean [315]. Then, with the Hubble Space Telescope's vision corrected in 1993, the first maps of bright and dark areas were made in 1994 (Figure 8.2), most notably at 940 nm (the wavelength of a typical TV remote control) but also in visible red light [316].[3] In addition to the near-infrared maps, the Hubble images at visible wavelengths showed Titan's atmosphere was "upside down" relative to what Voyager had seen in 1980 [317]. The reason was a seasonal shift in the amount of haze in the two hemispheres – 1980, where the north was about 20% darker in blue and green light, was northern spring equinox, and 1995 was

---

[1] The distinctions between an environment into which an extant organism could be introduced and not immediately die (which probably applies to Europa), and one in which organisms can flourish and evolve, and one in which an origin of life can and has taken place (which quite probably does not apply to Europa – its massive ocean may be starved of carbon and nitrogen) are crucial to keep in mind.

[2] A lightcurve is a plot of brightness against time or orbital phase (and thus, for a synchronously rotating object such as Titan, its longitude).

[3] This project was part of my first post-doc, at the University of Arizona.

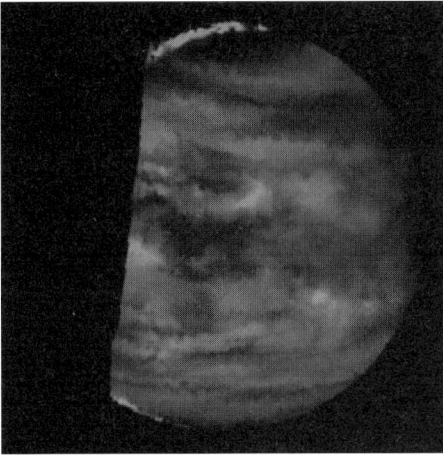

Figure 8.1 A near-infrared image at 2.3 μm, showing lower-level clouds on the nightside of Venus, obtained by the Near Infrared Mapping Spectrometer aboard the Galileo spacecraft in 1990. Bright slivers of sunlit high clouds are visible at left above and below the dark, glowing hemisphere. The contrasts on the dark side reveal differences in radiant heat welling up through varying cloud thickness 50–55 km above the surface, 10–16 km below the visible cloud tops. Near the equator, the clouds appear fluffy and blocky; farther north, they are stretched out into east–west filaments by winds estimated at more than 240 kph, while the poles are capped by thick clouds at this altitude. (Source: NASA.)

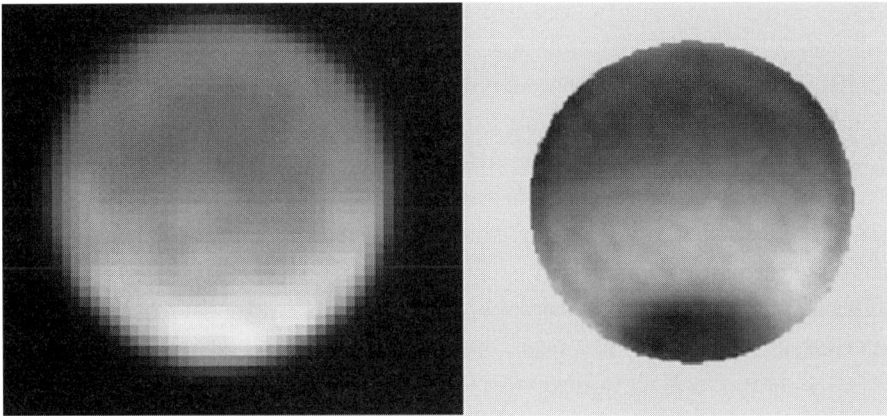

Figure 8.2 (Left) An image of Titan from HST in 1994, showing near-infrared light longward of 850 nm, including that at 940 nm which sneaks through the haze between methane absorption bands. The lower limb of Titan is bright, making a smile, due to thicker haze at that latitude and season. Irregular dark markings (about 10% of the total light) show bright and dark regions on the surface. Only features bigger than about 1/10 of Titan's diameter are distinguishable, so their nature remained unknown. (Right) A contrast-stretched image of Titan in blue/cyan light from 2003, showing the northern hemisphere darker than the south, but with a dark "polar hood". (Author, from data courtesy STScI/NASA.)

half a Titan year later. The ability of Hubble to image in light that was absorbed by methane (and thus only showed high-altitude haze above the methane-rich troposphere) showed that the difference was in the amount of haze, and suggested that the haze was being blown by north–south winds in the global (Hadley) circulation.

In 1998 came the revelation, from spectroscopic observations on the ground, that Titan had clouds [318]. Not the global-scale, high-altitude, near-uniform haze, but smaller, rapidly changing clouds in the troposphere, occupying only about 1% of Titan's disk [319]. These implied methane clouds, and perhaps an active hydrological cycle. While long speculated, there was little evidence of such clouds in Voyager data, and some argued that discrete clouds and convection might not happen. The new observations left little doubt, and were supported by some tentative indications of a transient bright region in Hubble images.[4] Unfortunately, Hubble observing time was seen as too precious to spend on monitoring for clouds that might or might not appear, so more systematic monitoring awaited the development of better near-infrared imaging from telescopes on the ground. But Cassini, with a camera and near-infrared mapping spectrometer, would be well-equipped to observe such clouds. Cassini flew past Venus twice (but being designed for the cold at Saturn, it was not able to use its optical instruments at Venus) in 1998 and 1999. After getting additional gravitational nudges from Earth and Jupiter in 1999 and 2000, it sailed on to Saturn. Meanwhile, by 2001, the giant 10-m Keck telescope and others, equipped with "adaptive optics" systems, fast-adjusting mirrors to compensate for the shimmering in our atmosphere, were able to show dramatic cloud systems (Figure 8.3) at Titan's south pole, where it was now mid-summer [321–323]. As these systems improved on several other telescopes, clouds began to be observed frequently: some vague patterns suggested that perhaps Titan's geography or topography might lead to preferential locations [324]. Cassini's arrival was eagerly awaited.

Planetary flybys such as Cassini's rely on exquisite navigation, as does the injection of spacecraft into orbit. In the hours leading up to Mars Climate Orbiter's arrival at Mars in November 1999, some discordant trajectory measurements became apparent, showing the close approach to Mars would be too close, but it was too late. The spacecraft burned up in the Martian atmosphere instead of braking into orbit. It was learned later that confusion over units had led to the accident, an issue that the headlong-rushed "faster, better, cheaper" Mars

---

[4] The HST evidence was presented at the DPS conference in October 1995; a map showing the transient bright region is published in Ref. [320]. The full details are still being written up formally, only half a Titan year later...

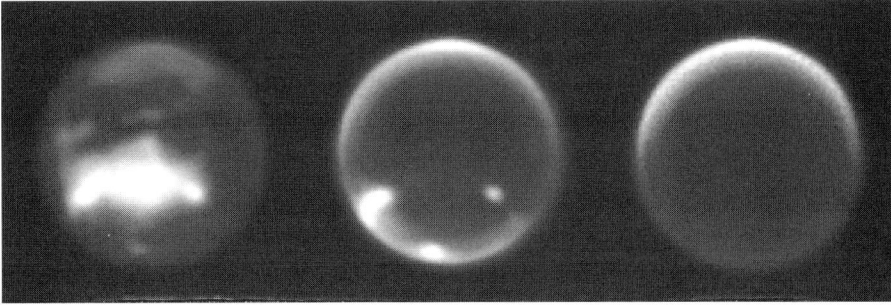

Figure 8.3 Titan observed in 2004 with the giant Keck II telescope with adaptive optics. At left an image at 2.0 μm (little affected by methane absorption) shows Titan's surface and troposphere. The middle panel at 2.12 μm blocks the surface, emphasizing the troposphere, showing prominent discrete clouds at southern high latitudes. At right, at 2.16 μm, methane absorbs light from the lower atmosphere, so this shows only haze in the stratosphere, which is now more abundant in the north. (Credit: W. M. Keck Observatory/H. Roe, Caltech and collaborators.)

program did not have the time to catch.[5] It was a particularly bitter blow for Mars climate scientists such as Dan McLeese of JPL and Fred Taylor at the University of Oxford, who had now seen their infrared investigation get to Mars twice (once on Mars Observer 6 years earlier), only to be lost at the last moment both times.[6]

The loss prompted close management scrutiny of the Mars Polar Lander project, already well on its way to the red planet. But reviews didn't save that mission. Another subtle failure, this time the spring-loaded bounce of a landing leg swinging into position, caused the descent engines to shut off while the lander was still tens of meters above the surface. In addition to cameras to survey its landing site on the layered terrain near the south pole, MPL carried a meteorology package and an instrument, TEGA (Thermal and Evolved Gas

---

[5] In fact the issue was quite subtle, relating not to the trajectory itself, but rather to predicted perturbations to the trajectory due to the operation of thrusters for pointing control. The spacecraft operators at Lockheed Martin reported the thruster torque in foot-pounds, whereas the navigators at JPL were expecting the numbers in the file to be in newton-meters. The quantity isn't an especially intuitive one, the torque is not measured directly, and the units differ by only a small factor, so the discrepancies weren't noticed. For other examples (sometimes subtle, sometimes horrifyingly obvious) of spacecraft accidents, see Ref. [325].

[6] Third time's a charm. Originally built for Mars Observer as the Pressure Modulated Infrared Radiometer (PMIRR), a redesigned instrument taking advantage of newer detectors and other developments flew as the Mars Climate Sounder on MRO (see next chapter) in 2005.

Analyzer) to "cook" soil samples scooped up by a robot arm [326].[7] This instrument was hoped not only to demonstrate the presence of ice and/or water bound in minerals, but also to detect whether carbonates were abundant, explaining the fate of a perhaps once-thicker CO_2 atmosphere. A workshop dedicated to Mars Polar Science and Exploration had been held in Texas in 1998, bringing together many terrestrial glaciologists with planetary scientists (this became a more or less regular meeting) and heightening interest in the Martian poles as the key to its past climate. Much of the debate at the meeting centered on whether (or how fast) the polar caps might be flowing. Glaciologist John Nye pointed out that if the south polar cap were solid CO_2, rather than merely being water ice with a persistent CO_2 frost cover, it should not retain the 3-km-thick shape it had (as recently measured by MOLA and stereo measurements on Viking images) but would ooze and flatten in a few million years [327]. But when MPL was lost in December 1999, attention at Mars shifted from glaciology to lower latitudes and mineralogy.

Around this time it began to be appreciated that Titan's polar regions might be distinctive. Although the 1-D post-Voyager perspective was rather that temperature gradients near the surface would (as for Venus) be rather small, just a couple of kelvin, the upper atmosphere was visibly different. As Titan's southern pole loomed out of winter, Hubble images showed an ultraviolet hood, presumably the counterpart of the northern one seen by Voyager 20 years earlier [328, 329].[8] Somewhat analogous to the high-altitude polar stratospheric clouds on Earth (which form inside the winter polar vortex and whose acidic surfaces catalyze ozone destruction) some materials were condensing, and Bob Samuelson's suspicion from examining Voyager infrared data fell on C_4N_2, dicyanoacetylene [331, 332]. If this material were condensing in a rather restricted latitude range, it might lead to an enhanced abundance of condensation nuclei, sweeping ethane and methane out of the atmosphere at the poles specifically.

Titan's couple of degree equator-to-pole gradient suggested by the Voyager data was in fact a little problematic: in part in anticipation of microwave radiometer observations we would make with Cassini's radar, and in part stimulated by thinking about the Martian polar caps and suddenly having time on my hands, it struck me that this was too big. Playing with a simple two-box energy balance

---

[7] I worked on calibrating this instrument at Arizona. I was also involved in a piggyback project to send two small penetrators to Mars along with MPL: these were never heard from again either. In the aftermath of these failures, when I suddenly had time on my hands since these missions returned no data, I explored ideas about equator-to-pole heat transport and entropy production.

[8] The role of winter shadow and nitriles in the hood formation was noted in Ref. [330].

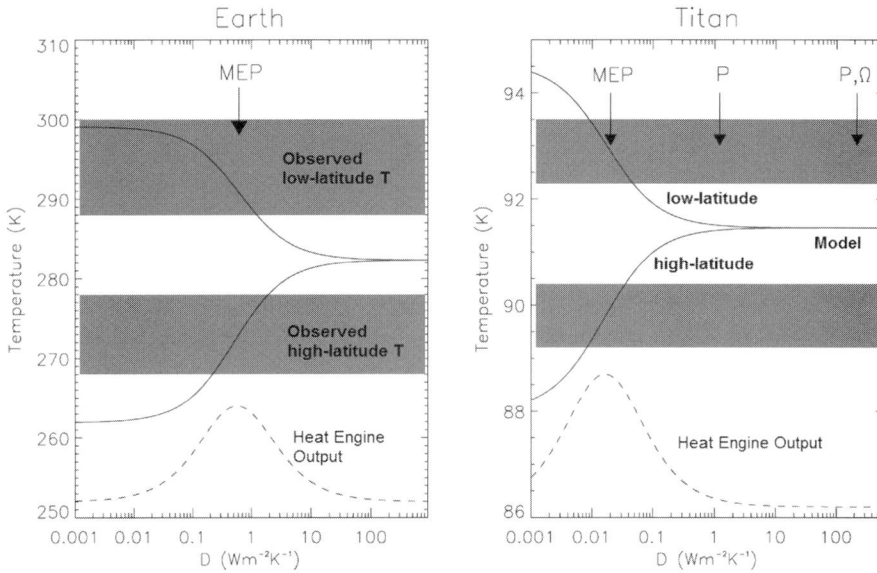

Figure 8.4 Equator-to-pole temperatures on Earth and Titan. The grey bars represent the estimated average equatorial and polar temperatures and the two lines are model temperatures in an energy balance model given the heat transport coefficient on the *x*-axis. Using the raw value for Earth (about $1\,\mathrm{W\,m^{-2}\,K^{-1}}$, roughly the MEP, there) at Titan or scaling the value used for Earth in Budyko–Sellers models by Titan's atmospheric pressure (P), or by pressure divided by rotation rate squared (P, Ω), yields temperature contrasts on Titan that are much too small. The observed temperatures there need a transport coefficient about 100 times smaller, which is where the entropy production or work output of the planetary heat engine (dashed curve, arbitrary axis) has a maximum (MEP). (Figure by author.)

model [333, 334], if I took the approach used by Mars energy balance modelers and just scaled a heat transport factor by the surface atmospheric pressure (with or without plausible corrections for planetary diameter and rotation rate), the temperature gradient should be only a hundredth of a kelvin. But when I chose the coefficient that yielded a maximum in entropy production (Figure 8.4), the couple of kelvin popped out. I became enchanted with the maximum in entropy production (MEP) idea. Interestingly, for Mars it suggested that the heat transport should not be too much less than Earth – and indeed, while the sensible heat transported by the atmosphere is small (allowing Leighton and Murray to get away without including it), the fact that tens of centimeters of CO_2 frost condenses at the poles in winter in fact means that the "pinning" of polar temperatures to the frost point, as done heuristically in most climate models, embodies a latent heat transport about equal to what MEP would require. Like the Gaia Hypothesis,[9] the idea had a seductively holistic aspect to it, and for planetary paleoclimate or exoplanet studies, where many of the details needed to constrain GCMs would be unavailable, it seemed like a useful hypothesis. The (at least superficial) relevance to observed heat transports on Earth, Mars and Titan, and some theoretical underpinnings for the concept laid out by physicist Roderick Dewar [335], addressed many of the objections raised against the idea when Paltridge suggested it in the 1970s. Significant challenges remain, however, particularly when both vertical and horizontal heat transports are considered.[10] I still think the idea may have some value,[11] and some progress continues to be made in oceanographic and atmospheric applications,[12] but so far there seem few settings in which we know it can be applied and yet tell us something we don't already know anyway.

As NASA scrambled to reformulate its program after the twin disasters of 1999, data continued to come in from Mars Global Surveyor. The gullies attracted closer attention, and discussion about what could be causing them. Could it be dry avalanches? Was it carbon dioxide? Or was it water? It was, in microcosm, a re-hash of the debates about the larger channels seen by Mariner 9 and Viking. Essentially the sometimes heated debate boiled down to thermodynamics (!). If the regolith were sufficiently insulating, and geothermal heat flow sufficiently high, then the subsurface temperature gradient would be quite strong, and temperatures able to support liquid water might be reached at suitably shallow depths. On the other hand, a cold subsurface (conductive regolith and/or low heat flow) might permit carbon dioxide to work. One extreme proponent of this scenario even cleverly named it "White Mars", as a riff on Robinson's books [339].

---

[9] Indeed, a workshop organized in Edinburgh in September 2002 by Tim Lenton and Inman Harvey, "Beyond Daisyworld", brought together most of the key individuals exploring the MEP idea and Gaia – Roderick Dewar, Hisashi Ozawa, Toni Pujol, Garth Paltridge and myself, as well as James Lovelock. In one of those "small-world" connections, although I worked as a planetary scientist in Arizona, and Dewar worked in a bioclimatology institute in France, it turned out our fathers lived only a few miles apart and played together at the Muckhart golf club in Scotland.

[10] A problem investigated by Toni Pujol and Gerald North (who had been a pioneer in energy balance modeling in the 1970s).

[11] despite objections from some august scientists: for example, the abstract of Richard Goody's 2007 paper [336] is more polite than its numbers are correct, saying his somewhat damning analysis "does not necessarily imply that MEP is incorrect, and inapplicable to atmospheres, but it does mean that the difficult and unexplored problem of dynamical constraints on the MEP solution must be understood if it is to be of value for climate research." Nonetheless, I repeat the health warning given when MEP was introduced in Chapter 4.

[12] One of the main objections was the apparent disregard of planetary rotation rate. This factor has been nicely explained in Ref. [337]. See also Ref. [338].

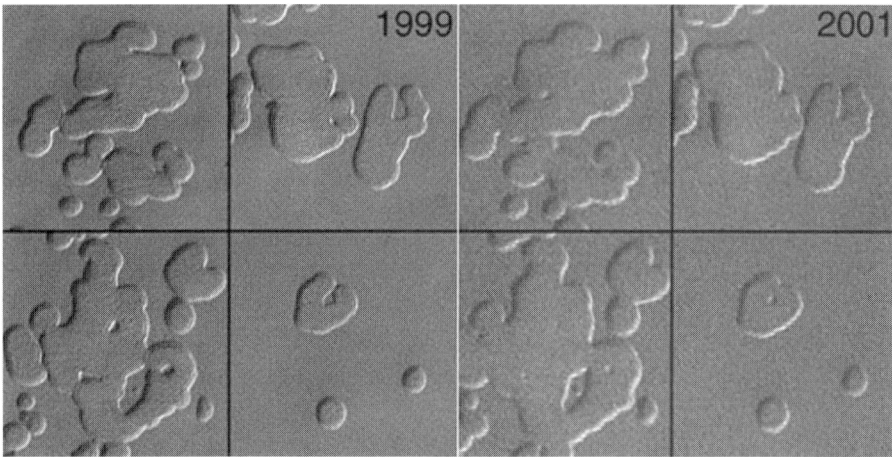

Figure 8.5 250-m-wide segments of images of the Mars south polar cap taken in 1999 and 2001, showing that pits had enlarged, mesas had shrunk, and small buttes had vanished: typically the scarps that enclose the pits and bound the mesas and buttes retreated about 3 m. (Source: NASA/JPL/MSSS.)

Another indication of near-surface ice was the detection of "rootless cones", basically small conical mounds on lava flows, interpreted to indicate that the lava flow had overrun wet or ice-bearing ground. The heat would boil off the ice, and the exploding steam would spray the lava up into a small cone. Such cones are found on Earth in Iceland, and were discovered on Mars in MGS images [340].

Meanwhile, the patterns of dust devil activity, dust storms and the polar frost cycle became better and better known as the years of MGS data accumulated. An intriguing observation was of "Swiss Cheese" terrain near the south pole (Figure 8.5), which appeared to show year-on-year changes, its CO_2 ice subliming away suggesting that, on top of the overall regular CO_2 frost cycle, there may be a systematic year-to-year change, which might be accompanied by a steady change in pressure on Mars over the past decade or few [341, 342].[13] Unfortunately, regional meteorological variations and the challenges of absolute calibration make it difficult to compare the various lander/rover pressure measurements to know for sure. MGS data from the TES instrument built up an important record of the seasonal cycle and interannual variations in water vapor, temperatures and dust (Figure 8.6).

An ongoing puzzle for Mars was that, despite the few-meter resolution of MGS/MOC, and the opportunity to make comparisons with Viking data from 20 years earlier, there was no evidence that Mars' ubiquitous sand dunes had actually

[13] An excellent review of Mars polar phenomena is by Byrne [343].

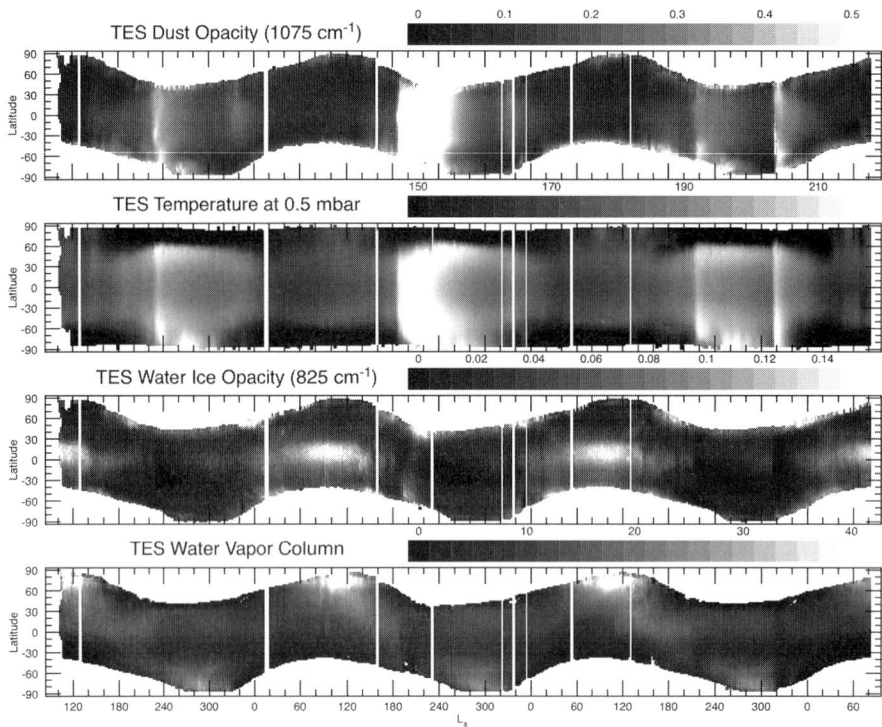

Figure 8.6 The Mars climate record 1997–2003 (3 Mars years) obtained by the Thermal Emission Spectrometer on Mars Global Surveyor. The white margins of three plots show winter darkness. Top to bottom – dust opacity, atmospheric temperature at 0.5 mbar, water clouds and water vapor. At top, the dust peaks in southern summer, near perihelion, but the second summer has a high sustained opacity due to a global dust storm. The heating of the atmosphere by the dust is evident in the second plot, which shows midlatitude warmth persisting after the dust. The third plot shows the regular equatorial aphelion cloud belt, while the last shows that the north polar cap is the strongest source of water vapor (during northern summer). (Image courtesy M.D. Smith/NASA GSFC.)

moved. Speculations were offered that perhaps the dunes were petrified (aeolian scientists like to say "indurated") by ice or evaporite minerals, or that maybe the dunes formed in a higher-pressure Mars epoch when sand moved more frequently than in the current climate.

After extensive review, it was determined that the 2001 orbiter, later named Mars Odyssey, should proceed more or less as planned. But the mechanics of landing were much less certain, and so a planned 2001 lander, which would have carried a small rover, was scrubbed. Instead, two larger rovers would be sent in 2003, resurrecting NASA's practice of the 1970s of sending pairs of probes, in part as insurance against random failure, but usually resulting in enhancement of the science.[14]

Mars Odyssey arrived at Mars, uneventfully, in October 2001, aerobraking into its final orbit by early the following year. It carried THEMIS, a moderate resolution thermal and visible camera, and a Gamma-Ray and Neutron Spectrometer (GRS). This latter instrument, which relies on natural cosmic rays to probe the near-subsurface, only works on worlds where the atmosphere is thin enough to allow the cosmic rays in and the nuclear shrapnel and flash of neutrons and gamma rays to come out. By patiently counting this radiation as the instrument repeatedly flies over given regions of the surface, a spectrum characteristic of certain elements down to a few tens of centimeters can be built up. By mid-2002, an unmistakeable signal had built up showing that the near-subsurface was abundant in hydrogen, and water was the only reasonable compound for that element to be in (Figure 8.7).[15] Hydrogen-rich deposits ranging between about 20% and 100% water-equivalent were found poleward of $\pm 50°$ latitude (meaning that perhaps if the Viking 2 lander had dug just a few centimeters deeper, it might have found permafrost). Surprisingly significant, albeit less rich (2–10%), deposits were found at near-equatorial latitudes as well. Given that the instrument probed down to about 1 m depth, these data suggested that the water reservoir, even ignoring the thick polar cap, had to be the equivalent of a layer about 14 cm thick. White Mars was dead.

Beyond the various evolutions of European instruments from Mars-96, Mars Express (MEx) also carried a small British-led lander Beagle-2. This rather entrepreneurial project had been conjured out of nothing by the efforts of Colin Pillinger, a meteorite scientist at the UK's Open University, and featured sensitive instrumentation to detect organics, as well as some rudimentary meteorological sensors [346]. The tiny resource envelope into which the lander had been squeezed left no room for backup systems to give the mission resilience, however, and after a parting photo of Beagle sailing off to Mars after separation from MEx, it was never heard from again. Some initial blame fell predictably on

---

[14] Mariner 9 was an example of this insurance policy: Mariners 8 and 9 were supposed to be in different orbits at Mars, addressing different science objectives. When Mariner 8 was lost at launch, Mariner 9 was put into a compromise orbit that partly addressed all the objectives. Vikings 1 and 2, and Voyagers 1 and 2 were examples of the successful multiplication of science return in the better-funded 1970s. Pioneer Venus, Galileo, Cassini, and indeed every other NASA mission since, apart from the post-crisis 2003 Mars rover, were one-offs.

[15] e.g. an early report by Boynton et al. [344]. In fact it was the Neutron Spectrometer part of this instrument, built in Russia, that was most informative about hydrogen specifically. Several more detailed analyses followed, including Feldman et al. [345].

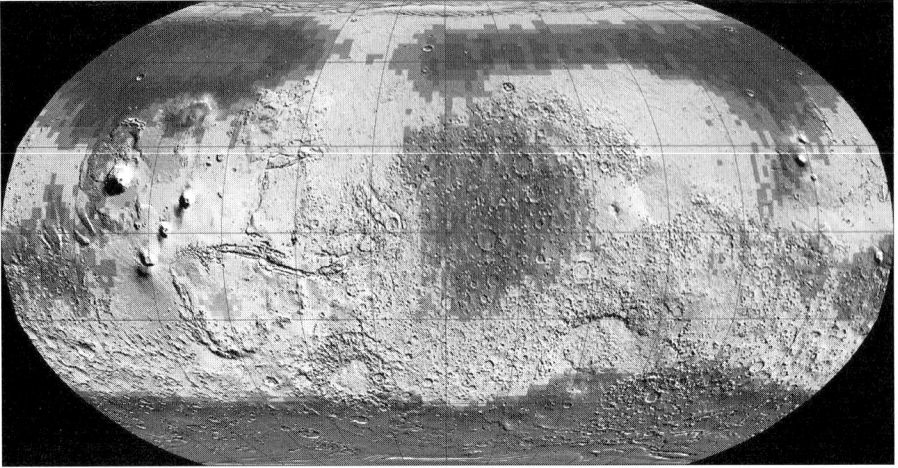

Figure 8.7 The abundance of near-surface water on Mars, inferred from the neutron spectrometer on Mars Odyssey (dark shade implies more water). This projection doesn't show the poles (the north polar cap is predominantly ice), but ice-rich permafrost extends down to 60 degrees latitude (and more in the north), and in fact most of Mars' surface has 5–10% water. (Source: NASA/JPL/University of Arizona/Los Alamos National Laboratories.)

Mars, but possibly a failure of Beagle's airbag landing system was the reason.[16] In any case, Mex's arrival was somewhat fraught. Further tensions arose with worries that the antenna of its large sounding radar might cause damage to the rest of the spacecraft on deployment.

In the end, however, the radar was deployed and was able to probe the structure and indeed the full extent of the Martian polar cap, revealing layers that could be traced across the whole structure (Figure 8.8). While the trauma of Beagle's loss was deep, it was a relatively inexpensive loss in the grander scheme of exploration, and many of the exquisitely miniaturized instrument developments would find use in future lander projects.

[16] Although I had no formal involvement in the project, I followed it closely. The meteorology team at the Open University included many of my Huygens colleagues, and indeed I had a test unit of the Beagle-2 ultraviolet sensors on my roof in Tucson for a year to get some familiarization data (there being more sunshine in Tucson than in Milton Keynes). Beagle's loss was a great disappointment. A bright patch in the post-separation image has been interpreted as a possible deposit of ammonia frost, leaked from the airbag inflation tank; the Philae comet lander similarly had a gas tank system for a hold-down thruster that failed to function. On the other hand, a HiRISE image in 2015 showed an apparently part-deployed lander: whether this means the last petal failed to unfold (thereby preventing the communication antenna from operating), or whether the tightly packed lander just sprang open on hard impact, is not clear.

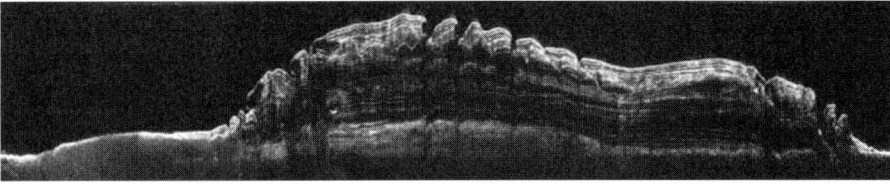

Figure 8.8 Cross-section of the Mars north polar cap, from the Shallow Radar (SHARAD) instrument on NASA's Mars Reconnaissance Orbiter reveals the layers of ice, sand and dust that make up the ~2-km thick north polar ice cap on Mars: similar data, at lower resolution, were obtained by the MARSIS radar on Mars Express. The image reveals four thick layers of ice and dust separated by layers of nearly pure ice, interpreted to correspond to one-million-year-long cycles of climate change on Mars caused by variations in the planet's tilted axis and its eccentric orbit around the Sun. (Source: NASA image PIA 10652.)

MEx's instruments opened new windows on Mars beyond its sounding radar. Its High Resolution Stereo Camera (HRSC), while comparable in resolution with MOC, gave the capability to make local stereo topographic models on a much smaller scale than MOLA data. One particularly intriguing observation was of the "hourglass" crater (Figure 8.9), which showed textures indicative of the surface flow of ice [347] – evidence of "rock glaciers" (flowing ice, armoured with rock which retards its melting or sublimation) was found elsewhere, at surprisingly low latitudes.

MEx also carried a near-infrared spectrometer, OMEGA,[17] to map minerals, most particularly to search for carbonates that might offer hope for a warmer, wetter past. Although it found volcanic rocks aplenty, and clays and sulfates attesting to water in the past, evidence of carbonates proved somewhat elusive (OMEGA's spatial resolution was only 0.3–2 km, so outcrops much smaller than this would have their spectral signature diluted into nondetection).

Another of MEx's spectrometers, the mid-infrared Planetary Fourier Spectrometer (PFS) grabbed headlines in March 2004 with the declaration that it had detected a whiff of methane on Mars (Figure 8.10) – a mere 10 parts per billion (ppb) [348]. While methane is a greenhouse gas (back in the 1970s, Ann Henderson-Sellers [176] calculated that 0.01 mbar of methane – what turns out to be a hundred thousand times more methane than the PFS observed – would be needed to warm Mars by 1 K), the excitement was not about the imminent prospects of terraforming, but rather that one possibility for a methane source

---

[17] A rather charming French acronym: Observatoire pour la Minéralogie, l'Eau, les Glaces et l'Activité.

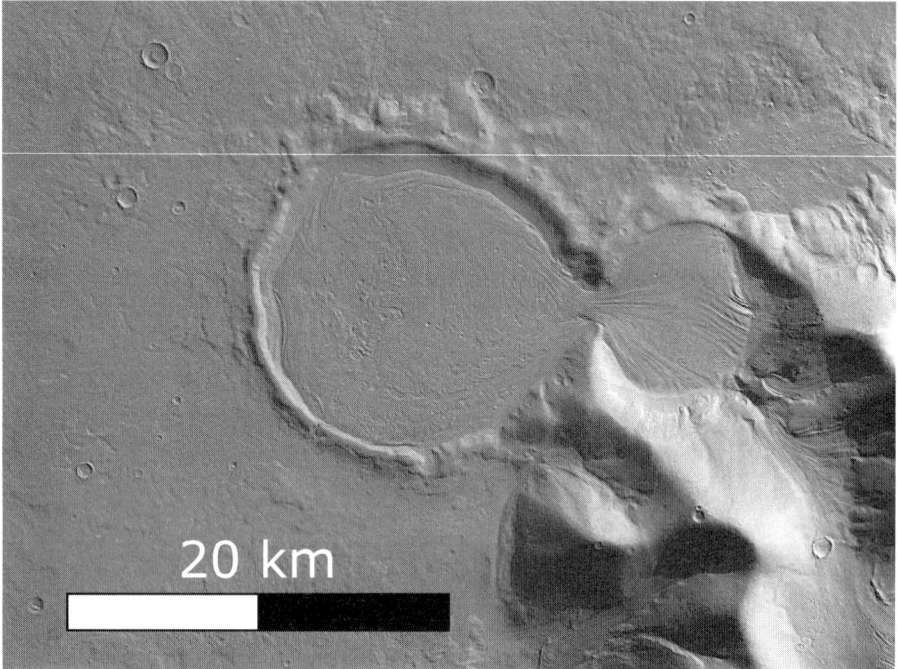

Figure 8.9 A Mars Express/HRSC image of the "hourglass" crater, showing evidence of glacial flow from the cirque at right. The feature is in highland terrain of Promethei Terra at the eastern rim of the Hellas Basin, at about latitude 38° S and longitude 104° E. (Image courtesy ESA/DLR.)

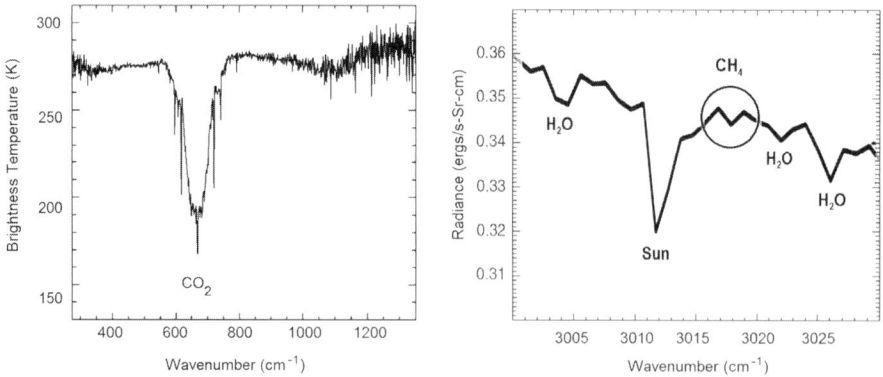

Figure 8.10 (Left) Rich detail is visible in the spectrum of Mars from PFS – the highly structured 15-μm CO_2 band is very prominent. (Right) A small blip in the PFS Mars spectrum is attributed to methane. I am not a spectroscopist, but I do not find this evidence overwhelming! (Figure by author.)

would be bacteria. Mars is an oxidative environment, so reducing species such as methane should not persist – as Lovelock had argued in the 1960s.

Groundbased observers, including Mike Mumma at Goddard Space Flight Center in Maryland, and Vladimir Krasnopolsky at the Catholic University of America (in nearby Washington, D.C.) had also reported detections [349, 350] (using the now-familiar technique of exploiting the Doppler shift to separate spectral lines of the Earth's and Mars' atmospheres); Mumma's observations showed that the methane amount varied substantially in space and time (and, made for good publicity, showed as a map with a garish color scheme to denote where the interpreted methane amount was largest). This unexpected variability might be taken as a warning sign, and indeed the observation was challenged by iconoclast Kevin Zahnle at NASA Ames [351], who noted that the Doppler shift put the Mars "regular" CH_4 line on top of a terrestrial line of $^{13}CH_4$, the trace of methane with isotopically heavy carbon.[18] Zahnle noted that in fact a spurious detection of methane on Mars had been announced in 1969 at a press conference by the Mariner 7 Infrared Spectrometer (IRS) team just two days after its flyby. The IRS had detected absorptions at 3.0 and 3.3 µm near the southern polar cap, and attributed them to ammonia and methane, before the team discovered that these bands could instead be due to CO_2 ice. Spectroscopic self-deception has a rich history at Mars and elsewhere.

But taking the twenty-first century observations at face value – three different investigators with independent methods and analyses – the problem with methane on Mars was in a sense opposite to that on Titan. Whereas the Copernican principle (i.e. that there is nothing special about our place and time) demanded that, since the lifetime of methane on Titan was short compared with the age of the solar system, there must be a source or reservoir, on Mars the estimated lifetime was too long. In other words, methane should not be confined to specific regions, but like CFCs on Earth should be well mixed. Even 5 years later, scientists noted that the abundance and variations of methane are not explainable with the chemistry as presently understood [352]. Perhaps some exotic process, such as chemical interaction with chlorine, or electrostatic discharge in dust devils, might be destroying the methane.

The question of where the methane might be coming from is less controversial, but not especially discriminating. While indeed the dominant source on Earth is bacterial action in anaerobic environments such as swamps and the guts

---

[18] Carbon 13 makes up about 1.1% of carbon on Earth: its amount varies slightly in materials of different biological origins. It is a stable isotope, not to be confused with the radioactive carbon 14, produced by the action of cosmic rays (and nuclear tests) whose 7000-year half-life allows its concentration to be used to date biological materials.

of ruminants (and indeed such strong sources would likely be needed to sustain enough methane for an early greenhouse on Mars or Earth), methane in modest amounts can be made simply by the chemical interaction of water with hot basaltic rock, in a process called serpentinization. The reaction yields minerals including talc and the greenish serpentine, as well as hydrogen and, in the presence of carbon dioxide, methane. So, in the Martian environment, this is a perfectly straightforward way to produce methane without biological activity. The question at this point remains unresolved – the debate stimulated the development in 2009 of a joint NASA–ESA Trace Gas Orbiter mission with very sensitive spectrometers to detect methane and many other gases. NASA withdrew from the effort in 2013, but the mission (reformulated as an ESA–Russia project) flew in 2016, albeit with less sensitive instruments. Methane continued to be considered as a solution to the faint young Sun paradox on Earth: Alex Pavlov, Jim Kasting and colleagues had found that methane-producing bacteria could indeed keep pace with photochemical destruction and maintain enough methane, about 100 ppm, to provide greenhouse warming [353]. Melissa Trainer and colleagues at the University of Colorado in 2006 made "tholin" generation experiments in candidate early Earth atmospheres (CO_2, N_2 and CH_4), and found that the haze generation rate depended in a complex way on the relative amounts of CO_2 and CH_4, but could certainly deliver a rich amount of organics to the surface [354]. The haze antigreenhouse difficulty noted by McKay and colleagues in 1997 may still be problematic, but the different characteristics of haze made under different conditions might yet allow methane to work.

Even as the methane controversy began, the Mars invasion continued apace. In early 2004, two new robots arrived. The Mars Exploration Rovers (MERs), Spirit and Opportunity, were designed to make explorations at a local scale, like a terrestrial field geologist. One rover, Spirit, was aimed at Gusev crater, a suspected former lake, while the other, Opportunity, was sent to Meridiani Planum. In fact, one factor in determining the landing sites was the predicted weather, specifically whether winds and turbulence would be severe enough to threaten the success of landing – the parachute/rocket/airbag system used for landing could encounter failures with winds above 10–25 m/s. Mesoscale atmospheric models were used to predict the winds as a function of the observed topography, thermal inertia, etc.[19] Indeed, some prospective landing sites (notably Melas Chasma, a deep trench, part of the Vallis Marineris canyon system) were ruled

---

[19] A mesoscale model simulates dynamics of a region, perhaps tens to hundreds of kilometers across – it is thus not a "global" circulation model, although in many respects its internal mechanics are similar. Usually GCMs are used to define the boundary conditions for a more detailed mesoscale simulation.

Figure 8.11 Scene near the Opportunity rover's landing site showing slabs of sedimentary rock, sprinkled with small blueish spheres a few millimeters in diameter, made of hematite. (Source: NASA/JPL.)

out on the basis of these meteorological predictions [355], a measure of the growing confidence in the ability to model the Martian atmosphere. Meridiani was known from MGS data to have a thermal emission spectrum consistent with hematite, an iron mineral that typically forms in wet environments.

Sure enough, hematite proved to be abundant, although not quite in the way scientists had expected [356]. In fact it formed little spherical concretions, which were nicknamed "blueberries" (Figure 8.11). These little balls,[20] which littered

---

[20] Similar hematite concretions are found in desert areas on Earth. I've seen some myself in Wadi Rum in Jordan and in southern Utah.

the Meridiani landing site, left little doubt that wet conditions had existed at some time in the past.

Meanwhile, although Spirit did not find evidence of a lake as such, many volcanic rocks were found to show evidence of weathering by acid fogs, and some other minerals such as goethite and silica required liquid water environments in which to form. Moreover, its small thermal spectrometer (mini-TES) found carbonate rocks that again required liquid water in which to form, and non-acidic conditions at that [357].

Although the Mars Exploration Rovers – in what seems an almost scandalous wasted opportunity in the history of planetary exploration[21] – carried no meteorological instrumentation, Spirit did make systematic surveys of dust devils with its cameras, and even made movies of their nearby passage (Figure 8.12). The statistics of dust devils observed this way were even good enough to show that, like earthquakes or forest fires on Earth, dust devil diameters follow a power law, with 100-m-diameter devils appearing four times less often than 50-m devils [359]. The dust devils were even kind enough to clean the dust off the rovers' solar panels on several occasions, a process that has likely been instrumental in the MERs' longevity [360]. At the Opportunity landing site, there seems to be less available dust on the ground and hardly any devils were seen (although its solar panels did see cleaning events, suggesting that dustless vortices are present). The MER's cameras are also, by taking images of the Sun, able to estimate the dust loading in the atmosphere (similar measurements were made by Viking and Pathfinder) and these data, spanning 5 Mars years, are a useful climate record (Figure 8.13) [361].

While Spirit eventually got stuck up to its axles in a pit of soft sand, in an orientation that did not give it enough solar power to survive the winter, Opportunity has continued to trundle around – for over a decade now – and its geological findings occasionally have climate implications. It traversed vast fields of ripples, and even detected some small movements. The major landmarks on its traverse (now some 45 km long as of mid-2018) included impact craters, which exposed layers of the Martian past. One such crater, Endeavour, showed intricate layers of cross-bedding (Figure 8.14), much like layers in sandstone on Earth, which can be decoded to infer something about the migration of sand in the past.

---

[21] Meteorological measurements seem to have been systematically short-changed in the Mars program. Perhaps the MER expected mission duration of 90 days influenced the noninclusion of meteorological instrumentation, although after a 2500+ sol mission of Spirit, and the ongoing mission of Opportunity, now 4200 sols in, the omission is particularly excruciating. Contrariwise, the European Space Agency ExoMars Schiaparelli lander in 2016 did carry a neat meteorology package, which was to operate for 2–4 days before its batteries ran out. But the lander crashed anyway (see Chapter 10).

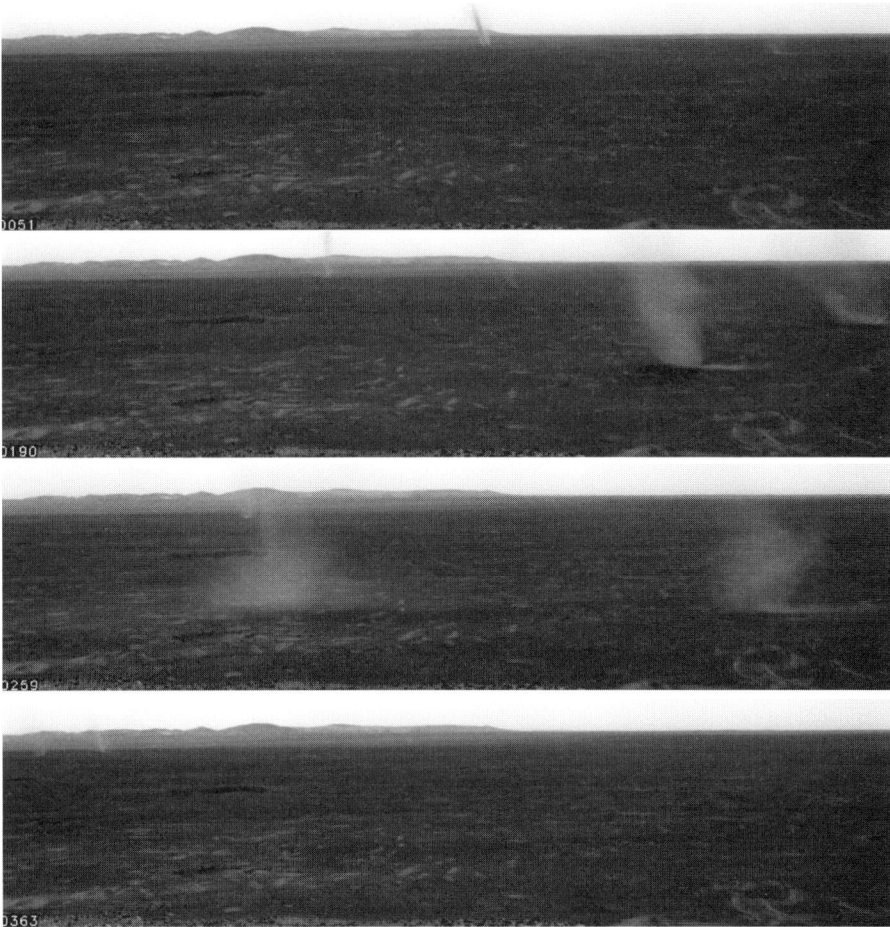

Figure 8.12 Dust devils seen on Sol 543 (July 13, 2005) of the Spirit rover mission at Gusev crater. Improvements due to the experience of scientist operators, to the better cameras, and to the higher bandwidth, led to much higher quality dust devil imaging than from Pathfinder. These are just four frames of a movie sequence, showing the dust devils evolving and migrating from right to left. Note that the dust devils in the foreground have a faster angular motion and are seen only in two frames, while the distant tall devil is seen in all four. (Source: NASA/JPL Texas A&M.)

Astronomical observations of Titan continued showing changes in the material suspended in that world's atmosphere: the north–south asymmetry and south polar hood evolved, the latter feature slowly disappearing just past midsummer [362]; the sustained observation at a range of wavelengths also showed that the upper levels of the haze (seen in the ultraviolet and in deep methane bands) responded more quickly than haze lower down (seen at green and red wavelengths) [363]. A millimeter-wave telescope in the (Spanish) Sierra Nevada

Opportunity Rover

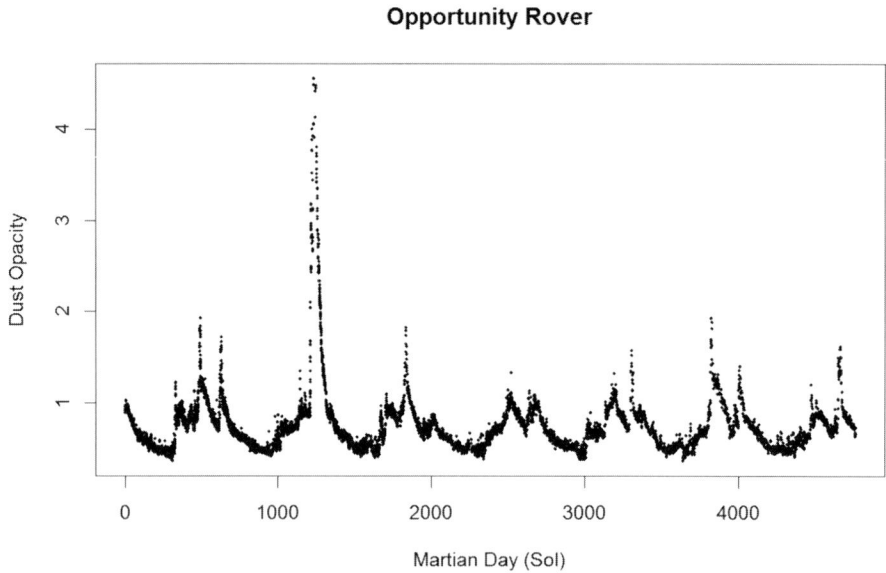

Figure 8.13 The Mars aerosol optical depth recorded by the Mars rover Opportunity, spanning 5 Mars years, some 3500 days from early 2004 to July 2013. There is a consistent seasonal pattern, with higher optical depth in the latter half of each Mars year (i.e. around southern summer). The dust opacity in the storms jumps up rapidly but decays slowly. (Author, data obtained by M. Lemmon.)

measured abundances of a number of nitrile compounds that were inconsistent with the photochemical models then available [364]. This observation also determined that the heavy nitrogen isotope ratio $^{15}N/^{14}N$ in hydrogen cyanide was enriched with respect to its value in terrestrial nitrogen, suggesting (somewhat analogously to the Venus D/H ratio) that Titan had lost significant atmospheric mass in its history.

Models grew in number and capability. Pascal Rannou in Paris succeeded in reproducing the "detached haze" structure seen in Voyager images with a GCM, showing that the same meridional circulation which shunted haze in the main layer to cause the north–south asymmetry cycle was responsible for creating the detached layer, which linked to the polar hood [355]. GCMs also attempted to achieve the observed superrotation inferred from Voyager data, stellar occultations, and a couple of challenging high-resolution spectroscopic observations that attempted to measure the stratospheric winds by the Doppler shift of ethane lines [366].[22] Interest in the zonal winds in particular was high, since the

---

[22] The wind speeds are large compared with the relative velocities being measured for extrasolar planet detection, but the challenge was to measure the difference in

Figure 8.14 Crossbedding on Mars. Opportunity mosaic of the Cape St. Vincent promontory on the edge of Victoria crater in Meridiani Planum, showing impressive layers in aeolian sands, reminiscent of that seen at Petra in Jordan, or in Zion National Park in the southwestern USA. These structures show that dunes once marched across this area, perhaps in a different climate epoch. (Image courtesy of Alexander Hayes.)

Huygens probe descent and possibly the quality of its radio link would be affected by them. However, GCM studies by Tetsuya Tokano and Fritz Neubauer in Köln, Germany, struggled to develop zonal winds that were as large (~100 m/s) as observations suggested [367]. They did, however, introduce a new feature into Titan GCMs, one that had not been needed for other objects simulated up to that point, the interesting possibility that Saturn's gravitational tides could cause a

Earth-relative velocity between the east half of Titan and its west half, given that the telescope could barely resolve Titan.

significant effect on Titan's winds.[23] They found that tidal winds might cause a noticeable effect on meridional winds in particular, perhaps a half a meter per second near the surface. They also experimented with a cloud model to simulate whether lightning should occur – finding that storms would develop over the course of a few hours, and perhaps deposit some tens of centimeters of rain [368]. This supported the speculation that, while the amount of solar power available to drive convection on Titan was small overall, it might be expressed in rare but violent storms [369].

Observations of Titan's clouds continued as several large telescopes with adaptive optics systems came on-line, with observers trying to see a pattern in Titan's clouds – were they associated with a particular location that might be a methane source? Some thermodynamical arguments were given to suggest why convective clouds on Titan were relatively rare – occupying only one percent or so on average of Titan's disk, compared with the ~30% cloudiness that prevails on Earth [370].[24]

Clouds on Earth came back to the fore (if, indeed, they'd ever really left) with ideas and paradoxes that are still being debated today. In particular, the balance between the short-wave and long-wave forcing by tropical clouds turns out to be critical in controlling the sensitivity of climate. As it gets warmer, there may or may not be more clouds. If there are more clouds, are they high cirrus that are warming, or low cumulus that are cooling (the high clouds have a lower short-wave reflectivity effect compared with their long-wave absorption effect)? If there are a lot more cumulus (or for that matter, a lot less cirrus) in response to warming, there is then a stabilizing feedback.

In fact, the potential strength of this feedback is powerful enough that Hsien-Wang Ou at Columbia University in New York proposed that it could solve the faint young Sun problem for the Earth, without invoking exotic methane or high-CO_2 paleoatmospheres [372]. The changing balance of clouds over time could compensate for the evolving solar constant. Some years later, Minik Rosing and colleagues in Copenhagen similarly suggested a lower Earth albedo in the deep past (with no land vegetation, and fewer clouds) was enough to avoid large greenhouse forcing [373], although others disagreed [374, 375].

But is this what happens? Ray Pierrehumbert in Chicago in 1994 had noted that the tropics by themselves are close to a runaway greenhouse condition,

---

[23] In fact, my first scientific (rather than engineering) publication had been a conference paper in 1992 noting the possibility of a gravitational tide in Titan's atmosphere, manifested as a pressure variation.

[24] The idea that Titan plumes might be energetic but rare can be traced to Awal and Lunine [371], which in turn related to ideas about Jupiter's clouds.

which they largely escape only because they export heat to higher latitudes [376]. He found that the main determinant of tropical heat budget is the area-averaged water vapor greenhouse effect, which is determined by the area fraction of dry downwelling air, which creates clear "radiator fins", allowing heat to escape to space. He suggested that determining the dryness and area of these subsiding regions were key unsolved problems in the tropical climate.

In 2001, Richard Lindzen at MIT suggested that the high cirrus might decrease in response to elevated temperatures, cleverly calling the stabilizing feedback an "infrared iris" [377, 378]. However, a closer study quickly dismissed this suggestion [379], and while this topic recurs in the literature, the claims of a negative feedback in the real world (and thus suggestions that current models lacking this feedback may overpredict global warming) seem to be a minority view. While scientific truth is not decided by vote, the deluge of data from the dozens of satellites and thousands of measurement platforms is able to provide some empirical test of these ideas, and the scientific process will eventually winnow the models that successfully describe data from those that do not. However, it is interesting in the planetary context where such abundant data are lacking that a few serious scientists are undeterred from proposing that even the sign (positive or negative) of some basic feedbacks is not what is widely assumed, even for Earth's present climate. Unless at least a few scientists challenge the dominant paradigm once in a while, the majority would never find out if they are wrong. The distraction in beating down outrageous ideas is probably worth the outside risk that, once in a while, they might be correct.

Another similar debate in climate science relates to pan evaporation. It is commonly assumed that as surface temperatures increase, the rate of evaporation of moisture from the surface will also increase as a natural consequence of the vapor-pressure curve of water (the Clausius–Clapeyron equation). However, a large body of data exists from evaporation pans[25] which seems to show that, in many places at least, the rate of evaporation has declined in recent decades [380]. This seems like a paradoxical result, and debate has ensued on its interpretation, especially following a report by Graham Farquar and Michael Roderick in Canberra, Australia, in 2002. Indeed, it had been recognized by Budyko in the 1960s (see citation in [381]) that evaporation was generally larger in winter (noticed too by anyone who dries their clothes outside) – the air is drier and conditions are windier in that season, which matters more than temperature itself. Taking the observation (and there are areas where it does not hold) to be

---

[25] There are various standard pans, a typical one is stainless steel, 53 mm deep and 1.2 m in diameter. It is placed away from obstructions on a wooden pallet, and the amount of water needed to fill it back to a fixed level is recorded every 24 hours.

true, there is the important question of whether it matters: to what extent does evaporation from a filled steel pan represent a relevant metric for the evaporation from real land surfaces [382], where the availability of near-surface moisture in the soil and the type of vegetation can significantly affect the transfer of moisture to the air?

Roderick and Farquar suggested that the decline in pan evaporation might reflect a decline in sunlight reaching the ground, due to increased clouds and/or aerosols. This would also show up as a drop in the day/night temperature difference or diurnal temperature range (DTR). Normally it would be difficult to disentangle such a "global dimming" effect from changes in greenhouse gas concentration – we cannot usually perform a controlled experiment on our climate. But in September 2001, a modest unintended experiment was performed, with the three-day grounding of commercial aviation in the aftermath of the 9/11 terrorist attacks. David Travis at Wisconsin and colleagues claimed

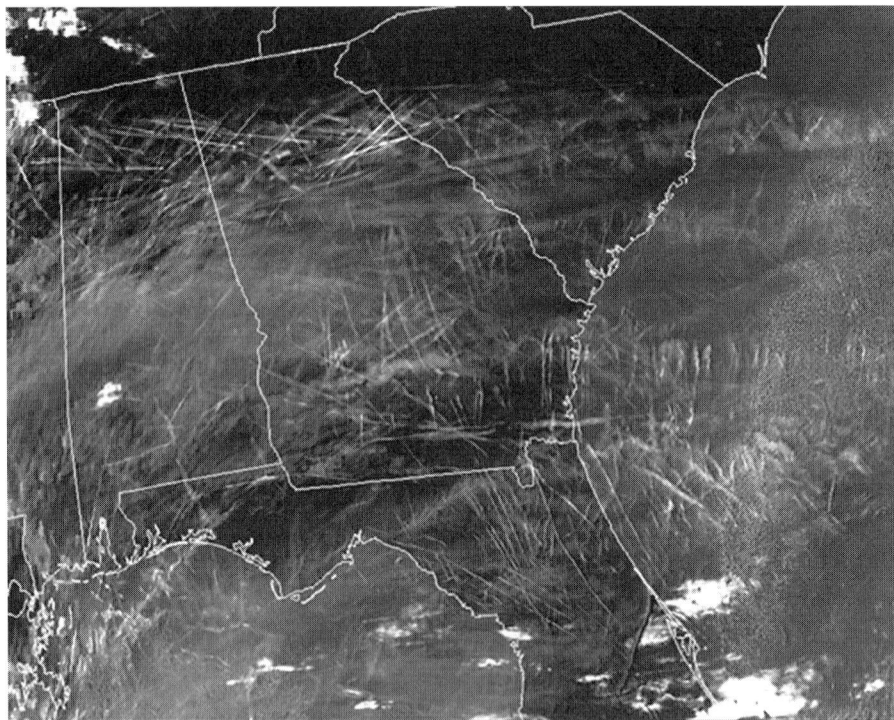

Figure 8.15 An infrared image from NASA's Terra satellite showing a widespread outbreak of contrails over the southeastern United States on the morning of January 29, 2004. The crisscrossing white lines are contrails that form from planes flying in different directions at different altitudes and trails drift in the wind after they form, so a fixed airlane may yield a parallel series of trails from successive aircraft. (Source: NASA.)

that the reduction in cirrus clouds produced by aircraft contrails produced a measurable increase in DTR [383], although others have suggested that they just happened to be clearer days overall [384, 385] (Figure 8.15).

While these fundamental features of Earth's hydrological cycle were and remain the subject of debate, rain continued to lash the surface of Titan. We would shortly learn much more about the solar system's other present-day hydrological cycle.

9

Dune Worlds

2004 to 2012

2004 – Cassini arrives at Saturn, observes convecting clouds around Titan's south pole

2005 – Huygens probe measures Titan conditions *in situ*, images surface close-up showing river valleys and tumbled cobbles

2006 – Cassini's radar discovers extensive dune fields on Titan, a key climate indicator

2006 – Venus Express arrives at Venus, shows sulfur dioxide variations

2007 – Lakes, then seas, of liquid methane discovered around Titan's north pole

2007 – Mars Reconnaissance Orbiter arrives; Mars Climate Sounder instrument observes atmosphere, high-resolution camera observes surface changes such as dune migration, formation of gullies

2008 – Phoenix lander explores north polar terrain on Mars, detecting subsurface ice, observes meteorology including "snowfall"

2009 – Cassini observes clouds and associated surface changes on Titan, showing rain reaches the surface today. Circulation models challenged to reproduce dune orientation

After its 7-year voyage from Earth, Cassini finally approached the Saturnian system. In Cassini's optical images (including the near-infrared at 940 nm) the views of Titan's south polar region on arrival at Saturn in July 2004 showed intriguing variegation (Figure 9.1). There were some lake-shaped dark spots (that is to say, dark spots shaped like not much in particular) which were certainly intriguing, as had been the much larger irregular dark regions seen by HST and other telescopes before. Thin dark tendrils appeared to link a couple of them, but

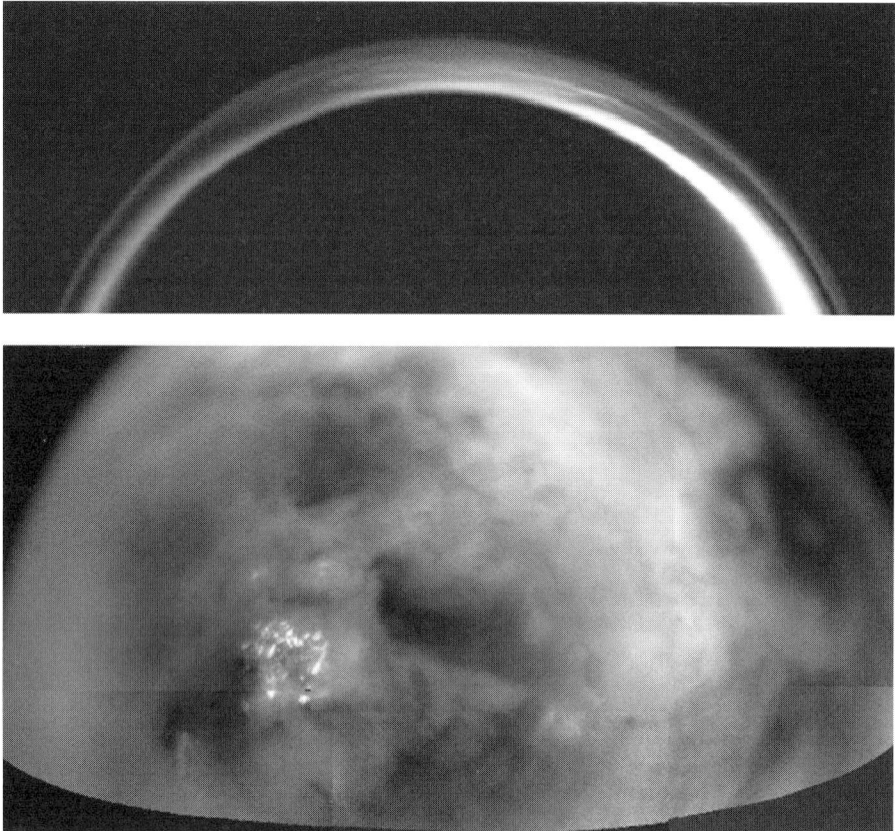

Figure 9.1 Views of Titan from Cassini in 2004. The upper panel is a UV image taken at a high phase angle, i.e. Titan is backlit and forward scattering shows multiple layers of haze over the north pole, with a single detached haze layer extending to lower latitudes. The lower panel is a mosaic of near-infrared images; the square patch of small bright splotches is methane clouds around the south pole, while the kidney-shaped dark spot to the lower left of those is now called Ontario Lacus and is an ethane-bearing lake, 250 × 70 km. On the other hand, the large dark area at upper right is Aztlan, later discovered to be an equatorial dunefield. (Source: NASA/JPL/Space Science Institute.)

without accurate measurements of how dark everything was, or topographic or other information, it was impossible to be certain what these spots meant.

The clouds, on the other hand, were vigorously churning around the south pole and Cassini had a grandstand view [386, 387]. Not only did the camera show the clouds evolving horizontally in short weather movies, being blown by the winds in some cases, but Cassini's near-infrared spectrometer could measure the height of the cloud tops via the methane absorption at different wavelengths [388]. Some of these cloud tops were ascending to about 40 km (i.e. the tropopause), at speeds of a few meters per second, just as models of methane cloud convection had predicted. But did methane rain reach the surface?

A hint to the positive was given in the first radar pictures taken by Cassini, which showed in October 2004 what could be bright channels winding in a dark highland, spreading out in a bright triangle – an alluvial fan [389, 390]. These form in predominantly dry areas on Earth, where occasional flash floods occur carrying cobbles and boulders. This debris is dropped abruptly as the flow spreads out from a mountain canyon and when the storm driving the flood dies down, leaving a rubbly surface that scatters radar radiation effectively – some beautiful examples on Earth are seen in China and in Death Valley, USA. But it was early days in understanding Titan and, again, there was more than one possible explanation.

But there was no such doubt in the pictures that came back from the Huygens probe as it descended by parachute through Titan's atmosphere in January 2005. At a stroke, its measurements solved many of the puzzles about Titan and provided definitive ground truth to help interpret Cassini's measurements. After coasting dormant for 22 days after its release from Cassini, Huygens started transmitting to jubilant controllers after its entry into Titan's atmosphere on January 14, around 10 a.m. That triumph in itself meant that the probe had survived entry, validating the models based on Voyager and other data.[1] And it kept transmitting after it hit the ground, as many hoped it would (but because of our ignorance of Titan's surface, could never be guaranteed).

Its pictures showed branching dark channels on a shallow bright highland that everyone could immediately recognize as river channels [392].[2] Rain must have indeed reached the surface, at least once. Moreover, the view from Huygens side-looking camera (Figure 9.2), about knee-high off the ground, showed a plain

---

[1] The story of Cassini's early findings, and the probe descent in particular, is told in Lorenz and Mitton [391].

[2] The paper was titled "Rain, winds and haze during the Huygens probe's descent to Titan's surface". *Nature* likes sensationalist titles – rain was not itself seen, only presumed to have occurred in the past.

Figure 9.2 (Left) Mosaic of three images taken during Huygens' descent, showing a network of river channels draining into a dark lowland. The image is about 7 km across. (Right) The view from Huygens' knee-high camera on the surface, showing rounded cobbles evidencing river flow in the past. (Source: NASA/ESA/U. Arizona.)

strewn with rounded cobbles, exactly as if "rocks" (made of ice or organics or something) had been tumbled in a fast-moving methane stream. These cobbles, 10–15 cm in diameter, sat atop a somewhat fine-grained material with the consistency of damp sand or packed snow,[3] and measurements with the probe's gas analyzer, which had a heated inlet that was jammed into the ground at impact, showed the ground "felt" cool and moist, and that methane, ethane and other vapors were sweated out by the probe's heat.[4] Although rain was not seen, perhaps one dewdrop may have been: a transient feature in one image taken after landing may be the sparkle from a 4-mm drop of methane that

---

[3] The penetrometer instrument that measured the consistency at impact was one that I designed and built (with much help from the SSP team) as part of my Ph.D. in 1991–4. After three years on Earth, and seven years in space, it worked for the 1/20 of a second it was supposed to.

[4] Reference [393] describes the composition measurements themselves. There were also indirect indications that the ground was damp by how slowly the inlet warmed up [394]. Another subtle indicator of material being sweated out of the ground by the heat of the probe was the suppression of ultrasound propagation by an instrument on the underside of the probe [395]. Extensive ground tests with the penetrometer, the subject of Karl Atkinson's Ph.D. thesis, showed better fits with damp material than dry stuff. How long ago it might have last rained at this location was an entirely different question, however, to which there is no compelling answer as yet.

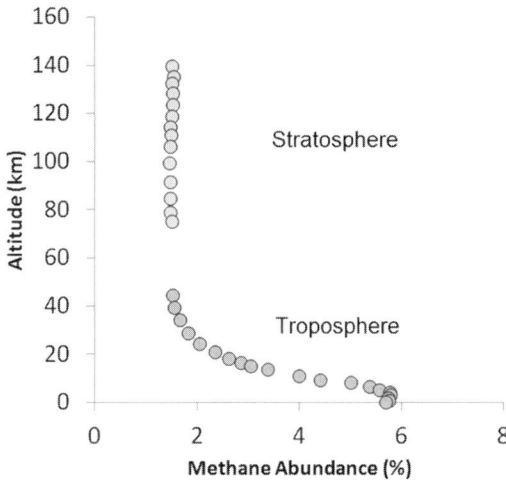

Figure 9.3 Titan's tropospheric profile of methane abundance, measured by the Huygens probe's mass spectrometer, resembles that of water vapor on Earth, increasing towards the warmer ground (where there is a reservoir of methane moisture). Unlike Earth, however, the tropopause is warm enough that Titan's stratosphere retains about one percent of methane, rather than the tiny amount of water vapor in Earth's stratosphere. (Author, data from Niemann et al., 2010 [393].)

formed on the camera baffle, probably from vapor generated by the heat from the probe's surface science lamp [396].

In addition to the cobbles and river channels, a couple of dark streaks could be seen in the distance, to the north. These in fact would prove to be one of the most powerful indicators of Titan's climate, about which more later. Huygens saw no rainbows, nor rain or clouds. A hint of a cloud layer was seen during its descent, but only by extremely close examination of the probe's side-looking images.

Whereas on Earth the profile of moisture in the troposphere is measured over a thousand times a day directly by weather balloons (as well as indirectly by satellite measurements), we have but one single profile of methane in Titan's equatorial troposphere (Figure 9.3). Interpreting it in terms of relative humidity and cloud formation, however, is not as straightforward as at Earth, in that nitrogen dissolves in methane (to about 30%) so that the thermodynamics of the binary mixture must be considered [397, 398]. Moreover, nitrogen is less soluble in the solid phase than in the liquid, so as cloud droplets ascend and grow in a cloud, they may freeze into hailstones at about 15 km altitude, and the nitrogen may come bubbling out as this happens.

In many respects the profile is similar to that of water vapor on Earth, where there is similarly a reservoir of the volatile on the surface, a cold trap at the tropopause, and photochemistry in the stratosphere that destroys the volatile. The typical profile on Earth has about one percent of water vapor near the surface (corresponding to about 50% relative humidity), declining to a tiny value (about 10 parts per million) above the tropopause. On Titan, the near-surface abundance is about 5% (also ~50% relative humidity), but the cold trap is much less severe. In this sense, Titan's methane may be an instructive analog for the water structure on early Venus, where the cold trap "leaked" moisture into the stratosphere where it was lost. As our Sun becomes a red giant, Earth's tropopause will become warmer and the stratosphere will become "wetter" like Titan's.

The wind and temperature profiles measured by Huygens were overall in remarkably good accord with the predictions based on the rather meagre Voyager data. But some details were notably different. A layer of wind shear in the stratosphere had been hinted at in models but was measured to be much stronger; prominent gravity waves were also detected. Something of a surprise was the deviation of the wind direction in the lowest kilometer of the atmosphere, and an initial report that the planetary boundary layer was 300 m thick was later revised to ~3 km, recognizing the subtle vestige of the previous day's layer, whereas the 300 m was only the mid-morning value.

Dark streaks seen in the second Cassini radar flyby were open to different interpretations, but on the fourth flyby a much better view resolved the question: they were giant sand dunes [399]. These massive (over 100 m high), radar-dark ridges were about a kilometer wide (Figure 9.4), separated by 2–3 km, and, in

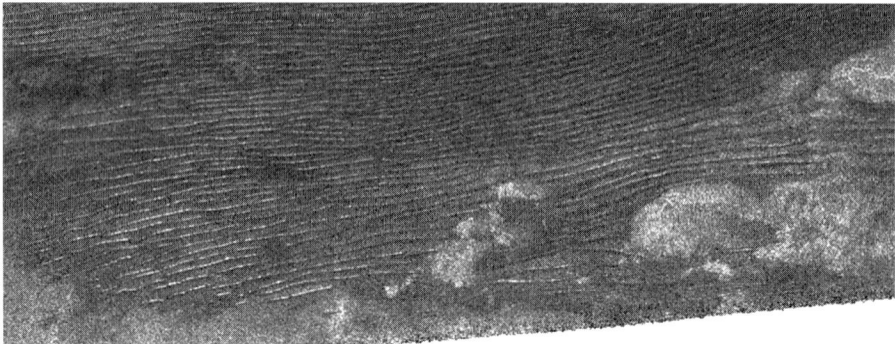

Figure 9.4 Dunes on Titan observed in October 2005 by the Cassini radar instrument. This image segment spans 100 km × 300 km, with radar illumination from above (north), showing very regularly spaced dark features, with glints on their north faces, confirming their positive relief – their length, height and spacing (and appearance overall) was noted to be very similar to the linear dunes in the Namib desert. Their streamline-like arrangement with respect to the bright hills at the right supported an association with a wind or liquid flow, although some scientists maintained that an erosive origin should still be considered. (Credit: NASA/JPL/R. Lorenz/Cassini Radar Team.)

many cases, hundreds of kilometers long. The way they parted and flowed around topographic obstacles suggested that they were linear or longitudinal dunes, which form in winds that switch between two dominant directions that converge at an obtuse angle (when winds flow at a nearly constant direction, dunes tend to form transverse or orthogonal to the winds; when the direction is strongly variable, star-shaped dunes arise). The 6000-km-long radar image also covered the Huygens landing site, and the two streaks seen in the distance in the probe images could be matched up with dark streaks in the radar image – they were shallow dunes.

These dunes formed a massive belt around Titan's equator, broken only by the large, bright mountainous Xanadu region on Titan's leading hemisphere. Indeed, the large equatorial dark areas speculated to be seas in Hubble images of the 1990s, and seen in more detail by Cassini's early camera images, all seemed to be seas of sand. But almost no dunes were found beyond 30 degrees latitude, immediately providing a good test for Titan GCMs. The transport of sand requires winds strong enough to overcome both the weight and cohesion of the sand, the latter property depending on how dry or damp it might be. Whether tholin-like organic sand on Titan made damp with liquid methane, or silicates on Earth made damp with water, wet sand is more difficult to move than dry – so was a dry equator consistent with Titan's meteorology?

In fact, GCMs had already predicted before the dunes were found that the low latitudes would become dessicated, and another GCM was soon formulated to address the question specifically. In essence, the slow rotation of Titan leads to a pole–pole Hadley circulation for much of the year, which leads to upwelling (and thus precipitation) in polar summer. Indeed, notwithstanding the river channels seen by Huygens, the equatorial region should dry out within a few Titan years, and methane moisture should accumulate at the poles [400, 401].

But what did the dune form and orientation tell us about the winds? At first it was speculated that the bimodal wind regime resulted from the diurnal tide in Titan's atmosphere,[5] but later it became clear that in fact seasonal changes in meridional circulation could yield the required alternation in direction, so the fact that the dunes appeared longitudinal in form was not too difficult to explain. But the arrangement with respect to obstacles and the way the dunes tailed off in places seemed to indicate that the net transport direction was to the east e.g.

---

[5] as suggested in Lorenz et al. [399], which at the time was the principal mechanism by which near-surface north–south winds had been suggested to occur – see for example, Tokano and Neubauer [402]. That seems to have been wrong, more recent models predict seasonally alternating north–south wind components at the surface due to the Hadley circulation.

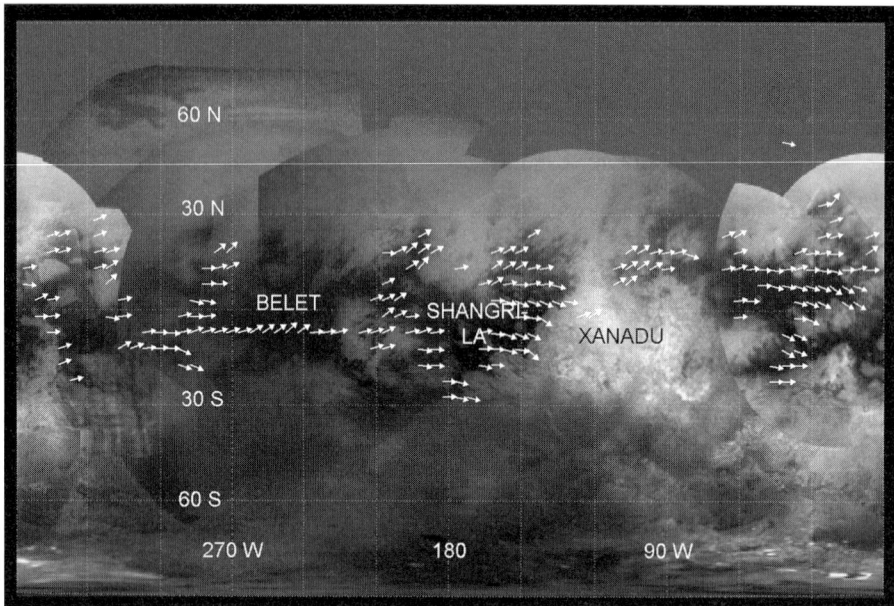

Figure 9.5 The average orientation of Titan's dunes observed in the prime mission in
5 × 5 degree boxes indicated with arrows. There is some hint of a poleward divergence,
and there are local deviations around bright areas. The background map is a near-
infrared one made with Cassini's camera – just visible to the upper left of the 60° N
label are the dark southernmost tentacles of Titan's largest sea, Kraken Mare. (Figure
created by author.)

Figure 9.5 [399, 403–405]. This in fact turned out to be a major difficulty for
modelers, causing operators of GCMs to initially question the geomorphological
interpretation of the radar images, since the large-scale surface flow at the
equator should generally be to the west, not the east as the sand dunes seemed
to indicate.[6] The matter took some years to resolve.

Meanwhile, after over a decade of neglect, Venus exploration resumed in April
2006 with the arrival of the European Space Agency's Venus Express spacecraft,
launched five months earlier. Originally intended to operate for 500 days, this
spacecraft lasted much longer, until November 2014. The mission had been
developed rapidly, most of its systems and instrumentation being modest adaptations of existing designs (notably, on Mars Express), and unfortunately one of
its most promising payloads (the Planetary Fourier Spectrometer) failed, with a
stuck mirror shortly after arrival.

---

[6] An early attempt to fit the dune pattern with a GCM was by Tokano [406]. The challenge
was highlighted by Wald [407].

While the mission yielded much more detailed insight into the chemistry and dynamics of the atmosphere at and above the cloud tops, and beautifully portrayed the figure-8-shaped south polar vortex, the mysteries of the lower atmosphere and surface were largely untouched (one exception being the detection of near-infrared surface "color" on a volcano, suggesting perhaps it had lavas fresh enough to be unaltered by the atmosphere[7]).

VEx also gave some of the most robust evidence yet of lightning on Venus. Venus has atmospheric convection that is vigorous at some altitudes, and clouds made of suitable liquids,[8] so it seems reasonable that some sort of electrostatic effects might occur, but various Pioneer Venus and Venera observations had positive and negative results, making the overall perspective rather equivocal. VEx's magnetometer detected signals that were interpreted as originating with lightning discharges, but could only be picked up occasionally because their detection relied on propagation along magnetic field lines linking the source region to the spacecraft. Venus' thick clouds, well above the surface, would likely only produce cloud–cloud discharges, rather than cloud–ground strokes, and the thick layers might obscure flashes from observation optically. The presence of lightning might also explain the abundance of nitrogen oxides, which is otherwise difficult to explain.

Venus Express data showed a dramatic up–down variation in the amount of sulfur dioxide in the Venus stratosphere [409] (Figure 9.6): the changes seen by Esposito in the 1980s were not an isolated occurrence. Whether these changes indeed indicate volcanism, or merely a change in circulation that dredges SO_2 up from the lower levels of the atmosphere to altitudes where it can be measured, remains to be seen, however. Meanwhile, SO_2 variations have been contemplated as a possible means of detecting volcanism on exoplanets, a long shot if ever there was one [410].

Sulfur dioxide has continued to be a temptingly pungent spice with which to heat up the Martian greenhouse in the past. Unlike that early favourite, ammonia, SO_2 has the advantage of geochemical plausibility since it is a common volcanic gas on Earth, and sulfates are found in abundance on Mars both by landed measurements and in spectra measured from orbit. In addition to its own greenhouse warming potential, acidic surface conditions would tend to suppress or reverse CO_2 weathering into carbonates (after all, the standard

---

[7] Smrekar et al. [408] described the features as indicating hotspot vocanism. Note that "hotspot" should not be interpreted to mean the surfaces were hotter than their surrounds by virtue of having been freshly erupted – although there have also been rather unreliable claims of such emission in Venus Express data.

[8] specifically, high polarizability. Liquid water and sulfuric acid are much better in this respect than methane.

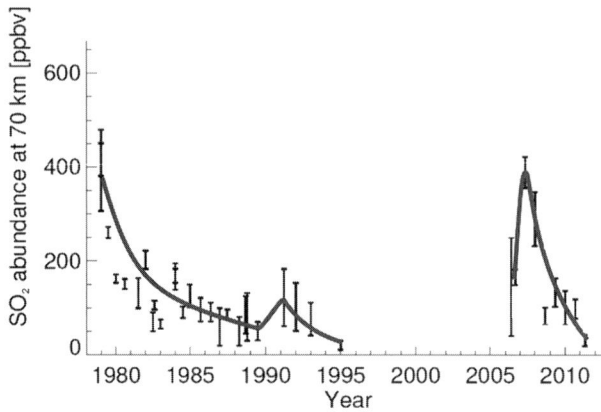

Figure 9.6 The slow decline of Venus' atmospheric SO_2 abundance, punctuated by sudden increases. Data at left were from the Pioneer Venus orbiter, data at right from Venus Express. (Credit: ESA.)

chemistry laboratory method to generate CO_2 is to add acid to carbonates). The question was (re)examined by several authors in the 2000s. Feng Tian at the University of Colorado explored the effect of SO_2 with 1-D photochemical and radiative-convective modeling, but found that, within weeks of adding SO_2 to the atmosphere, the formation of sulfate aerosols would increase the planetary albedo, overcoming the enhanced greenhouse effect [411], while Itay Halevy at MIT postulated that since the removal of SO_2 from the atmosphere is accelerated by liquid water on the surface, there may have been a climate feedback stabilizing the Martian paleoclimate around some temperature near the freezing point [412]. Proliferating Mars GCMs were also brought to bear, first considering only the enhanced greenhouse [413], but later the formation of aerosols. The conclusion of all this work, summarized in the 3-D GCM study by Laura Kerber in Paris in 2015 [414], is that "*mildly warm conditions require many improbable factors, while cooling is achieved for a wide range of model parameters*". Nonetheless, sulfur gases are still appealed to for solving the Earth's faint young Sun problem too [415].

Another means to provide the early Earth with a stronger greenhouse is simply to make the atmosphere thicker with nitrogen. Colin Goldblatt speculated that some of the nitrogen now thought to be in mantle rocks may have once been in the atmosphere, and suggested that if the atmosphere were twice as thick 2.5 billion years ago as today, a warming by a little over 4 °C would have resulted [416] (mostly by pressure-broadening of the CO_2 absorptions, blocking more of the infrared window; as Kasting had noted for CO_2 Mars' atmospheres, the stronger Rayleigh scattering of sunlight by the thicker atmosphere causes a cooling effect, but, in the case of nitrogen on Earth, the greenhouse warming is dominant). Such a drastic evolution of the Earth's atmosphere seems implausible, but is difficult to assert as impossible. Whereas there are some chemical signatures in rock deposits that place limits on the amounts of oxygen and greenhouse gases in the deep past, nitrogen is difficult to estimate. One of the few constraints that does exist is a physical one – if the atmosphere were much denser, raindrops would fall more slowly and hit the ground less hard. Sanjay Som and colleagues in 2012 documented the imprints of raindrops[9] in South African rocks from 2.7 billion years ago, and argued that their size required that the atmosphere be less than twice as thick as today [417]. On the other hand, Kavenagh and Goldblatt have demonstrated that the interpretation of such imprints is deeply challenging and that the limit provided by Som's example is not nearly so constraining [418]. Som and colleagues later interpreted the size distribution of gas bubbles that solidified in a 2.7-billion-year-old lava flow as indicating a surprisingly low total atmospheric pressure then of 0.23 bar [419]. This indeed would make the faint young Sun problem even worse, demanding other strong greenhouse gases: further examples and interpretation will likely be needed before such a profound proposed evolution in such a fundamental parameter of our planet is widely accepted.

Nitrogen is responsible for much of Titan's greenhouse effect too: although, as Tyndall noted, nitrogen is not a radiatively active gas, there is enough of it on Titan that the long-wavelength end of the thermal spectrum is largely blocked by collision-induced absorption of nitrogen molecules. Moreover, the methane and nitrogen greenhouse effects are similarly due to nitrogen–methane and nitrogen–hydrogen collisions. Benjamin Charnay in Paris examined Titan's climate with a GCM [420], throwing nitrogen clouds into the mix as well as modeling latitudinal gradients (not considered in the 1-D climate simulations of the 1990s). Like that earlier work, Charnay found that, especially if the surface albedo were set at high values, and the faint young Sun was considered, the nitrogen atmosphere would partly collapse, but that reaching a Triton-like frozen state demanded exceptional parameter choices. But it seemed that if the methane had run out in the past, nitrogen would largely take its place, forming clouds and rain: such a

[9] Charles Lyell (a Scottish geologist who, in addition to his own work, popularized Hutton's writings and helped arrange publication of Darwin and Russell's work on evolution) documented the fossil imprints of raindrops in Carboniferous rocks in 1851.

nitrogen-based hydrological cycle may have played a role in eroding Titan's surface for much of its history.

When raindrops fall, their speed is a balance between their weight and aerodynamic drag. And just like the heat generated on the shoes of a brake, this drag generates a tiny amount of heat. Olivier Pauluis and colleagues in New York calculated [421], and later verified with satellite radar measurements [422], that over the tropics this dissipation amounts to about 2 W m^{-2}.[10] This isn't much compared with the flux of sunlight, but is in fact quite comparable with the greenhouse forcing by minor gases such as ozone and nitrogen oxides, which climate modelers do worry about, yet this dissipation is typically not included in models. Carnot's vision of the atmosphere as an engine bears recollection here – the engine is not perfect because rain introduces some friction. Much of the rest of the ideal output of the engine is expended in dehumidifying the atmosphere, concentrating well-mixed water vapor into raindrops themselves. Just as desalination requires useful energy to unmix salt and water, so reducing the mixing entropy of the atmosphere soaks up some of the potential work that the atmosphere could otherwise do. This thermodynamic perspective on atmospheres has yet to be fully exploited (especially for the other climates), although it has found some use in estimating the intensity of hurricanes as a relatively simple function of sea surface temperature [424], and Kerry Emanuel of MIT has calculated the dissipation in tropical cyclones, finding that the total has increased substantially in recent decades [425]. While one can never confidently attribute a single weather system to global warming, some patterns do seem to be emerging, and there is some fundamental physics behind the link between storms and surface temperature.

Mixing and distillation have a bearing on a possibly unlikely window on Martian polar meteorology that was opened by the Mars Odyssey GRS. The GRS could detect gamma rays of a particular energy (1294 keV) emitted by the noble gas argon. As measured by Viking, argon is present in the Mars atmosphere at about 1.6%. However, if a parcel of CO_2-rich atmosphere freezes out on the ground in the polar winter, the argon is left behind as a gas. Thus the condensing atmosphere becomes more argon rich by freeze distillation – like making apple-jack from cider. The GRS was able to observe this process in action, specifically over the Martian south pole, where in winter it seemed that the argon abundance increased by a factor of 6 [426]. This process had been hinted at by Viking data, where infrared brightness temperatures seemed to be lower than the CO_2 frost point for the expected pressure: this could be explained if the CO_2 in contact with

---

[10] In fact the drag/dissipation question has been raised as a potential issue in exoplanet circulation too [423].

the frost were more dilute than the bulk atmosphere. Other ideas had been that CO_2 snow was reducing the thermal emission somehow.

This factor of 6 enhancement, seen by 2007 in two successive southern winters on Mars, was larger than GCMs predicted. One study with special model adaptations (e.g. a cube grid to better model polar processes) managed to get an enhancement of about 4.7 [427], but even this fell short, suggesting there is still work to do. But overall, with growing amounts of observational data against which to test them, Mars GCMs have been improving all the time and growing in predictive capability. A critical challenge has been the treatment of airborne dust and the condensation of water and CO_2 vapor: in the post-Viking era these three types of cloud tended to be treated as entirely separate entities, but improved microphysical schemes now allow dust to serve as nucleation sites for the condensation of volatiles, while such accretion of ice around dust can help sweep the dust clear. Further, while it was tempting at first to model accumulation and ablation of water and CO_2 ices on the surface as separate processes, it emerged that they are often tightly coupled. For example, water ice near its melting point will sublime into the atmosphere rather quickly, so (counterintuitively) gullies would not be expected to form in the warmest parts of Mars, since any ice there would disappear before it could melt. What may be needed is a colder region where CO_2 ice can also accumulate – then the CO_2 frost will protect the water ice, keeping it at the CO_2 frost point temperature until the frost burns off later in the summer. At that point, the water ice warms up rapidly, possibly melting before it can sublime away. However, dry avalanche, CO_2-driven and water-driven ideas about gully formation continued to be debated.

Tony Colaprete at NASA Ames and colleagues were able to show in 2005 that the south polar cap, whose offset from the geographic south pole had always been a puzzle, results from a topographic perturbation to atmospheric circulation caused by the deep Argyre and Hellas impact basins [428]. This perturbation in turn affected the style of CO_2 deposition – the stormy hemisphere west of Hellas tends to have a higher albedo, which Colaprete attributes to precipitation (i.e. CO_2 snow), whereas east of Hellas is the dark "Cryptic" region where CO_2 apparently forms in near-transparent slabs, condensing directly onto the surface.

Similarly, François Forget in Paris showed that at high obliquity, when the Martian polar tilt is up to 40 degrees and the poles see much stronger peak insolation that can burn off more water vapor, windward slopes of low-latitude high volcanoes can see accumulation of snow. The regions of predicted snow accumulation [429, 430] match up well with observations of surface textures indicative of glacial action [431] (Figure 9.7). Of course, any given model might just happen to be tuned right to get some observed result, but these types of results signify a growing capability to predict site-specific weather conditions on Mars.

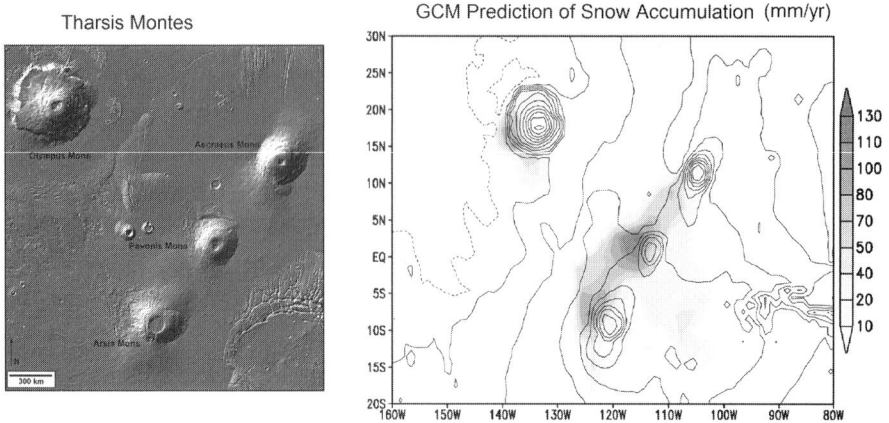

Figure 9.7 Close examination of the Martian surface revealed what looked like glacial deposits, even near the Martian equator, on the western slopes of the Tharsis volcanoes (approximately the white areas in this shaded relief view from MOLA data). Circulation model experiments by François Forget and the Paris group (right) found that these specific areas indeed should see snow accumulation during periods of high obliquity (but not in present conditions). (Image courtesy François Forget.)

Mars was subjected to even more intense orbital scrutiny in 2006, when a new NASA spacecraft, the Mars Reconnaissance Orbiter (MRO) arrived in March, and after aerobraking its way down to a low science orbit, began observations in autumn. It carried MARCI, a meteorology-oriented camera, able to image the entire globe, daily and in color.

Also among MRO's payloads, at last, was the infrared sounding instrument that Dan McLeese at JPL had waited so long for. (It had been launched as PMIRR on Mars Observer in 1993, and then a replacement had flown on Mars Climate Orbiter in 1998, only to be lost on that mission too). This iteration, the Mars Climate Sounder, was of a more compact design than its predecessors but would still yield frequent regular vertical profiles of the temperature, pressure, dust and clouds of the lower 80 km of the atmosphere.

The Compact Reconnaissance Imaging Spectrometer for Mars (CRISM) would complement OMEGA by offering higher spatial resolution (~100 m), to map mineral and other deposits. Also bringing higher resolution in a different dimension was SHARAD, a shallow-sounding radar.

Whereas the Mars Orbiter Camera (MOC) on Mars Global Surveyor in 1997 brought a significant improvement in image resolution, from the tens to hundreds of meters typical for Viking to several meters, the further jump to the ~half-meter pixel scale of the High Resolution Imaging Science Experiment (HiRISE) – essentially a big spy-satellite telescope on a low-orbiting Mars orbiter – transformed our understanding of Mars. Not only was the camera of high resolution, but MRO had a very high capability downlink; indeed, over its first 10 years of operation it sent back 250 Tb of data.

The enhanced resolution of HiRISE exposed many new phenomena on Mars. Not least was the movement of ripples and even dunes. It had been a puzzle for many years why Mars' abundant sand dunes did not observably move. Perhaps the dunes were all fossils of a past climate, when the atmosphere was thicker, making it easier to move sand. Another idea was that the sand was cemented, either by salts or ice. Even uncemented, though, in present conditions, with an atmospheric density 50 times smaller than on Earth, exceptionally strong winds are required to provide enough lift and drag to pluck sand into the air. However, it became recognized at the end of the 2000s that – unlike on Titan and Venus, and much more than on Earth – there is a big difference between the wind speeds needed to keep sand moving once it starts and that needed to get the first grains moving. Like the climate system, there is hysteresis in sand transport. But the improved resolution of the HiRISE camera allowed closer observation, and sure enough, after some tentative indications of sand movement in a few locations, definite observations of dune migration (Figure 9.8) were obtained [432, 433].

HiRISE gave a closer look at intriguing features first noticed in MOC images – the "spiders" of Mars. These are groups of bright cracks radiating from a central point that appear to form in the polar regions in spring (Figure 9.9). Often they are associated with dark spots or streaks. It seems that they form in much the same way that Triton's plumes operate – a transparent glaze of CO_2 ice, perhaps a meter thick, gets heated at its base once the Sun comes up [434, 435]. This causes evaporation, which builds up pressure that is kept confined under the ice slab, until either mechanical or thermal stress causes the ice to crack and gas erupts through, dragging with it dark dust, which is then winnowed to form a streak downwind of the crack.

Whereas on Earth there are seafloor and ice-core records with reasonably well-estimated age profiles, and which contain records of oxygen isotopes that are closely related to the ice volume, decoding Mars' climate history is much less straightforward. Layers seen in orbital images are exposed near-horizontally (at least until we have a rover rapelling down a crevasse or gazing sideways as it trundles down a trough in the ice cap) and stereo images or MOLA data are used to estimate the elevation of the layers. The layers as seen in Mariner, Viking and MGS images are presumably dust-rich and dust-poor, although many also seem in higher-resolution HiRISE data to have different textures, some being more rubbly than others. The largest layers, seen in sounding radar data, similarly must have some compositional or textural contrast to give a radar reflection.

Jacques Laskar, a French astrodynamicist who calculated Mars orbital and rotational histories (revising the 1970s work of Bill Ward, and pointing out that

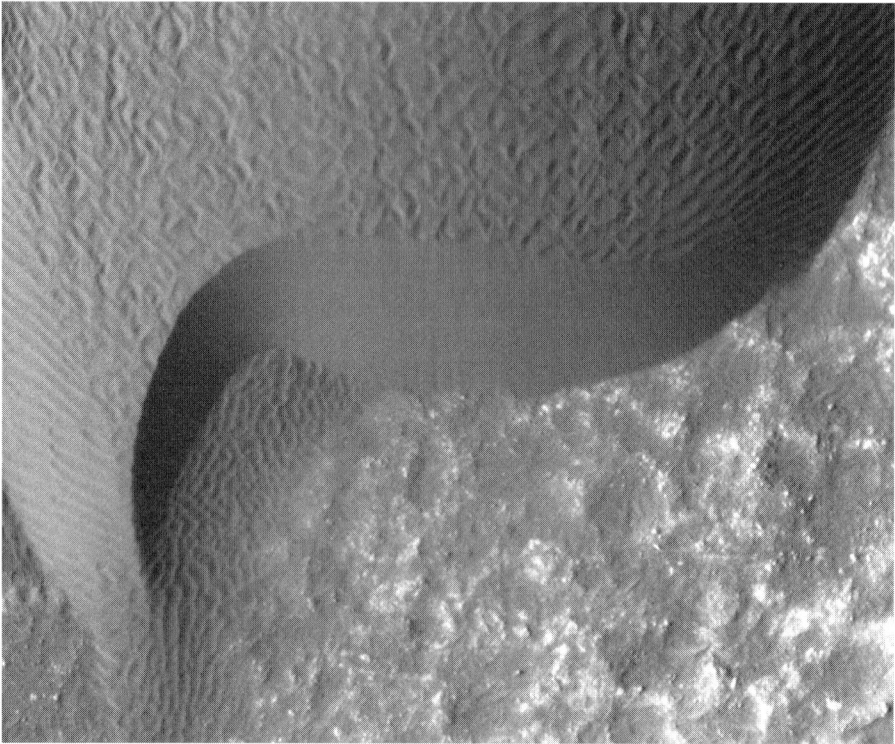

Figure 9.8 A dune on Mars seen to move. The textured ground underneath the dune helps to provide a reference against which to measure the small (meter-scale) migration seen in images months or years apart with HiRISE. The ripples on the surface of the dune have also been measured to migrate. (Source: NASA/JPL/U.Arizona.)

Mars' obliquity may have varied chaotically in the past since, unlike Earth, it lacks the anchor of a large moon to keep its pole straight), ventured a correlation between the north pole insolation history and the observed layers using MGS image data and MOLA elevations (Figure 9.10) [436]. Arbitrarily, he assumed a change in the depth/age relationship part-way through the record.

More elaborate analyses of the layer stratigraphy have been made by various American, French and Danish scientists [437–439], but the question of what the layers actually correspond to in terms of climate process (a common assumption is that dusty material accumulates at a constant rate, and ice is added at some rate that relates to insolation) is still rather open. There was, furthermore (as at Earth), the question of which insolation is relevant – Norbert Schorghofer at Hawaii has pointed out that the insolation at different latitudes responds differently to the eccentricity and obliquity cycles [440]. Thus, although the appeal of the layered deposits remained compelling as a climate record, the message had not yet been decoded [441–443].

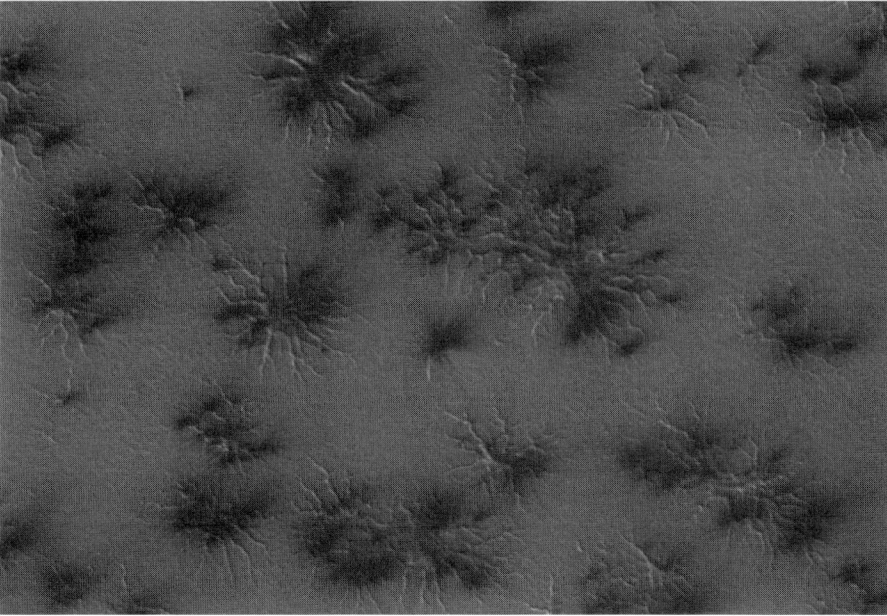

Figure 9.9 The spiders of Mars. Bright cracks in the near-transparent CO_2 slab ice radiate from vents where gas from sublimation at the base has broken through, hosing dark dust and sand into the air, where it is blown downwind to form streaks. (Source: NASA/JPL/U.Arizona.)

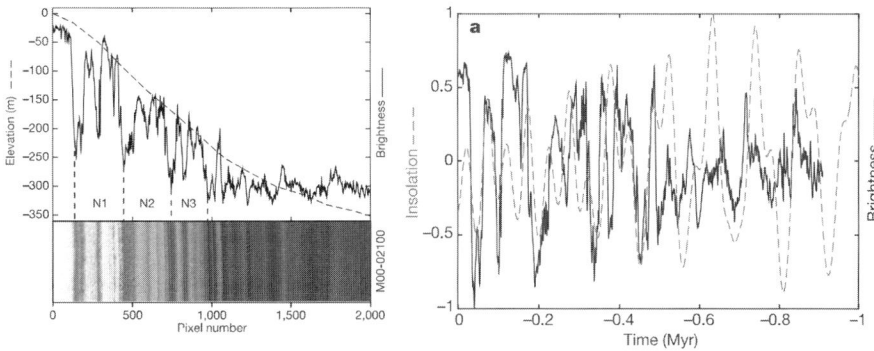

Figure 9.10 Laskar's attempt to reconcile layered texture in the north layered terrains (image at bottom left, with height profile and image brightness above) with the insolation history. (Reprinted by permission of SpringerNature from Laskar, J., Levrard, B. and Mustard, J.F., Orbital forcing of the Martian polar layered deposits. *Nature*, **419**(6905), p. 375 © 2002.)

In fact, although the Shackleton work in the 1970s was convincing that a 100-kyr eccentricity signal was prominent in the last 400 kyr of deep-sea cores of Earth climate, the signal rather disappears going further back, and a shorter-period signal dominates. Core records are often discontinuous, and accumulation rates (i.e. the age/depth function) differ from place to place, but computational methods – not unlike those used to splice sequences of base pairs in genome mapping – were brought to bear by Lorraine Lisiecki at Brown University in 2004 to generate a long, high-quality climate record from 57 cores [444, 445]. This "stack" (Figure 9.11) shows strong coherence with the obliquity and precession cycles of insolation at 41- and 23-kyr periods over the last 5 million years.

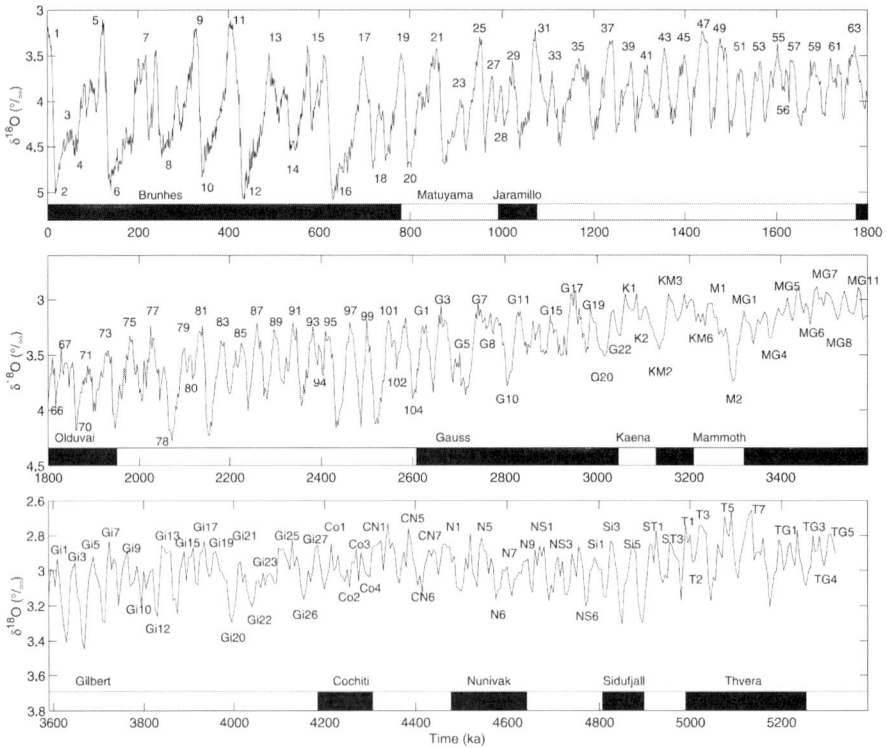

Figure 9.11 The oxygen isotope history over the last 5 million years, synthesized by kneading and splicing 57 different seafloor mud-core records together. Referred to as the LR04 stack, after Lorraine Lisiecki and Maureen Raymo of Boston University, who co-authored the study, it shows that the prominent 100-kyr signal disappears in the early Pleistocene (i.e. before about 500 million years ago) and instead, shorter-period signals dominate. (From Lisiecki, L.E. and M.E. Raymo (2005), A Pliocene-Pleistocene stack of 57 globally distributed benthic $\delta^{18}O$ records, *Paleoceanography*, **20**, PA1003, doi:10.1029/2004PA001071, with permission from Wiley.)

What this means, of course, is not easy to determine. The astronomical forcing of climate is weak, and so the climate system must amplify the signal – the oceans or ice sheets somehow resonating with it [446–438]. But in the last million years or so, the organ pipes of the Earth's climate have changed their tune, once whistling to obliquity and precession but more recently booming to the deeper notes of eccentricity. The earlier sensitivity to ~40-kyr fluctuations has been attributed to the behavior of the West Antarctic Ice Sheet, but doubtless other mechanisms are possible. For example, going back only a little further into the Pliocene three million years ago, the planet overall was a couple of degrees warmer than present and sea level 20–30 m higher; although the continents were all basically where they are now, North and South America were not connected by the Isthmus of Panama, and currents could surge between Pacific and Atlantic, so ocean heat transport would have been quite different from the present.[11] Perhaps the Martian climate may be intrinsically simpler to understand, but for now the record we have to work with on that world is rather meagre compared with the terrestrial climate reconstruction.

Many of the objectives of the daring 1998 Mars Polar Lander mission were embraced by the Phoenix lander. This reached Mars in 2008, when planetary alignments favored delivery to the northern polar regions, rather than the south. Phoenix was to use much of the lander hardware that had already been developed for the cancelled 2001 lander, but with instruments largely derived from those of the ill-fated MPL.

The lander was destined for the Vastitas Borealis formation, the large low-lying plains around the Martian north polar cap, and what had presumably been the seafloor at some point, if Oceanus Borealis had ever been a sea. Remote sensing showed patterned ground in this area (regular polygonal cells, indicated by slight troughs at the corners, often with concentrations of coarser rocks), suggestive of near-surface ice. And, of course, the Odyssey neutron data indicated that hydrogen was abundant in the near-subsurface at these latitudes. So a prime goal of Phoenix was to measure *in situ* the amount and nature of near-surface ice, using soil samples dug with a digging arm, cooked in the TEGA (the Thermal and Evolved Gas Analyzer, reflown from MPL). As well as measuring ice, this instrument would detect water of crystallization in salts, and the presence of a small percentage of carbonates [449], long speculated to be a sink for the Mars atmosphere. Phoenix would also measure weather conditions,[12] including a LIDAR to

---

[11] We see that Thomas Jefferson's thought experiment in climate modification has in fact been performed by the changing Earth already. But humans weren't around to observe it.

[12] Phoenix at least had a decent set of pressure and temperature sensors and a humidity/soil conductivity probe. Its wind measurement capability was somewhat rudimentary,

Figure 9.12 The "Holy Cow" region underneath the Phoenix lander, where the rocket exhaust (the nozzles are just visible at the top of the image) had blasted away regolith, exposing a layer of hard white material, presumed to be water ice. (Source: NASA/University of Arizona.)

profile clouds and dust, and would examine the soil with a wet chemistry experiment and even with an atomic force microscope.

In many ways, the most persuasive indication of near-surface ice was an entirely unplanned one (Figure 9.12). Phoenix's descent system used rocket engines down to the last couple of meters, and the blast of the exhaust from these (even though deliberately diffused through small nozzles to minimize this effect) scoured the regolith away under the lander [450]. When a camera mounted on the robot arm was used to peek under the lander, it found a flat sheet of white material under about 8 cm of regolith, most probably ice.

But proving the material was ice became an overriding concern for impatient NASA officials.[13] As the Phoenix team struggled with challenges such as short circuits and learned-as-they-went how to control the arm, the pressure to get a successful TEGA sample showing ice dominated other science priorities. This was frustrated by difficulties in getting a sample into the analysis oven – it turned out that the soil was actually rather sticky, forming clods that were reluctant to

---

however, essentially a small "telltale" windsock, which had to be imaged with the camera to get a measurement – yet another in the compromises that Mars-landed meteorology has suffered. Even so, useful records of some ~7000 measurements were made.

[13] It didn't help that some of the TEGA doors failed to open, and operations were also plagued by short circuits and late development of the sampling arm software. For an interesting embedded account of Phoenix science operations, see Kessler [451].

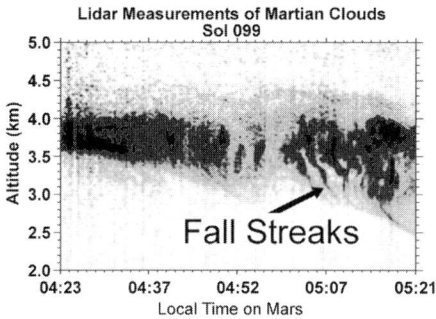

Figure 9.13 Clouds and streaks of snow falling towards the Martian surface (but subliming away before they reach the ground) observed by the Canadian LIDAR on the Phoenix lander. (Source: NASA/U. Arizona/JPL/Canadian Space Agency.)

drizzle through the wire mesh covering the oven loading mechanism.[14] The wet chemistry experiment indicated a possible reason, the presence of perchlorates in the soil, deliquescent salts that kept the regolith moist even with very low humidity in the air [452, 453]. This low humidity (measured by a probe on the sampling arm to be less than 5% [454]) allowed white frost exposed by arm digging to sublimate away, and indeed some samples may have dried out on their way from the ground to the TEGA ovens. But eventually success was declared, ice was present (just as everyone expected anyway) and the hydrogen signal seen by Odyssey could be attributed to permafrost.

Winds were measured up to 16 m/s [455], and in fact particles of a calibration sample drizzled from the robot arm into TEGA were seen to be swept sideways as they fell. Temperatures ranged typically from 190 K to 243 K [456], and a few dust devil vortices were detected per day by Phoenix's pressure sensor [457].

As the mission went on, and the Sun sank lower and lower in the Martian sky, the lander became colder and could develop less and less energy from its solar panels. But before it succumbed to the Martian winter, its camera spotted a few dust devils as well as clouds, and its lidar showed streaks of snow falling towards the ground (Figure 9.13) [458] (in fact, later simulation work would suggest these

---

[14] In fact, in my last year in Arizona in 2005/6, I helped perform soil loading tests on mockups of the TEGA ovens, drizzling ice–sand mixtures chilled with dry ice through the mesh, which I never liked (it was intended to prevent blockage of the oven by large grains). By holding and shaking the scoop just right, dust and sand tended to go in OK, although when things warmed up in the lab things did get more sticky. It didn't occur to me (or anyone else) that the soil on Mars would behave that way – it was supposed to be too cold (and indeed the highest air temperature recorded during the mission was −20 °C). A lesson, perhaps, in the utility of tests to show that something works versus tests to find in what ways something may fail.

are not passively descending particles, but, rather like microbursts, plumes of negatively buoyant air cooled by radiation from the snow itself [459]!).[15] After 150 sols, the lander became silent. Few expected it to survive the winter (unlike Viking, which had power and heat from a radioisotope power source, and which had been at a somewhat lower latitude) and indeed pictures from the HiRISE camera in the spring showed evidence that Phoenix's solar panels had broken off, probably weighed down by something like a 30-cm thickness of CO_2 frost in the winter.

Titan's polar regions had revelations of their own. In 2006, the first flybys at high northern latitudes on which the radar instrument observed the surface (at the time, the north pole was in winter darkness) showed hundreds of lake-shaped dark features, typically 20 km across. The radar reflectivity measured was much darker than the equatorial dunes, and just what one would expect for smooth, flat surfaces. The fact that these surfaces were made of a low-dielectric-constant material consistent with liquid methane was confirmed by the microwave emission, measured separately by the radar instrument. This combination of quantitative data was all but definitive that Titan's north polar regions, indeed perhaps both poles, had extant bodies of liquid.

Cassini, in orbit around Saturn, flew by Titan at intervals of between 16 and 90 or so days, typically once a month. Sometimes its observations were directed at measuring the interior structure, sometimes at upper atmospheric measurements, and sometimes its surface via radar or near-infrared measurements, but rarely at the same time. As geographical coverage by radar slowly built up, it became clear that, while there was a large kidney-shaped lake in Titan's southern hemisphere, Ontario Lacus, about 250×75 km in extent, there were otherwise relatively few bodies of liquid in the south. In contrast, in addition to the hundreds of small lakes in the north, three much bigger bodies of liquid were present in that hemisphere (Figure 9.14), large enough to merit designation as seas rather than just lakes. These three seas, named Punga, Ligeia and Kraken Mare, were ~300, ~400 and ~1000 km across. Thus, by 2009 scientists were confronted with the impression that the north had a much larger inventory of liquid than the south.

[15] The idea is that, once the cloud particles form, they allow the air to cool further by radiation into space, and plumes plunge down from the cloud layer to the surface, although they may only reach it if the cloud layer is low enough. Adiabatic heating of the descending air causes the ice particles to sublime away, and thus the plumes are a little like virga on Earth. Convection caused by radiative cooling does occur sometimes in marine boundary layer clouds on Earth, and has been recognized as possibly a significant factor in Venus' clouds, e.g. [460].

Figure 9.14 Radar mosaic of Titan's north polar region, showing its seas as dark areas. The three seas (Mare) are named after mythical sea monsters, straits (Freta) after characters in the Asimov Foundation novels, and bays and inlets (Sinus) after inlets on Earth. (Author/Cassini Radar Team.)

This apparent dichotomy in some ways mirrored that of Mars decades before – why was the Martian residual south polar cap so much smaller than the north polar cap? That mystery had two factors at work, the quite different elevations of the two caps (the north being in the lowlands, where the atmospheric pressure and thus the CO_2 frost point temperature was higher) and the orbital eccentricity of Mars which meant the northern summer was longer and cooler than that in the south.

Topographic information derived from Cassini's radar measurements showed that while both poles were in fact rather lower (~1 km) in elevation than the equator, the two poles were not substantially different in that respect, and there was nothing to suggest any surface or subsurface properties differentiated the two hemispheres. That left Titan's seasons: like Mars, Saturn and Titan's present spin–orbit configuration is such that its eccentric solar orbit has its periapsis close to southern summer solstice, such that southern summer is shorter but more intense than the northern. This means that summer (when the methane rain occurs, perhaps counterintuitively the season in which liquid accumulates) is longer in the north than the south, and the strongest sunshine and highest temperatures, which might best cook volatiles off the surface, occurred in the south, drying out the lakes. As Caltech scientist Oded Aharonson explained [461], this arrangement, as on Earth or Mars, is only temporary, depending on the precession of the poles, and on the variations in Saturn's orbital eccentricity in response to gravitational perturbations from Jupiter. In other words, the Croll–Milankovitch cycles that may drive some climate change on Earth and Mars may occur on Titan too, and perhaps 30,000 years ago, the seasonal asymmetry on Titan would have been reversed.

Evidence that Titan's climate today was different from how it may have been in the recent past came from analysis of the near-infrared images of Ontario Lacus, suggesting it might in fact be rather muddy, and that a bright margin around its edge was suggestive of a "bathtub ring" of evaporite deposits [462, 463]. Of course, these are not salts familiar as solutes in terrestrial waters, but some organic analog where differential solubility in an evaporating basin has been preferentially deposited at the shrinking margins. Such a bathtub ring is evident at terrestrial lakes that have seen recent drops in lake level, notably Lake Mead near Las Vegas.

In fact, a comparison between an optically measured outline in 2004 and the margins in a radar image (Figure 9.15) some years later suggested that Ontario may have shrunk in extent, due to seasonal evaporation [464]. Such an observable shoreline retreat relied on very shallow slopes (less than one part in one thousand), such that a drop in lake level by only a meter or two could yield kilometers of shoreline movement. In fact, such shallow slopes were measured by Cassini's

Figure 9.15 Ontario Lacus as imaged by Cassini's radar (left). At 235 × 70 km, Ontario Lacus is the fifth-largest body of liquid on Titan (after the three northern seas and Hammar Lacus) and is the only large body of liquid in the southern hemisphere. Its morphology resembles that of some playa lakes on Earth (notably the much smaller – 4 × 2.5 km – Racetrack Playa in Death Valley, USA, seen from the air at right in author's photo), consistent with it slowly drying up in the present climate epoch. A delta-like feature part way up suggests transport of sediment by the dark river channel to the right.

radar altimeter at the margins of Ontario, and are seen in some terrestrial dry lake beds, such as Great Salt Lake in the USA or Australia's Lake Eyre. Some scientists had other favourites, such as the Etosha Pan in Ethiopia or Racetrack Playa in Death Valley, USA. These morphological and topographic comparisons supported the notion that the southern hemisphere might be drying out (as the astronomical climate picture of Aharonson suggested).

On the other hand, the possibility that liquids were accumulating in the northern hemisphere was supported by the appearance of some of Titan's northern seas, Ligeia Mare in particular, which had some coastlines, inlets and islands that resembled those where lake levels have been rising on Earth (Figure 9.16), such as Lake Mead or Powell, USA (where damming decades ago led to flooding of river valleys, the more recent drop in level and formation of bathtub rings notwithstanding) or the Musandam peninsula where the Oman mountains on the Arabia plate are plunging under the Eurasian plate and the Arabian gulf at ~1 cm/year.

As Cassini's radar coverage of Titan slowly built up by 2008, more and more of the images overlapped. Images were mapped onto the surface assuming that Titan's pole was perfectly perpendicular (normal) to its orbit plane, and that Titan

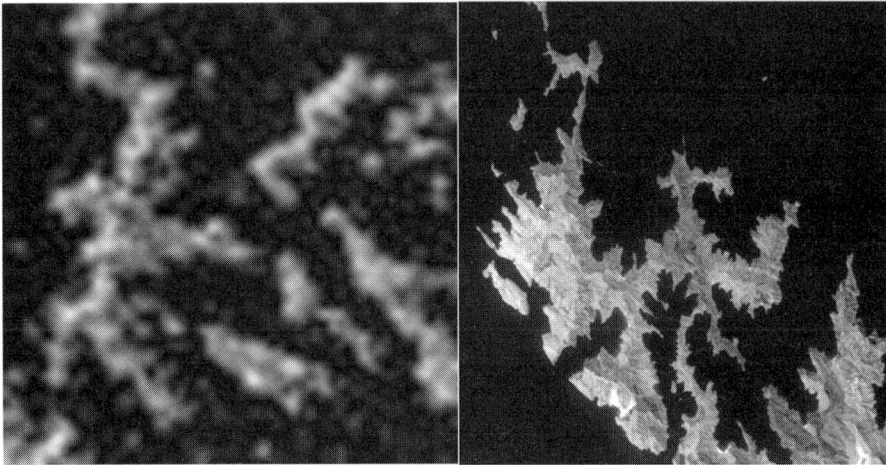

Figure 9.16 Seas rising around mountains, or mountains sinking into the sea, lead to island networks and flooded valleys. On the left is a radar image of the coastline of Ligeia Mare near the north pole of Titan; on the right is a (rather higher-quality) image in visible light of the Musandam peninsula on the north coast of Oman, where the Arabian tectonic plate is migrating northwards and diving under the Eurasian plate, such that the mountains are sinking into the Arabian Gulf at about 1 cm/year. It is tempting to speculate that similar evolution is happening or has happened on Titan. (Author, from NASA data.)

rotated perfectly synchronously with its rotation period exactly equal to its orbit period around Saturn. But with these assumptions, the images didn't match up – some mountains and other features seemed several kilometers out of position in successive views. JPL engineer Brian Stiles adjusted the assumed parameters of Titan's spin state to match features up, and this suggested the rotational pole was 0.3 degrees off the orbit normal, and that the spin rate was not quite synchronous with its orbit period. This was an exciting result, as such an effect had been predicted by Tetsuya Tokano and Fritz Neubauer: seasonal changes in zonal winds predicted in a GCM should exert stress on the solid surface [465, 466]. But the only way this seasonal angular momentum change would result in an observable rotation change (effectively a change in the length of the day) would be if the ice crust were detached from the massive core, by floating on an internal ocean of water much like that on Europa. At first look, the Cassini radar rotation rate seemed to match up nearly exactly with the predictions.

The Earth's length of day changes dramatically with time, although this is mainly an oceanographic, rather than meteorological effect. The Moon, via the tides it raises in the sea, saps angular momentum from the rotating Earth. In so doing, it boosts its own orbit, with the result that the Moon is slowly receding from us (something Halley recognized from eclipse timing) and the Earth's rotation is slowing down. From tidal rhythmites, muds and sands laid down in estuaries with large tides which reflect the daily, monthly and annual cyles in the tides, we know that 600 million years ago a day was about 22 hours long (so there were about 400 days a year). In fact, the Moon's slow climb (or "secular acceleration") can be measured directly by timing the bounce of a laser pulse off reflectors left on the Moon by the Apollo astronauts – it is receding from us at about 4 cm/yr. On top of this variation in the length of the day, of about 2 ms/yr, there are shorter-term signals (detectable astronomically) due to the slow change in the Earth's shape as the crust bounces back from being unloaded of the weight of the ice sheets in the last 20,000 years, and there are annual and semi-annual cycles with amplitudes of about a third of a millisecond, as seasonal changes in the zonal winds exchange angular momentum with the solid Earth [467]. On Mars, there are also small changes, due in part to winds but mostly to the transport of a large mass of CO_2 onto and off the polar caps in the seasonal frost cycle [468] – in fact, the rotation rate change was measured by precise radio tracking of the Mars Pathfinder lander [469].[16]

---

[16] Despite many landings on Mars since, this remains the best measurement for now, as the other landers (rovers) inconveniently do not stay in the same place. Without knowing its traverse to ~cm accuracy, even Opportunity's decade on Mars does not improve on Pathfinder's observation.

The possibility that Titan's length of day might be changing in an observable way due to seasonal variation in the winds inevitably led modelers to explore the effect, and several GCMs were adjusted or tuned (as, indeed was a model of the currents in Titan's seas) to determine the effects on the length of day. Unfortunately, it emerged that in fact Titan's rotation might be quite complicated,[17] and precession of the pole in a sort of cycloidal motion might be masquerading as a nonsynchronous rotation. Many more years of observation would be needed to tease apart this effect from any seasonal change. Despite the (my) initial report being something of a red herring, it stimulated much new work in atmospheric angular momentum budgets.

Rich Achterberg at GSFC, analyzing data from Cassini's thermal infrared spectrometer, found in 2008 that the contours of stratospheric temperatures, which one would expect to be circularly symmetric around the pole, weren't [470]. Specifically, the center of rotation of the atmosphere seemed to be displaced towards the Sun by a few degrees, an effect explored by the prolific Tetsuya Tokano in a GCM [471].

Titan's linear dunes stimulated much work on the relationship of dune morphology (shape) to winds. Beyond their near-uniform orientation, another striking feature of Titan's dunes was their regular appearance: while their height did seem to vary from place to place, reflecting perhaps the availability of sand, their spacing was almost perfectly uniform at 2–3 km (there being a slight variation with latitude). Equally striking, perhaps, was that this spacing seemed to be the same as the large linear dunes seen deep in the Arabian and Namib deserts on Earth. Despite Titan's dunes being made of different stuff, in a gravity one-sixth that of Earth, and transported in air four times denser in winds perhaps an order of magnitude weaker, whatever factor controlled the size of the dunes seemed to be the same.

As it turns out, there is a minimum dune size (related to the distance over which a particle will accelerate in the wind) below which dunes cannot form.[18] But above that scale, a perturbation to a sand sheet will grow in amplitude, and the dune pattern can be complex, depending on the history of wind speeds and directions. Ultimately, however, a dune may grow in height to be about 1/12 of

---

[17] for example, even though mechanically isolated from the crust by a water layer, a nonspherical core could be gravitationally coupled to the crust.

[18] Strictly, the minimum dune scale is close to the saturation length, the distance over which wind over a sand sheet will pick up more and more sand until the sand flux reaches a limiting value. Patches of sand smaller than this length will just tend to disappear, but patches larger than this will grow. This length is about 50 times the drag length, itself the product of the particle size and the ratio of particle density to atmospheric density. For much more on all this, see [433].

the height of the planetary boundary layer: at this point, the airflow above the dune becomes squeezed between the top of the dune and the top of the layer, which resists upward displacement much like the free surface of a liquid. The enhanced wind stress at the crest as the growing dune "feels" the top of the layer removes material from the crest, and so the growth becomes self-limiting at this height. Since large dunes invariably are 10–12 times wider than they are high, the ultimate limit on dune spacing is roughly equal to the planetary boundary layer height. On Earth, this height can be a few hundred meters over or near the ocean, where the thermal inertia of the sea limits the growth of the layer, but inland it can be several kilometers, and this gradient is seen reflected in dune size in coastal sand seas such as the Namib. At Titan's equator, lacking large seas, the boundary layer thickness is essentially uniform at about 3 km [472],[19] a measure supported by the Huygens descent measurement, and this is reflected in the uniform dune spacing. Benjamin Charnay found a similar boundary layer height emerging in a GCM [474].

All else being equal, one would expect that winds near the surface at the equator would generally be easterlies (i.e. to the west), since this westwards drag should balance out the eastwards drag at higher latitude. As George Hadley had pointed out in the 1740s, if such a balance did not occur, the angular momentum of the atmosphere would undergo a secular change, and it does not. It was therefore a matter of some consternation to the operators of Titan global circulation models when, by the late 2000s, it became clear that the geomorphological interpretation of the vast equatorial dunefields was that dunes appeared to have grown towards the east, in exactly the opposite direction.

The answer is fundamentally simple, although debate about the finer details is still ongoing. The sand in general is not moving, but only moves in response to the strongest winds, when those winds exceed a threshold of about 1 m/s. Close examination of the seasonal cycle of surface winds in at least one GCM found that there is a brief period around equinox when the upwelling Intertropical Convergence Zone crosses the equator (Figure 9.17). This leads to more mixing, which brings the momentum from higher altitudes (where the wind is always westerly) down to the surface for a short period.

Note that, while this discussion has focussed on the east–west component of the winds, it should be borne in mind that on each of these sand movement events the wind is actually blowing to the NE/NNE or SE/SSE, rather than purely

---

[19] Note that an earlier analysis had suggested the boundary layer measured by Huygens was only 300 m deep [473], but this was in fact the growing boundary layer that morning (the Huygens descent was at 9 a.m.), whereas close examination showed another step in the potential temperature profile at about 3 km, the vestige of yesterday's boundary layer.

2003 - Southern Summer

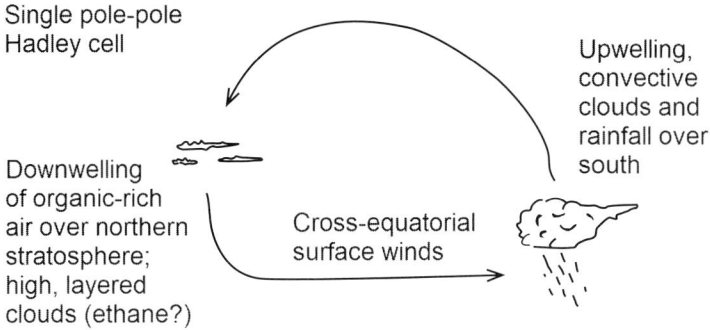

Single pole-pole
Hadley cell

Upwelling,
convective
clouds and
rainfall over
south

Downwelling
of organic-rich
air over northern
stratosphere;
high, layered
clouds (ethane?)

Cross-equatorial
surface winds

2009 - Northern Spring

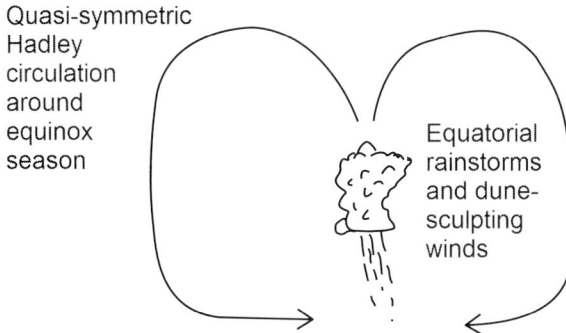

Quasi-symmetric
Hadley
circulation
around
equinox
season

Equatorial
rainstorms
and dune-
sculpting
winds

Figure 9.17 Mean Meridional Circulation, less formally termed "Hadley circulation" on Titan. Titan's slow rotation and long year gives a single hemisphere to hemisphere circulation cell (top), except briefly at equinox (below). Earth's conditions, however, give a more symmetric pattern with the Intertropical Convergence Zone merely moving north and south by 20 degrees or so. The inner cells are termed the Hadley cell: the outer two are the Ferrel cell. Mars' circulation is more like Titan's than Earth's.

east. If the wind blew consistently east, we would see transverse dunes marching slowly in that direction: the longitudinal and rather symmetric form of the dunes implies an alternating but slightly convergent set of sand-moving winds. It isn't necessary for the sand to move every equinox, only that the probability of it moving NE is about the same as it moving SE, so that the net sand transport averages out to be eastwards.

Although the dune pattern overall seems to be somewhat symmetric about the equator, and the dunes are mostly themselves rather symmetric in form, closer examination (using image-enhancement techniques that attempt to model out

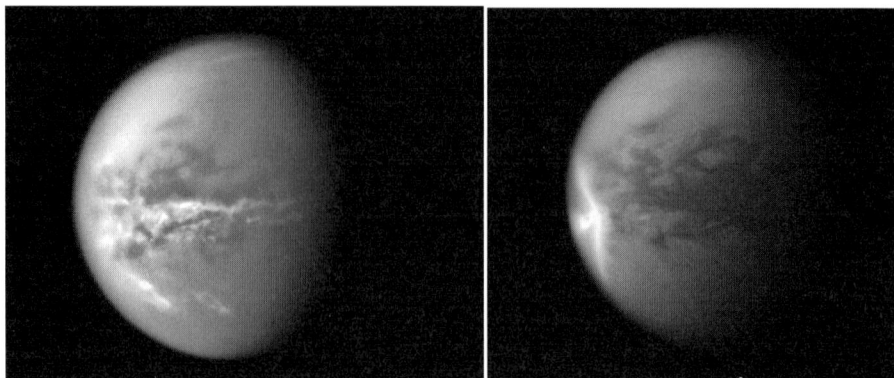

Figure 9.18 Cassini near-infrared images of equatorial clouds on Titan near equinox. (Source: NASA/JPL/SSI.)

the intrinsic "speckly" appearance of radar data) has indicated that some of the dunes have a somewhat crescentic or barchan shape, with horns pointed north-east [475]. This may be a signature of the dunes re-orienting in response to the Croll–Milankovitch cycles, and paleoclimate modeling seems to suggest the winds indeed may have changed in the way suggested by the dunes [476], the change also reflected in the distribution of lakes and seas [477]. A nice consistency between models and observations is emerging.

The long Cassini mission permitted study of Titan's seasonal cycles. The south polar clouds disappeared soon after Cassini's arrival as autumn began, but a flurry of cloud detections were made near the equator around equinox[20] in 2009 (Figure 9.18). Elizabeth Turtle of the Cassini imaging team reported that nearby surface areas were seen to darken after the storm's passage, and then progressively brighten in subsequent months, as if the ground were darkened by moisture and then progressively dried out [478]: the rain evidently reached the ground and the hydrological cycle on Titan is active today as we watch.

One prominent cloud had a distinctive arrow-shaped appearance (Figure 9.18), which Jonathan Mitchell suggested may be due to the propagation of a Kelvin wave in the atmosphere [479]. Indeed, these large waves in the atmosphere may allow one storm on Titan to trigger others, hundreds or thousands of kilometers away. Titan promises to be a laboratory for the all-important cloud convection processes that are so hard to understand on Earth alone, and so crucial to our

---

[20] Since 2009 is northern spring equinox on Titan, it corresponds seasonally to April on Earth. Thus these rain storms were April showers on Titan.

climate. Cassini and groundbased observers have been keeping Titan under close surveillance for some years, yet large cumulus cloud systems had not been observed in the northern polar regions, even by the 2017 end of Cassini's mission in midsummer.

A range of Cassini data, including radio occultations, thermal infrared measurements, and microwave radiometry, have now built up at least a rudimentary picture of seasonal temperature changes at the surface. Temperatures have see-sawed (Figure 9.19): while the deep south was about 1 K cooler than the equator in 2006–10, and the north about 3 K cooler [480, 481] the north has warmed by 1 K and the south has cooled by 2 K. These results, and the temperature and composition evolution in the stratosphere, give GCMs a new target to aim at.

On the other hand, at higher levels, where the thinner atmosphere responds more quickly to the changing season, dramatic changes have been seen. The detached haze layer, hovering at 500 km altitude when Cassini arrived, dropped suddenly in 2010 or so to the 350 km altitude at which it had been seen by Voyager at a comparable season (just after equinox) [482]. Presumably the Hadley circulation, which must reverse in direction over a few years (as indicated in GCMs as well as by the north–south asymmetry in the main haze deck) weakens or otherwise changes character during this reversal. The northern polar hood

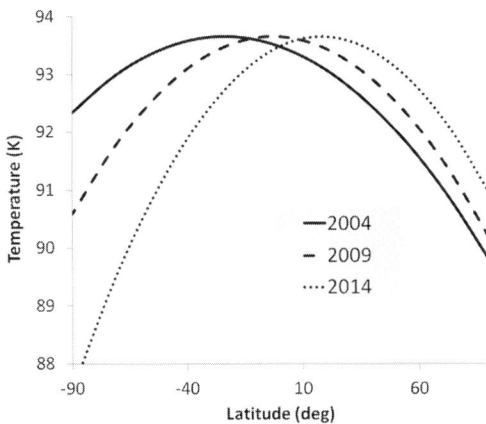

Figure 9.19 Titan surface temperatures indicated by Cassini infrared (CIRS) measurements – the equator stays more or less constant in temperature, but the poles swing up and down relative to their seasonal average. Microwave measurements, sensing a little way into the subsurface, show a slightly muted version of this evolution. The plot, of analytic fits to the data, probably overestimates the south pole cooling in 2014. (Author, from data in Jennings et al., 2016 [481].)

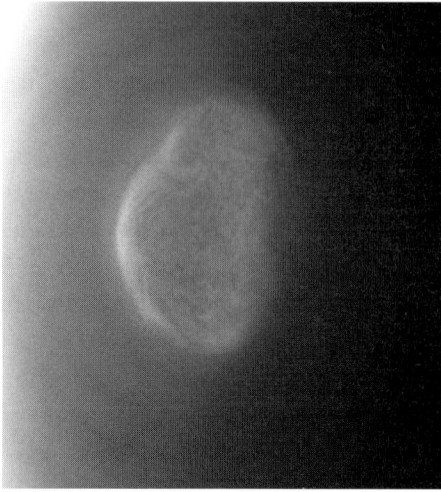

Figure 9.20 South polar vortex cloud on Titan as those latitudes enter winter darkness in 2012. The bi-lobed structure in many ways resembles that of Venus' polar vortex. (Source: NASA/JPL–Caltech/Space Science Institute.)

progressively decayed, and the build-up of nitrile gases there peaked and then declined. Meanwhile, a striking bi-lobed cloud system in the polar vortex, which spectroscopic data showed to be made of hydrogen cyanide ice crystals [483, 484], formed over the south pole (Figure 9.20). Even though these dramatic changes occur hundreds of kilometers above the surface, it may be that by influencing the number of condensation nuclei, they affect how organic material is deposited onto the ground. The seasonal polar condensation in the stratosphere may also affect the amount of sunlight reaching the surface, muting or augmenting the seasonal cycle.

Just as fossil raindrops were used on Earth to place bounds on the thickness of an early atmosphere, so imprints of another scale seen in HiRISE data were used to place similar limits on the thickness of the Martian atmosphere. Edwin Kite and colleagues at Caltech examined images to look at a heavily cratered (and thus, old – indeed Aeolis Dorsa was dated at 3.6 billion years) terrain that was nonetheless dissected with river channels [485]. By looking with the unprecedented resolution afforded by HiRISE, they could count small craters, noting that if the atmosphere were thick (about 1 bar) they would not form, since the smallish meteorites needed would burn or break up in a thick atmosphere before they hit the ground. Similar analyses had once been proposed (but at a rather larger scale – 10 km craters, rather than the few tens of meters for HiRISE at Mars) to constrain paleoatmospheres on Titan and Venus [486–488], but, in the event, geological resurfacing on both worlds rather confounded the exercises. Kite's work did appear, however, to exclude thick CO_2–H_2O greenhouses for extended periods, suggesting that either Martian rivers were transient phenomena or a more exotic greenhouse was needed.[21] Ramses Ramirez at Penn State, working with Jim Kasting and others, even tried spicing up a CO_2 greenhouse with molecular hydrogen [490], and found, unsurprisingly, that it helped, although a CO_2 background of a bar or more was still needed.

On the terrestrial planets, with their upper atmospheres warmed by the Sun, a hydrogen greenhouse is a short-lived affair, as this light gas escapes rapidly to space [491]. However, Dave Stevenson at Caltech playfully suggested that if you put a big enough hydrogen greenhouse on a planet, even rogue worlds roaming in interstellar space, with only a tiny amount of radiogenic heat and no sunlight at all, could have surface temperatures high enough to support liquid water [492]. Life on such dark, isothermal bodies would not have much excitement, however, conditions only changing over aeons with the slow decay of radioactive elements in the rock towards chilled oblivion. Closer to home, however, there is much going on, and much to look forward to, the subject of the next chapter.

---

[21] But those hoping for a warmer, wetter, thicker Mars will not just grasp at straws, they'll grasp at rocks too. A rather fun analysis by Michael Manga of Berkeley [489], also working with Kite, of a "bomb sag", downward dragging of sediment strata by a rock thrown presumably from a nearby volcano, was interpreted to imply an atmospheric pressure of >0.4 bar. On the other hand, the provenance of this rock, seen in Spirit rover images, is not known – the slow impact implied by the sag pattern may just have been because the rock had bounced to its final resting place.

10

Looking Ahead

2012 to 2020

2010s – Titan circulation models capture key aspects of surface interaction –
dune orientation, distribution of surface liquids

2012 – Curiosity rover explores Gale Crater on Mars, allowing more detailed
decoding of paleoclimate and paleohydrology

2014 – MAVEN spacecraft explores loss of Mars atmosphere to space

2015 – New Horizons encounters Pluto, observing haze layers, mobile nitrogen
ice deposits

2015 – Akatsuki returns to Venus, observes bow wave linking atmospheric
circulation to surface topography

2016 – ESA Trace Gas Orbiter arrives at Mars; Schiaparelli lander lost

2017 – Cassini mission ends, after almost half a Saturn year exploring that planet
and Titan

While this book has focussed on Titan as the center of climatological interest in the Saturnian system, two other objects there are also of interest. First is Saturn itself. From a climate standpoint, it has several features that distinguish it from Jupiter – there are modest differences in size, rotation rate and composition, but, unlike Jupiter, Saturn has internal heating that is comparable with the amount of heat it receives from the Sun. Furthermore, its obliquity (which is essentially the same as Titan's) gives Saturn seasons which Jupiter (broadly) lacks. But Saturn's seasons are not the same as Titan's, exactly, because the rings get in the way. As Huygens visualized in the 1600s,[1] when Saturn is at equinox the rings are edge-on as seen from the Earth and Sun, but towards the solstice, the rings cast great curved filigree shadows onto Saturn, and so the tropical seasonal history of solar illumination is not a smooth function of time, but has abrupt steps down to zero as those latitudes are eclipsed for years on end. The dynamics at the shadow's edge will doubtless be the subject of many interesting model studies. One prominent feature of Saturn's dynamics that has been examined in some depth is its northern polar vortex, which has a strikingly hexagonal appearance. This was spotted in Voyager images [493], and the connection with wave stuctures seen in dishpan experiments (see Chapter 3) was noted, but Cassini has had a grandstand view of this distinctive jetstream circulation system. Raúl Morales-Juberías, Kunio Sayanagi and colleagues have modeled this feature numerically as a meandering jet stream [494], matching the hexagon's rotation relative to the rest of Saturn. Curiously, the south polar vortex does not have the same hexagonal shape, instead resembling a "simple" hurricane structure, perhaps slightly dipolar like the Venus polar vortex.

While the hexagon was a familiar feature, Saturn's relatively placid appearance has occasionally been punctuated by enormous storms. One was seen shortly after the Hubble Space Telescope came on-line in 1990, and historical records suggest these "great white spots" may appear every 30 years.

But in 2010 (nowhere near the expected 30-year period) a bright storm erupted in Saturn's northern midlatitudes (Figure 10.1) [495]. First noticed, in fact, by amateur astronomers, its evolution was tracked closely by larger telescopes and of course by Cassini, where the storm system had observable signatures not only in visible and infrared, but even to Cassini's microwave radiometer (the storm alters the distribution of ammonia gas, which is a strong microwave absorber), and to its radio. Saturn generates a radio "crackle" due to lightning discharges, and this storm brought a crescendo of activity [496].

[1] Because the Earth and Sun can be 6 degrees apart as seen from Saturn, astronomers can sometimes sneak a view of the shadow of Saturn on the rings. Robert Hooke did so in 1666.

Figure 10.1 A beautiful Cassini image of Saturn on Christmas Eve, 2010, showing the bright "tadpole" storm in the northern midlatitudes. The rings themselves are the faint perfectly straight horizontal line (the Cassini spacecraft was in the ring plane, i.e. right over Saturn's equator, so the rings are edge-on) but the Sun was already at 12 degrees north, and so the rings case the curved banded shadow onto Saturn. (Source: NASA/JPL-Caltech/Space Science Institute; image PIA12824.)

Three-and-a-half million miles away, all was quiet on airless Iapetus. This 1500-km-diameter moon, discovered in 1671 by Jean-Dominique Cassini, would not normally merit discussion in a book on climate. However, it has the odd feature that its leading face is much darker than the trailing hemisphere (so dark, in fact, that Cassini at first "lost" the satellite – it was invisibly dark when he first looked for it on the eastern side of Saturn). It was once thought that the dark color, forming an ellipsoidal patch (now named Cassini Regio), was the result of dust infalling from Phoebe, Saturn's outermost major moon. But the composition and distribution don't quite work. When Cassini (the spacecraft) flew past Iapetus in 2007, its close-up images revealed details of the distribution of bright and dark material (Figure 10.2), and its thermal spectrometer showed that, at 129 K, the dark terrain is the warmest surface in the Saturnian system.[2] This results in large measure from the slow rotation of Iapetus on its long (80-day) orbit, so that, unlike the more even roasting received by the inner satellites, terrains have longer to warm up.

John Spencer and Tilman Denk showed that the resultant temperatures are just high enough that water ice can sublimate on timescales of billions of years [497], and so the observed variegation is not from the deposition of dark stuff

---

[2] The surfaces of Titan's haze particles are actually warmer, but don't count in this comparison!

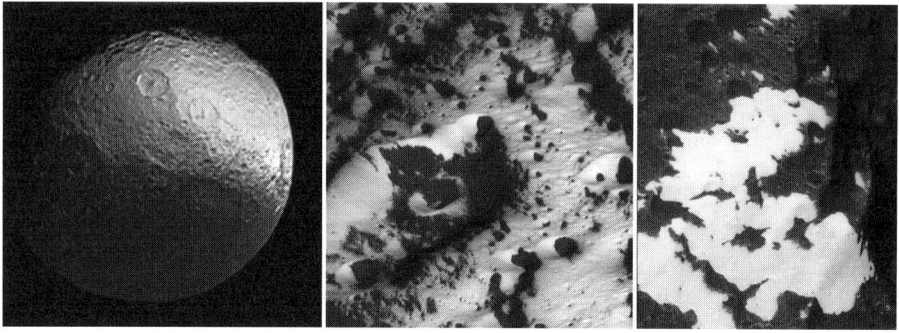

Figure 10.2 Iapetus as Daisyworld. Iapetus' yin-yang appearance (left), with dark reddish Cassini Regio towards bottom. Close examination by Cassini (middle and right) shows that the bright material at the edges forms irregular patches, not unlike the patterning on a cow that emerges from mathematically similar feedbacks. (Source: NASA/JPL/Space Science Institute.)

(in fact believed to be carbonaceous material, not too dissimilar in composition from Titan haze or dune sands) but rather from the migration of ice away from the dark stuff (in much the same way ice may have crept into dark, cold hiding places in polar craters on the Moon or Mercury – bright patches are also seen in shadowed areas on Callisto, attesting to this sort of migration on airless surfaces, albeit rather faster than on Iapetus).

While Iapetus doesn't really have a climate, there is a climate-related feedback here, in that once a patch gets a little darker than its neighbors, it will be a bit warmer and therefore ice will migrate faster out than it moves in, and so the contrast gets amplified. The large-scale segregation (i.e. the formation of a dark hemisphere) may well have been triggered by an external source such as Phoebe, but the distribution of bright and dark observed today has been effected by thermal migration. Indeed, the small-scale patterning clearly has little to do with orbital mechanics, but is quite characteristic of a feedback system and emerges naturally in simple two-dimensional Daisyworld models [498]. Similar feedbacks occur in the autosegregation of human and other populations, and indeed the emergence ("morphogenesis") of irregular patches of bright and dark via simple chemical feedbacks ("reaction–diffusion" equations) was first studied by the mathematician Alan Turing [499]. Later studies of Pluto would show that similar volatile-segregation processes are likely at work on that world too [500].

The Cassini mission operated until northern summer solstice in 2017, making a fiery plunge into Saturn's atmosphere (to safely dispose of its plutonium generators). But Cassini took data to the last, sniffing Saturn's upper atmosphere even as the spacecraft lost contact. Cassini's entry took place at the end of a series of especially close orbits, giving grandstand views of Saturn's atmosphere and probing its interior with gravity measurements. This was at the same time that another spacecraft, Juno, was at Jupiter (it arrived in July 2016) to study that planet's meteorology and interior structure in a similar swooping set of orbits. Comparative planetology of the gas giants is set for busy times.

But before all this, Cassini continued to study Titan, which continues to present modelers with challenges. It had been expected that cloud convection should begin and intensify in Titan's northern polar regions as we approached northern summer solstice, to mirror the clouds seen in abundance in the south in 2001 and onwards. But in fact rather little cloud activity had been seen. Is this just bad luck (in observation timing), is it stochastic variability, or is there a north–south asymmetry in meteorology, with the gentler northern summer heating taking longer to get activity going compared with the south?

This is an obvious question to explore with GCMs, and indeed experiments by Jonathan Mitchell, Ray Pierrehumbert and colleagues [501] showed that the amount of methane in the climate system on Titan can influence how far north and south the solar-driven convective upwelling (on Earth, this is called the Intertropical Convergence Zone or ITCZ) that drives cloud formation can migrate (Figure 10.3).

At first, Titan GCMs each seemed to get only bits of the Titan story right. Some can get the superrotation, but many cannot. But the ones that get the superrotation have other issues. Over time, as at Earth, the models are improving. Whereas at Mars good global measurements of elevation, albedo and thermal inertia exist, Titan so far lacks such surface constraints, but slowly they are being put together.

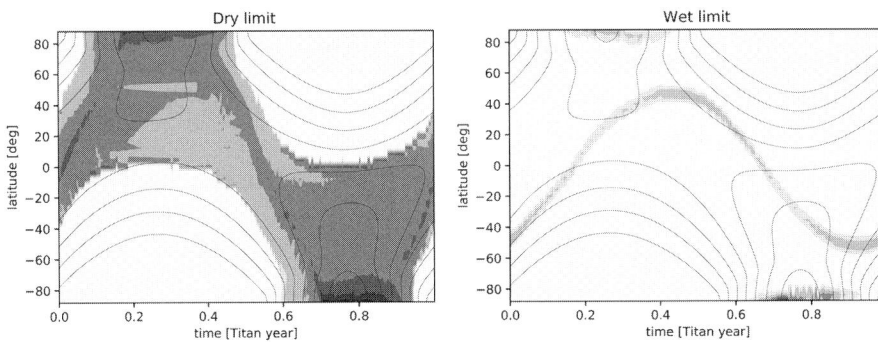

Figure 10.3 Circulation model experiments by Mitchell et al. [501], showing the latitude–season pattern of convection, cloud formation and thus precipitation for a dry Titan (left) and a moist Titan (right) as a function of time since Southern Summer Solstice (SSS) – October 2002. In the dry case, the upwelling swings from pole to pole, whereas in the moist case the pattern is more restricted in latitude. (Figure courtesy of Jonathan Mitchell.)

As Harlow Shapley (the astronomer who edited a 1953 book on climate change) once remarked "No one trusts a model except the man who wrote it; everyone trusts an observation, except the man who made it." And indeed, model deficiencies are known – for example Mitchell's GCM above has a simple grey radiative scheme and does not consider the stratospheric haze. Juan Lora in Arizona[3] noted that the thick haze layer on Titan absorbs and scatters light such that, while the maximum daily insolation at the top of the atmosphere is indeed at the summer pole (as Halley calculated for the Earth), the sunlight has to filter through such a long path of hazy atmosphere at the highest latitudes that the energy is attenuated. Thus the strongest heating at the surface is at summer midlatitudes, not at the poles [502].

In fact, the latitudinal extent of tropical upwelling is a subject of much current interest in terrestrial climate studies – it seems to be expanding [503]. Since 1990 the tropical belt has expanded by about 1 degree per decade, nudging storm tracks poleward. The widening can be reproduced in models by (and thus has been attributed to) increase in troposphere sunlight absorption by ozone and black carbon aerosols [504, 505]. What this might mean for our climate isn't clear, but there are evident parallels and synergies in exploring the same problem for Earth and Titan.

Titan has a smorgasbord of different cloud flavors, served at different seasons and altitudes, including not only the polar vortex cloud of hydrogen cyanide up at 300 km but also broad polar cloud layers associated with the Hadley circulation – one at about 80 km apparently of ethane seen in late northern winter [506], and another of cyanoacetylene at 140 km [507]. But the methane convective clouds in the troposphere, most closely resembling cumulus convection on Earth, complete with rain and quite possibly hail and haboobs, are the most exciting kind to observe, but are a rather rare delicacy.

Early work on Titan rain suggested methane drops could be larger than Earth raindrops, perhaps forming wobbly globules a centimeter across, and would fall slowly in Titan's low gravity and dense atmosphere, at about 1.6 m/s, the speed snowflakes fall on Earth [508].[4] Falling so slowly, isolated raindrops would have time to evaporate if they fell through unsaturated air, a phenomenon called virga, quite common in desert areas on Earth (less so in damp Britain, where I calculated these raindrop properties). However, within a storm, successive waves of drops

---

[3] At the time of writing Lora had moved to UCLA to work with Mitchell, so all this will get straightened out soon. Thus is scientific progress achieved.

[4] This paper, which I wrote as a Ph.D. student, was in part motivated by the engineering question of whether the Huygens probe might encounter supercooled droplets that might form ice deposits as it descended. It is probably one of the better papers I've ever written.

would moisten a column of air, forming a "rain shaft", and so it would rain on the surface after all – cloud convection models suggest storms might dump tens of centimeters in a few hours [509, 510], and it seems likely that just such storms carved Titan's abundant river valleys. Recent modeling by Sean Faulk and colleagues at UCLA has suggested that, while the highest average rainfall occurs near Titan's poles, the most intense rainfall (i.e. the largest amount of rain in a single day) seems to occur in the midlatitudes, consistent with the observed distribution of alluvial fans in Cassini radar data [511]. Alluvial fans, seen prominently on Earth in e.g. Death Valley or the Himalayas, are triangular deposits of cobbles and boulders that literally fan out from narrow valleys in mountainous regions, and result from suddenly sharper rates of sediment transport due to intense storms. While the records of Titan's climate in its geology, and their interpretation by models, are coarse and far behind those of Mars, progress is being made.

Much work remains on the microphysics of cloud condensation – for example, is the supersaturation of vapor needed to start droplet or ice-grain growth the same for nonpolar compounds such as methane forming on cyanoacetylene or tholin particles as it is for water condensing on sea-salt aerosol on Earth, or for CO_2 or H_2O on dust on Mars? At one time it was thought that Titan's lower atmosphere might be supersaturated, but the Huygens profile suggests it was not. Preliminary experiments by Daniel Curtis in Colorado found methane freezes onto particles at lower supersaturations than ethane needs [512]: in both cases it turns out the situation isn't too different from water on Earth. Thermodynamic subtleties complicate the picture at Titan – nitrogen dissolves in liquid methane, but is less soluble in the solid phase, so a methane–nitrogen drop carried up in a cloud updraft may freeze when it reaches an altitude of about 14 km. Like the carbon dioxide in a can of soda put in the freezer, the nitrogen may come out of solution, forming a frothy hailstone. Sonia Graves, working with Chris McKay at NASA Ames, updated my raindrop calculations [513], finding that the evaporative cooling of raindrops or hailstones (an effect I hadn't included) might allow them to reach the surface after all, although they didn't include any frothiness. As on Earth, hailstones littering the surface should be a very transient phenomenon – they are typically far out of equilibrium with their surrounds, and thus should quickly melt. Note that, although Titan's surface conditions are only a few degrees above methane's freezing point, nitrogen and ethane act as antifreeze, so persistent methane ice is not expected in the present climate, although conceivably evaporative cooling of lakes might cause such freezing. Jason Barnes of the University of Idaho has noted in Cassini VIMS data some evidence of bright precipitation deposits on Titan, which may be somehow freezing-related, or perhaps related to dissolved material [514] – there is much we still do not understand about the exotic possibilities of Titan's fluids.

A haboob is a dust storm, typically driven by cold winds that flow out beneath rainstorms: the air chilled by evaporation of raindrops and accelerated by their drag can cause violent downdrafts (also called "microbursts", occasionally responsible for aircraft accidents) that blast outwards when they hit the ground. A recent interpretation by Sebastien Rodriguez in Paris of Cassini VIMS data taken near equinox indicated a patch of opacity in the lowest few kilometers of Titan's atmosphere that seems most simply interpretable as a haboob, kicking dust up from the dunefields (indeed, the Huygens probe may have kicked up some dust at impact [515, 516]). It seems outflows from equinoctial rainstorms appear to be one of the most plausible ways to generate winds strong enough to sculpt sand into the dunes [517]: what was guessed at as a general regional mixing down of eastwards momentum in the last chapter at equinox is most probably performed in a very localized and infrequent, but violent, way (Figure 10.4).

Dust is emerging as a player on the terrestrial climate stage. Aerosols in general (such as smog, as well as volcanic aerosols and smoke from forest fires especially in South-East Asia) are an important climate forcing factor, and

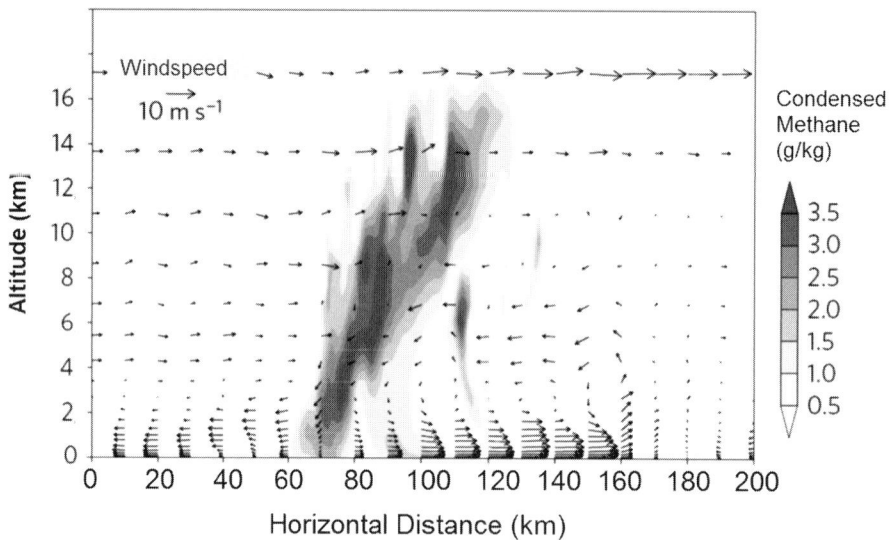

Figure 10.4 Mesoscale simulations of a Titan rainstorm and the outflowing winds. The grey contours indicate the condensed methane content (i.e. cloud particles, rain and hail). In this simulation the main storm at 160 km is dissipating, but its violent outflow winds (driving eastwards, having picked up momentum from the zonal flow at higher altitudes) have triggered a secondary storm (with updraft visible) at 250 km. The outflow winds may sculpt the dunes and create haboobs. (Adapted with permission from SpringerNature from Charnay, B., et al., Methane storms as a driver of Titan's dune orientation. *Nature Geoscience*, 8(5), p. 362. © 2015.)

Figure 10.5 A great plume of dust blasting out across the Atlantic from a dust storm in North Africa in Spring 2000, observed by NASA's MODIS instrument. (Credit: Jacques Descloitres, MODIS Rapid Response Team, NASA/GSFC.)

wind-blown dust can be a significant contributor. This is the case particularly in China, and in the Sahara. In fact, Saharan dust can blow all the way across the Atlantic (Figure 10.5), and the cooling effect of the dust may lower sea-surface temperature in the tropical belt where hurricanes form, reducing their intensity [518].[5] During the Earth's ice ages, it was likely a much dustier place – indeed, vast hills of packed wind-blown dust ("loess") were laid down in Europe, North America and China, where it slowly erodes out, making the Yellow River yellow. Generally drier and windier conditions, and vast flats of silt and sand exposed by sea level drop (it was about 100 m lower at the peak of the glaciation) made material much more movable by winds, and many dunefields that are now vegetated (such as the Nebraska sand hills, or the Kalahari in southern Africa) were barren and active.

Modern climate models must book-keep not only the behavior of the atmosphere, oceans and cryosphere (ice caps, sea ice etc.) but also the biosphere.

---

[5] The long-range transport of Saharan dust was noticed much earlier than this, e.g. Ref. [519].

Each grid cell in a coupled climate model may track the relative amounts of different types of vegetation – forest versus grass, for example – because these different types not only have different albedo, but also different aerodynamic roughness (affecting wind friction) and different moisture transport. Book-keeping different amounts of different vegetation types allows the generation of synthetic pollen records that can be compared with the records in lake sediments etc., allowing models to be calibrated and tested against, for example the growth and decay of the Last Glacial Maximum 20,000 years ago.

In such coupled climate–land-cover models, feedbacks proliferate and effects become harder to anticipate. For example, while the greenhouse effect of carbon dioxide on temperatures is straightforward, some effects in coupled models (and, one assumes, reality) are less so. Plants are honed by evolution to ingest just the right amount of CO_2 for photosynthesis and growth, which they do through tiny pores called stomata which open during the day. As CO_2 rises, plants can manage with smaller or fewer stomata. But plants lose moisture through the same apertures, and so plants will transpire less water vapor than they used to, affecting atmospheric humidity, not least in the rain forest. How will that affect clouds and precipitation?

Plant growth is often nutrient-limited. In some settings the limiting nutrient is iron – and mineral dust can supply it, so the settling of mineral dust may drive plant growth. In fact, Sahara dust may be a significant iron contributor to the Amazon basin [520], coupling the climates of two continents in a rather exotic way. It was once asked [6] *"Does the flap of a butterfly's wings in Brazil set off a tornado in Texas?"* via purely fluid-dynamical effects, but we are now attempting to ask broader biogeochemical questions too: Did a rainstorm in Chad cause a haboob which lofted the dust that nourished the Brazilian tree to let it grow the flower on which the butterfly flapped?

Iron fertilization of the oceans has even been proposed intentionally (as early as 1988 [521]), as the effects of climate change begin to be recognized to the point where deliberate mitigations are contemplated. What was once the "terra-forming" of other planets in science fiction is gaining a little respectability as "geoengineering" a solution to a crisis here on Earth. Ideas as exotic as reflectors in space to reduce the effective solar constant have been considered, to fleets of ships that might generate low clouds to increase the planetary albedo, to adding iron to the seas to stimulate plankton growth and thus the uptake of CO_2 from

---

[6] The sentiment is attributed to Edward Lorenz, although his 1953 paper suggests that the *"flap of a seagull's wings might alter the course of weather forever"* before he later adopted the more poetic butterfly.

the atmosphere.[7] It would be prudent to understand the full effects of such manipulations of the climate system before attempting them. In fact, some of these experiments happen naturally – while the sulfate aerosol was the most obvious climate effect of the Pinatubo eruption, in fact Andrew Watson and others have noted that the iron delivered to the ocean from Pinatubo's ash may have led to a perceptible (barely, and temporary) downward blip in the CO_2 rise [522]. These biogeochemical effects, with wide-ranging timescales, make it challenging to predict our climate.

On Mars, the coupled cycles of dust, CO_2 and water are being better and better understood, as a wide range of datasets are brought to bear. Dust-lifting into the atmosphere seems to be a combination of dust devil activity (or at least, some meteorological parameters that correlate with dust devil activity) and the runaway lifting that generates the dust storms [523].

Radar observations of Mars' south polar region showed that embedded in the layered dust/ice layers are in fact large deposits of CO_2 ice after all [524], probably enough (~10,000 cubic kilometers) to add ~5 mbar to the atmospheric pressure, an 80% increase. More CO_2 may be distributed in undetectably small pockets, enabling pressure cycles in response to orbital changes. But it seems nowhere near enough to permit hundreds of millibars of CO_2 in the past. Similarly, examination of Niili Fossae carbonate deposits by Christopher Edwards and Bethany Ehlmann of Caltech suggests that the observed carbonates similarly can't be the graveyard of a thick Mars paleoatmosphere [525].

In fact, radar sounding has yielded estimates of volatile inventories on Titan too. Cassini's radar – never designed as a subsurface sounder – unexpectedly gave a direct measure of the depth of Ligeia Mare on Titan, with Marco Mastrogiuseppe and colleagues finding an echo from the sea bed to be 160 m deep [526], within a factor of 2 or so of what one might guess knowing that most large lakes on Earth are roughly as deep in meters as they are wide in kilometers. Detecting such an echo with this type of radar required that the sea be very radar-transparent, suggesting a rather clean, methane-rich composition (more analogous to a fresh, clear mountain lake than an organic-saturated ethane-rich composition analogous to a salty sea). The implications for Titan's climate are just beginning to be considered, but the discovery reinforces the previously long-standing puzzle of where the photochemically produced ethane went. Speculations have ranged from sequestration in clathrate ices (a process which, if concentrated at high latitudes, might even explain Titan's somewhat pole-flattened shape), to incorporation in the haze [527–530].

---

[7] Oceanographer John Martin once famously said "Give me a half tanker of iron, and I will give you an ice age."

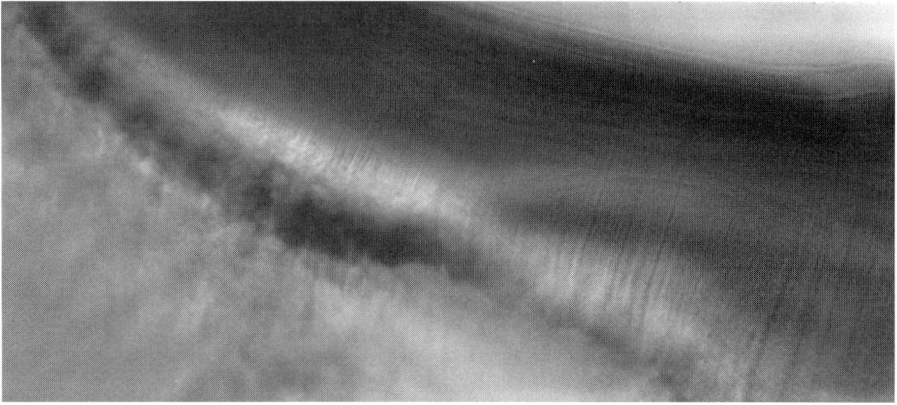

Figure 10.6 An image (20 km tall, at 40 m/pixel) of katabatic winds, made visible by cloud and snow formation, streaming downwards off the Martian north polar cap, taken by the THEMIS camera on Mars Odyssey in 2004. The layers exposed by wind-driven sublimation in the equator-facing slope of a polar trough are seen at upper right, and at lower right the laminar streams of wind are seen widening towards the south. When they hit the trough floor, there is a hydraulic jump into a slower, thicker flow (with enhanced formation of snow) evident in the turbulent patchiness towards the bottom left. (Source: ASU/JPL/NASA THEMIS image V12295001.)

The curious shape of the Martian north polar cap, with its spiral troughs, may also have been explained. Isaac Smith and John Holt of the University of Texas, with a combination of radar and imaging data, have elucidated a mechanism by which the troughs form and migrate [531, 532]. The troughs (Figure 10.6) are in fact formed by sublimation erosion by katabatic winds blowing off the polar cap,[8] and their spacing may be controlled by the thickness of the cold layer. A hydrodynamic instability, the hydraulic jump (familiar at the base of weirs on rivers, or the circular shock that forms when you pour a stream of water onto a horizontal plate) triggers a flow transition which enhances, thickens and slows the layer, promoting redeposition of the sublimed material as snow (Figure 10.6).[9]

---

[8] Katabatic winds are slope winds: cold air is dense and flows downhill, which is why it is sometimes uncomfortable to camp in mountain valleys – what seems like shelter is in fact a duct for cold, dense air at night. Katabatic winds are a prominent feature of Greenland and Antarctic meteorology.

[9] There are in fact related features in Antarctica called snow megadunes (they look like large stripes in satellite optical and radar images, see Lorenz and Zimbelman, *Dune Worlds* [433]) and in fact were thought at first to be a good analog for similar stripes seen on Titan, but as the earlier chapters noted, those turned out to be real dunes. The Antarctic features have almost no topographic expression at all, but seem to result from an ice texture change, perhaps the snow sinters or erodes differently as a result of some kind of wave in the boundary layer flow.

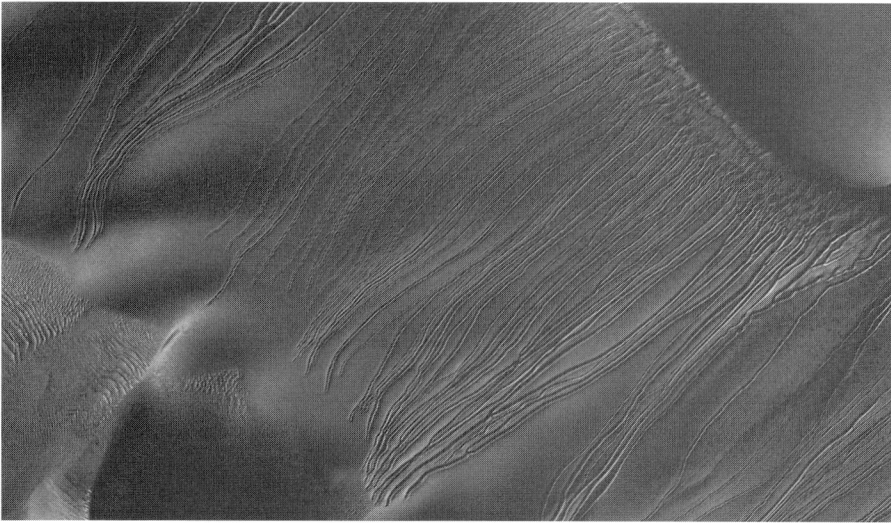

Figure 10.7 This HiRISE image shows "linear gullies" on a large sand dune inside Russell Crater. These grooves, 2 km long, with relatively constant width and raised banks or levees along the sides, seem unlike gullies caused by water-lubricated flows on Earth and possibly on Mars as they lack aprons of debris at the downhill end of the channel. These are known to form in spring, and are suspected of being made by slabs of dry ice sliding downhill as the sunlight hits this volatile material. (Image PSP_001440_1255, courtesy of NASA/JPL/University of Arizona.)

Thus the troughs migrate polewards over time, as evidenced by their topography and the exposed layers on the equator-facing slopes.

A rather entertaining solution to the question of a distinct set of slope gullies that appeared on sand dunes was found (Figure 10.7). These particular gullies are fairly straight and of uniform width, with slight ramparts, and form on polar dunes that see winter conditions cold enough for CO_2 frost. It seems slabs of CO_2 ice may break off and slide down, the CO_2 sublimating on the warming sands, forming an air cushion like a hovercraft. Serina Diniega and colleagues noted the seasonality of gully formation [533] and demonstrated the process with slabs of dry ice on the Coral Pink Sand dunes in Utah [534].[10]

On the other hand, another set of transient slope features (Figure 10.8), dark linear stains that seemed to track the Sun, named recurring slope lineae (RSLs), were eventually confirmed to involve water, but not apparently the groundwater breakouts first speculated for the MGS gullies. In particular, Luju Ojha, a Nepalese student at Georgia Tech working with the HiRISE team and with CRISM data, showed that the spectral signature of perchlorates was associated with RSL

---

[10] The dry ice field experiments are enormous fun to try yourself.

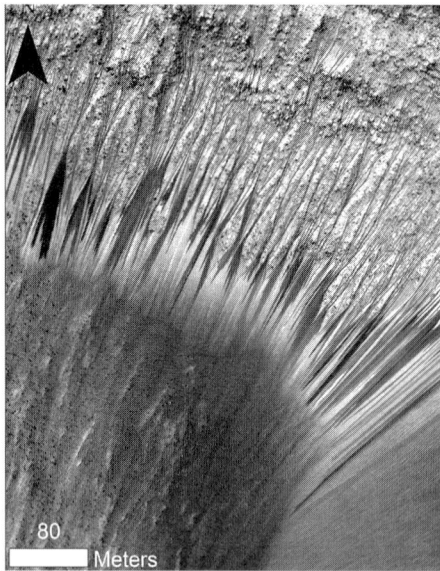

Figure 10.8 Recurring slope lineae on Mars. These dark streaks develop at a specific season, apparently from dampness wicked from the atmosphere by deliquescent salts such as perchlorates. (Source: NASA/JPL/University of Arizona.)

appearance [535]. It seems this deliquescent material was able to abstract enough moisture from the atmosphere in some seasons to cause surface flows. As so often in the history of planetary science, a binary choice between explanations seems to have been false, in that both CO_2 and water are involved in different examples of slope streaks/gullies.

Somewhat ironically, as evidence has mounted of at least small, transient exposures of liquid water at the present day, the weight of opinion seems to be moving from the idea that Mars had a much thicker, warmer atmosphere over an extended period in the past to permit formation of the fluvial and lacustrine (lake) features, towards favouring episodic warming with often only locally wet conditions.[11] The analogy of ice-covered lakes in the Dry Valleys of Antarctica shows that annual average temperatures need not be above freezing to permit liquid water to appear. Of course, transient climates beg the question of what might drive them, and gas release from the interior, volcanic heat, and the supreme "Deus Ex Machina" of planetary science, impact cratering, have all been invoked. These challenges of attribution aside, the fluvial, lacustrine and glacial landforms from the scale of Oceanus Borealis seen from orbit to the small ripples

---

[11] This is in principle quite an old idea, but some recent incarnations are Refs. [536] and [537]. For a starting point on the Dry Valleys, see Ref. [538].

and ice polygons seen by landers and rovers, together with the geophysical indications of buried ice and spectra of hydrated minerals, all point to a rich history of aqueous episodes on Mars.[12]

The year 2014 heralded no less than three new objects in the Martian firmament. The first to be anticipated was MAVEN, a NASA orbiter to understand the upper atmosphere and the atmospheric loss processes at Mars, key for understanding its climate evolution (much of this science was the target of the ill-fated Nozomi mission in 2002).

Something of a new space race has developed between India and China, both of which flew orbiter missions to the Moon in the mid-2000s. China had built a small satellite to be delivered into Mars orbit by the Russian Phobos-Grunt mission: when Phobos-Grunt followed the now-traditional fate of Russian Mars projects, crashing back to Earth shortly after launch in 2011, this created an opportunity for India. If they could build a mission in time for the 2014 launch window, they could be the first Asian country to reach orbit around another planet. The Mars Orbiter Mission (MOM) was put together in record time, albeit with a very modest payload, and, after launch on an Indian PSLV rocket, reached Mars in September 2014. Although in principle a technology demonstration, the mission has at least returned some spectacular views of the red planet (Figure 10.9).

The third Mars visitor was also the oldest – a comet first detected at the Siding Spring Observatory.[13] The comet, observed by rovers on the surface, sailed past Mars on October 19, 2014, at a distance of about 140,000 km. Its effects on the Martian atmosphere, modest in a climatic sense, were nonetheless quite apparent. In particular, iron and sodium from particles of comet dust that burned up as meteoroids were detected in the Martian upper atmosphere by MAVEN. (In fact, traces of these elements from meteoroid ablation are a notable feature of the Earth's atmosphere at altitudes of 80–100 km – it is possible that such meteoric material may play a role in the nucleation of cloud particles at lower altitudes.)

At the time of writing, the data from these spacecraft are only just beginning to be interpreted. The initial indications from MAVEN are that present-day atmospheric escape from Mars is still rather significant, and thus that Mars may have lost much of its atmosphere this way. MOM's camera tells us relatively little that isn't known from the armada already at Mars, but from its highly

---

[12] For a summary of results from lander and rover missions, see Ref. [539]. New results and interpretations of Curiosity rover results at different stratigraphic levels (ages) are still coming in thick and fast and defy succinct summary.

[13] Formally comet C/2013 A1. It was discovered only in January 2013, some 7 AU from Earth (beyond the orbit of Jupiter).

Figure 10.9 Faint water-ice clouds are visible over both polar regions in this full-disk image of gibbous Mars taken from the Indian Mars Orbiter Mission. The Tharsis volcanoes are visible in the center, Olympus Mons at left and Vallis Marineris at right. The albedo variations across Mars are evident, and can change with time due to dust redistribution. (Source: Indian Space Research Organization.)

elliptical orbit gives some spectacular global color views, which dramatically portray some features of the climate. MOM also carries a spectrometer designed to detect methane – but so far no results have been reported.

Approaches to studying Mars have started to converge with those used to document the Earth's climate. In particular, observations anywhere tend to have gaps (when Callendar looked through 200 records in a Smithsonian compilation of World Weather, he found 18 going back a century, only two of which were continuous) and sometimes instruments or practices change. Thus most modern work uses so-called reanalysis datasets, where a GCM has been used to assimilate observations (i.e. like weather forecasting models, these historical simulations continuously fold in whatever observations are available, anchoring them to reality). The resulting dataset, while only partly grounded in actual observations, is a more complete and consistent record.[14] Such a reanalysis has been developed for Mars by Luca Montabone and colleagues at Oxford and the Open University [540]. The MACDA (Mars Analysis Correction Data Assimilation) project used Mars Global Surveyor data to guide a GCM over three Mars years and thereby generate convenient and consistent fields of atmospheric and surface variables.

Some progress has also been made in decoding the long-term astronomical cycles of climate recorded in the Martian polar deposits. Isaac Smith and colleagues, exploiting the high-resolution SHARAD profiles of radar bacskcatter in the north polar region, identified a consistent widespread horizon, a sequence of recent accumulation layers corresponding to Croll–Milankovitch cycles, which they attributed to the polar insolation change 370,000 years ago [541]. They suggested 87,000 km^3 of ice had been deposited in the north polar cap over this period, an amount equivalent to a global layer about 60 cm thick.

The major feature of the present Martian climate that is not yet well understood is the interannual variability, specifically in the global dust storms. These have been recorded throughout the history of telescopic observation, and there now exists a robust, detailed record obtained from the continuous robotic presence at Mars itself since 1997 (Mars year 23[15]). One can imagine feedbacks wherein lofted dust intensifies winds and lofts more dust, that seasonally there may be a runaway effect. But why don't dust storms happen every year?

It seems there is a dust distribution cycle, with regions that are particularly susceptible to strong winds and/or dust devils becoming darker with time. Then, once in a while, a local dust storm (Figure 10.10), often triggered at the edge of the northern polar cap [543], usually in southern summer, grows to become global, and the dust distribution is re-set. The albedo changes are in fact enough to cause appreciable changes in temperatures and wind patterns: Lori Fenton in 2007 noted that regions over 50 million km^2 have been observed to change their reflectivity by 10% or more, and showed with a GCM that such changes could nudge global temperatures by 0.65 K and increase the likelihood of dust devil formation;[16] and Mars Global Surveyor observations show that local

---

[14] Popular ones are the ERA-15 1978–1994 dataset by the European Centre for Medium-Range Weather Forecasting (ECMWF) and the NCAR NCEP reanalysis (1948–present).

[15] by ad-hoc convention, Mars years (MY) are enumerated from MY 1, starting at the Mars spring equinox in April 1955 [542].

[16] See the paper by Fenton et al. [544]. Newman et al. (see below [547]) have suggested, however, that Fenton's work reported albedo changes that were misinterpreted as long-term changes, but were actually only short-term excursions.

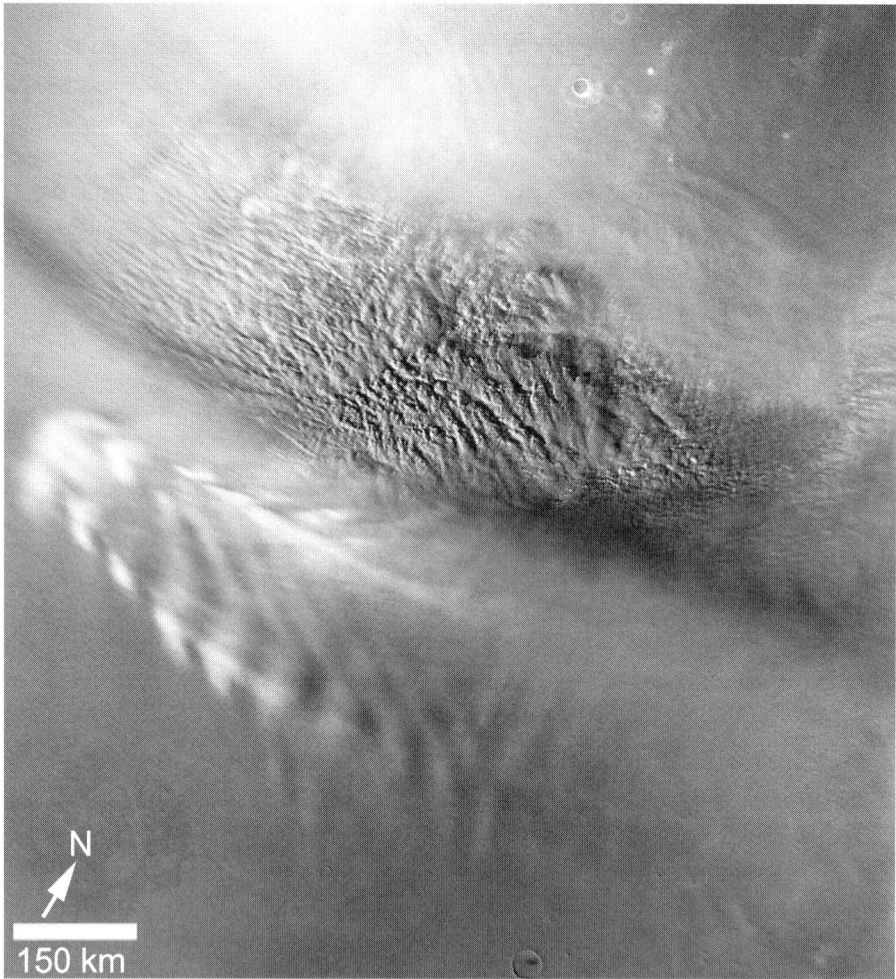

Figure 10.10 A dust storm over Utopia Planitia (53.6° N, 147.9° E), along the north seasonal polar cap edge in late northern winter snapped by the Mars Color Imager instrument on NASA's Mars Reconnaissance Orbiter on November 7, 2007. The dust storm pictured here (irregular patch, upper center) was short lived, lasting less than 24 hours. The image also shows the seasonal north polar cap (at top of figure) and the ribbed texture of gravity-wave (lee-wave) water ice clouds forming in the wake of Mie crater, just south of the storm. (Image Credit: NASA/JPL-Caltech/MSSS.)

temperatures can change by several degrees kelvin due to dust-driven albedo changes [545].

If Mars really works this way then the cycle may be too chaotic to predict: we may never know the thickness of the dust layer in different places well enough to guess whether one storm will happen to grow to become global. The system behavior is very susceptible to just how finely tuned the threshold wind speed to lift dust happens to be: Alexey Pankine and Andy Ingersoll have suggested that the system may be an example of self-organized criticality (SOC) [546]. Like the individually unpredictable avalanches on a sandpile, a chaotic sequence of big and small events emerges from a steady trickle of sand at its apex. So, if the Mars dust climate teeters on the edge of stability, perhaps the history of dust storms may be essentially random, but with a power-law distribution of event sizes.[17] Various schemes of dust lifting (including the effects of dust devils as well as large-scale winds) have been explored in GCMs, including those with a finite dust reservoir that just gets pushed around ([547] is one of the latest in a series of experiments).

However, it may be that some external forcing paces the dust storms. As was the case for early weather forecasting on Earth, empirical correlations can offer a useful basis for prediction, even when the underlying mechanisms are not understood. A rather surprising but intriguing correlation was noted recently by James Shirley [548] and Michael Mischna [549] of the Jet Propulsion Laboratory, who pointed out that in the last 20 Mars years, the 9 years (specifically the perihelion half of the Mars year) in which major dust storms have occurred have been associated with a particular phase of the Mars angular momentum about the solar system barycenter.[18] Shirley's paper freely confesses that a physical mechanism for this effect – if not some statistical fluke – has not yet been elucidated, but this of course is often how science makes headway. They even venture a speculation, that the phase of the angular momentum in the 2016–17 perihelion season was similar to that of the 1971 and 2007 seasons, and that the angular momentum amplitude is even larger for the 2018 dust storm season, suggesting that (especially if no dust storm occurs in 2014–15 [it didn't]) that a global dust storm is likely for 2016 or 2018 (Mars years 33 or 34), or both. We shall see.[19]

Titan too, it seems, may have interannual variations. Remarkably, one of the most powerful recent insights into this fundamental question on Titan's climate may derive from some of the most modest instrumentation. Climate science often relies on patient, sustained and well-calibrated measurements (Keeling's CO_2 measurements being the most prominent example), even if the

---

[17] Self-organized criticality is a useful paradigm for considering economic shocks, earthquakes, forest fires and a host of other events – see P. Bak, *How Nature Works*, Copernicus, 1996. These events tend, like Martian dust devils, to have a power-law size distribution.

[18] Mars does not orbit the Sun, but rather the Sun and Mars and the other planets orbit the barycenter of the solar system. The Sun's motion about the barycenter – much like the reflex motion used to detect exoplanets by Doppler spectroscopy – is largely due to Jupiter and Saturn, and the Sun's motion modulates the Mars angular momentum.

[19] As this book went to press in July 2018, Mars had been in the grip of an intense planet-encircling dust storm for several weeks.

measurements themselves are not especially sophisticated. Astronomer Wes Lockwood at Lowell Observatory in Flagstaff has recorded the disk-integrated brightness of Titan (as well as Uranus and Neptune) at blue and yellow wavelengths, since 1974 [550–552].[20] The observations have used the same 21-inch telescope, and the same filters and photomultiplier detector – and the same observer. They are thus unusually homogenous in quality. Over the mid–late 1970s, all three objects had brightened, suggesting a common cause, which was thought to be solar variability: not in the total heat output of the Sun, but the flux of ultraviolet light, which in turn affects the rate at which photochemical haze is produced in their upper atmospheres.

When the Voyager encounter in 1980 revealed the north–south asymmetry in Titan's haze, this seasonal effect explained much (but importantly, not all) of the total brightness variation seen in Lockwood's photometer data. The brightening and darkening of the two hemispheres by ~20% in blue light over the ~30 year Saturn seasonal cycle, combined with the changing exposure of the two hemispheres, led to a ~10% variation in brightness with a ~15 year period,[21] and indeed the north–south asymmetry observed by Hubble in the late 1990s was consistent with the cycle extrapolated from Voyager. However, these important observations have continued, and show that the cycle does not in fact perfectly repeat (Figure 10.11). This means that there is not just a seasonal cycle, but there must be an additional effect. Perhaps it is some external forcing, like the solar activity originally thought of, or perhaps even cosmic-ray variations (which by changing the amount of ionization in the lower atmosphere may influence cloud condensation – an effect occasionally suggested to occur on Earth). Or perhaps it is some chaotic internal variability like El Niño on Earth or Mars' dust storms – Titan's atmosphere is massive enough, as the laborious spin-up required in numerically modeling its winds attests,[22] to have a "memory" to carry information from one year to the next.

When I started writing this book, I was not expecting to discuss Pluto. Indeed, its atmosphere is breathtakingly thin,[23] yet the findings of the New Horizons mission

---

[20] Lockwood's photometry data proved to be a useful adjunct to understanding Titan's seasonal change in the late 1990s as Hubble data emerged, e.g. Ref. [553].

[21] With only the data in the late 1970s, it was difficult to discriminate this ~15 year cycle from any effects of the ~11 year solar activity cycle.

[22] When you run a GCM with the atmosphere at rest, it may take many years (and thus much computer time) to reach a steady-state. This issue is particularly inconvenient for Venus and Titan with their massive atmospheres.

[23] Literally. On Pluto, or Mars, for that matter, the air would be instantly sucked out of an unprotected human's lungs, they would lose consciousness within seconds, and die soon thereafter. By contrast, on Titan a human could hold their breath for a minute or so; with only an oxygen mask and warm clothes, they could function for some time. Venus is too unpleasant to even contemplate.

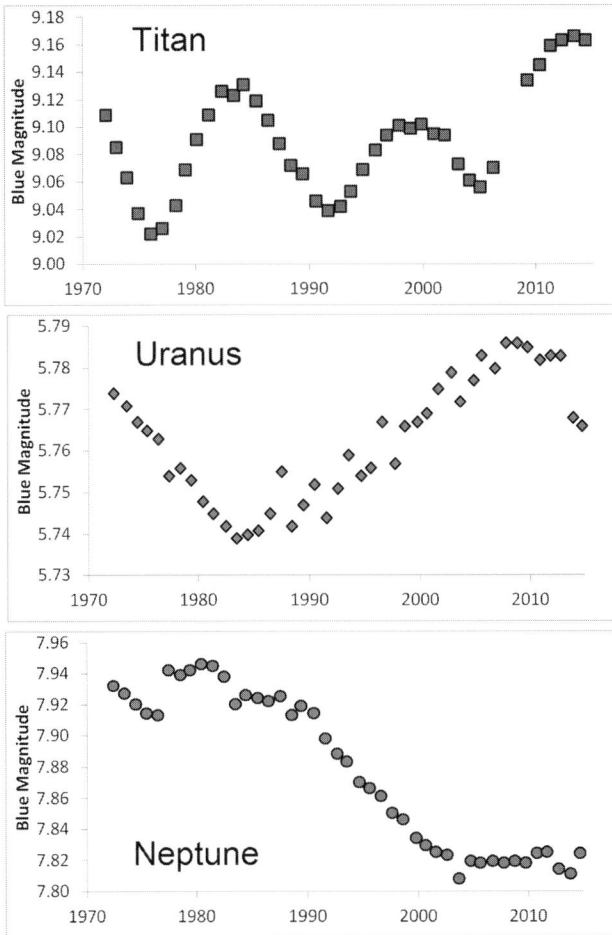

Figure 10.11 The Keeling Curve for Titan? Lockwood's carefully sustained measurements of Titan's brightness in blue light over a whole Titan year – most of a professional lifetime – shows a 14.5 year cycle.[24] Like the Mars pressure cycle, the two halves of the 29.5 annual variation are not equal, perhaps for a similar reason – the eccentric orbit. But notice that the annual cycle does not repeat – in the 2009 equinox Titan was darker than it had ever been recorded to be – evidence of climate variability? The brightness is reported in astronomical magnitudes, so brighter actually has a lower numerical value, hence the scale is inverted. (Figure by author, from data courtesy W. Lockwood.)

---

[24] Lockwood's Neptune data have been argued by British scientists Karen Aplin and Giles Harrison to support a combined solar ultraviolet/cosmic-ray influence on its cloudiness. Their paper is forced to use laborious statistics to support the asserted effect [554]. Nonetheless, the idea that Neptune may be more susceptible to cosmic-ray effects than is Titan or the Earth is a reasonable one, as indeed Lockwood and Thompson's original 1979 paper suggested [551]. If we are to understand Earth's clouds to percent precision, or perhaps Neptune's to even 10 percent precision, it may be necessary to include such subtle effects – indeed the idea that cosmic-ray flux could influence the cloudiness of the Earth goes back at

Figure 10.12 Just 15 minutes after its closest approach to Pluto on July 14, 2015, NASA's New Horizons spacecraft looked back toward the Sun and captured this near-sunset view of the rugged, icy mountains and flat ice plains extending to Pluto's horizon. The smooth expanse of the informally named icy plain Sputnik Planum (right) is flanked to the west (left) by rugged mountains up to 3500 m high, including the informally named Norgay Montes in the foreground and Hillary Montes on the skyline. To the right, east of Sputnik, rougher terrain is cut by apparent nitrogen glaciers. The backlighting highlights more than a dozen layers of haze in Pluto's tenuous but distended atmosphere. The scene is 1250 km wide. (Source: NASA/Johns Hopkins University Applied Physics Laboratory/Southwest Research Institute.)

show that Pluto seems to support a staggering array of surface–atmosphere interactions, as well as striking layers of Titan-like haze (Figure 10.12).

Interpretations of the New Horizons encounter with Pluto are still emerging. The presence of glacial textures in the surface suggests that soft nitrogen ice can accumulate at such rates that it flows (Figure 10.13). There may even be river valleys attesting to rain. This time, more than at any other occasion in planetary exploration, models of rather high fidelity are ready. François Forget and the LMD team in Paris showed with a GCM that a depression on an otherwise featureless Pluto would tend to fill with nitrogen ice, providing a rather straightforward

---

least to Ney in 1959 [555]. Edward Ney participated in early satellite observations, including those that first mapped lightning from space. One challenge in understanding the role of cosmic rays is that their flux at Earth and the planets varies with the solar cycle, which in turn also affects the flux of ultraviolet light from the Sun, which influences upper atmospheric temperatures and chemistry. Ney's paper suggests that the cosmic-ray effect, anti-correlated with the solar cycle, could explain the apparent paradox pointed out by Humphrey in 1936 of "the hot sun and cool earth".

[25] As in other early stages of planetary exploration, informal names are adopted for surface features: by convention, the IAU Committee on Planetary Nomenclature approves names to ensure that they are pronounceable, representative of different cultures, avoid offensive terms, etc.

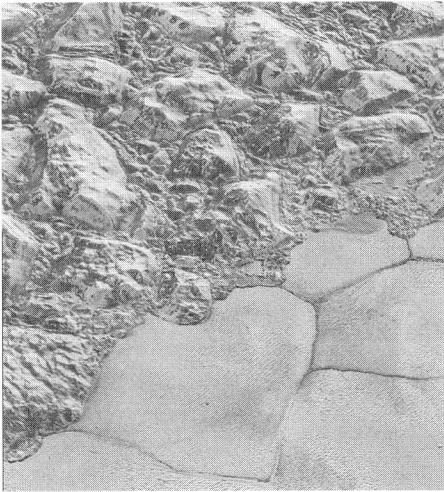

Figure 10.13 In the northern region of Pluto's Sputnik Planum, swirl-shaped patterns of light and dark suggest that a surface layer of exotic ices has flowed around obstacles and into depressions, much like glaciers on Earth. (Source: NASA/Johns Hopkins University Applied Physics Laboratory/Southwest Research Institute.)

scenario for the heart-shaped bright smooth patch informally named Tombaugh Regio,[25] after the astronomer who discovered Pluto (at Lowell Observatory).

Hints of the atmospheric structure of Pluto have been given over the years by a handful of stellar occultation measurements, even suggesting seasonal change [556]. However, it was not clear if these probe all the way to the surface – and indeed this debate still goes on. A temperature inversion, or a layer of haze, could block the light ray in a manner indistinguishable from the solid surface. That ambiguity at least was resolved by the New Horizons data, in that the diameter of the solid surface could be imaged directly. And, of course, a radio-occultation experiment avoided the haze issue and provided a pair of definitive temperature profiles. Pluto's atmosphere is basically a pure stratosphere, with a warm layer overlying a cold region that is in contact with the (cold) surface – not unlike one of the pre-Voyager concepts for Titan.

Venus at last is receiving new attention. A Japanese spacecraft, the Venus Climate Orbiter, renamed Akatsuki after its launch in May 2010, reached Venus in December of that year. It fired its engine to brake into an orbit that would periodically match the rotation of the cloud tops, ideal for making weather movies. But, like Mars Observer, a problem with the propulsion system led to a catastrophic failure – not only was the burn cut short, but in fact part of the rocket nozzle broke off! Akatsuki helplessly swept past Venus in an orbit around the Sun that would bring it closer and hotter than it had been designed for, but,

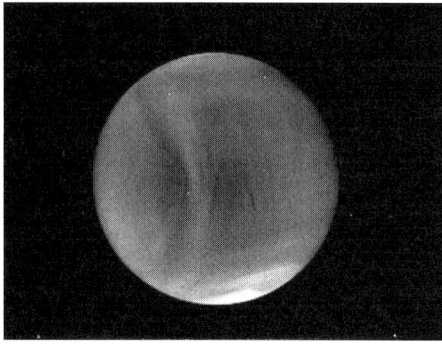

Figure 10.14 Akatsuki Image of Venus at 10 μm, taken within a few hours of its miraculous orbit insertion in December 2015. A chevron-shaped feature is visible, as well as light banding and a bright south polar feature. Given that Venus has a tiny obliquity, the interhemispheric difference is particularly interesting. (Source: JAXA.)

unlike Mars Observer, contact and control was maintained and the problem diagnosed. Its trajectory was adjusted to re-encounter Venus nine solar orbits later in December 2015, where it used its small control thrusters to limp into a highly elliptical orbit around Venus. This new orbit is poorer for observations than the original intended one, but, equipped with three infrared and one ultraviolet camera, as well as a dedicated lightning detection instrument, Akatsuki promises new insights on Venus' cloud dynamics (Figure 10.14) – notably in discovering large wave-like features forced by mountainous terrain, even though the clouds are 60 km above the ground. Meanwhile, Russia is also contemplating a return to that world and there is even talk of a possible Indian mission to that planet, but NASA has yet to rejoin the party directly.

As discussed in the next chapter, now that astronomers are confronted with growing numbers of Earth-sized (i.e. Venus-sized) planets at different distances from their parent star, the question of how the climates of the terrestrial planets diverged assumes new prominence. In particular, it seems likely that early Venus – before it lost its oceans – was a much more clement, and perhaps habitable, world. GCM simulations of progressively higher fidelity have been built and can explore the sensitivity of Venus conditions to parameters such as its rotation rate and the distribution of land surface. As ever, clouds are the key. Michael Way of GISS and colleagues [557] determined that a sufficiently slowly rotating Venus could have been habitable even until less than a billion years ago, in part because thick clouds form on the dayside, reflecting away the solar heat, whereas the abundant elevated (and thus dry) terrain at low latitudes (in contrast to Earth, where low latitudes are dominated by ocean) limits evaporation and allows the nightside to radiate heat away. Indeed, as noted by Yukata Abe and colleagues in 2011 [558], in climates forced by strong solar heating, stability and habitability seem to be enhanced on dry worlds because the greenhouse feedback cannot so readily run away.

Mars continues to be the darling of the space agencies. The European Space Agency, with Russia, flew a dual mission to Mars in 2016. This featured a Trace Gas Orbiter, aimed at monitoring methane and other gases with spectrometers, together with a camera and an improved neutron spectrometer to more tightly map subsurface water. Riding along was a lander,[26] intended (much like MOM) as a technology demonstrator – to show that ESA could shake off its Beagle demons and land safely. The lander, named Schiaparelli, carried a small meteorology package, but, being battery-powered, was limited to only a few days of operation. Part of the package was an electric field sensor, intended to make the first atmospheric electricity observations on Mars – it is expected that dust storms or devils may have interesting signatures. Sadly, Schiaparelli was lost.

A NASA mission not designed to study the climate nonetheless promises to yield much important climate information. InSight (NASA missions selected in a competitive process often have rather contrived names) is intended to study the Martian crust and interior using a sensitive European seismometer (derived from French developments originally for Mars-96). Such a sensitive instrument is also affected by the weather, notably pressure and wind. Thus, InSight, planned to operate for at least one Martian year, will make measurements of the surface meteorology conditions with higher precision than Viking and with better continuity and frequency than the landers and rovers since; high-accuracy tracking to measure the planetary rotation state is also planned.

The mission also features a "mole" or self-hammering drill, which will deploy a string of temperature sensors down to 5 m in the shallow subsurface. In principle such a temperature profile, whose principal goal is to measure the geothermal heat flow, bears a memory of the past climate [559] (for example, temperatures in the Greenland ice sheet show a gentle dip at depths of a few tens of meters corresponding to the Little Ice Age). Whether a single such profile can be usefully generalized to the rest of Mars remains to be seen, but we recall that such heat flow was an important consideration in assessing the stability of water (or CO_2) in the subsurface at gully-forming locations and at the base of the polar caps. Originally supposed to launch in March 2016, development difficulties with the seismometer deferred the mission, which was launched in May 2018.

ESA plans a Mars rover equipped with a subsurface drill (originally for 2018, now for 2020), and NASA also plans one in 2020. While both are geared towards

---

[26] formally "EDM" – Entry and Descent Module. Since its prime function is only to demonstrate safe landing, it carries no solar panels.

Figure 10.15 The proposed Titan Mare Explorer, intended for launch in 2016, would have observed surface meteorology and air–sea exchange processes on Titan's second largest sea, Ligeia Mare, and would have sent its data by radio directly back to Earth. However, prospects for such an efficient mission have faded as in the 2030s (northern winter) the Sun and Earth will be below the horizon as seen from Titan's northern seas. (Source: Johns Hopkins University Applied Physics Laboratory.)

geology/mineralogy and astrobiology, as on their predecessor rovers these investigations will help understand past conditions, and the NASA rover at least has some rudimentary meteorology capabilities. Remarkably, a new player may be stepping into the Mars climate ring. The United Arab Emirates, aiming to move to a more knowledge-based economy in the post-petroleum era, decided to celebrate its 50th birthday with a Mars mission.

The Titan environment is one to which exploration systems are closely coupled. For example, hot air balloons will be carried in the wind, and several simulations have been performed to see how many circumnavigations a balloon might make in a year, for example [560]. Similarly, wind results from no less than four different GCMs were evaluated for Ligeia Mare in 2023 [561, 562], the planned destination for a mission proposal (the Titan Mare Explorer – TiME [563] – Figure 10.15). The winds are of interest for several reasons. First, the thick Titan atmosphere forces a parachute descent of a couple of hours for typical probe designs, giving winds time to displace the capsule downwind, and so the size of the landing ellipse,[27] and the ability of a capsule to splash down safely in the sea, depends on the winds [564]. Second, the drift of the floating capsule would depend primarily on the wind drag on its superstructure (in fact it was Edmond Halley that first calculated the force balance on a wind-blown floating object[28]). And thirdly, a floating capsule is at the mercy of the waves [567] – pitching and rolling affect the allowable exposure times to avoid motion smear in camera images, and how fast an antenna drive might need to operate to compensate; in the limiting case, large waves might cause capsize.

Titan remains of great interest, both as an "Ocean World" like Europa and Enceladus, offering the prospects of understanding prebiotic chemistry and habitability of icy satellites, and as a quasi-terrestrial planet with a rich climate and landscape, showing processes familiar to us on Earth in different conditions and with different working materials. Titan's environment, benign apart from its deep chill, lends itself to a range of exploration platforms, and at the time of writing NASA has funded study of Dragonfly [568],[29] a rotorcraft lander (basically looking like a Mars-rover sized quadcopter, able to fly in Titan's dense atmosphere and low gravity to visit different sites tens of kilometers apart. It would also be able to fly repeatedly up and down through Titan's planetary boundary layer to observe atmospheric interactions with the surface.)

Heading further out, New Horizons will pass a Kuiper Belt object in 2019 – maybe it will be as surprising and dynamic as Pluto (possibly it will not – it is about a hundred times smaller than Pluto so there seem slim prospects for an atmosphere). But beyond the bounds of our solar system lie countless stars, planets and climates.

---

[27] The scatter of possible landing points on Mars missions usually depends only on the entry angle and the somewhat circular dispersion of delivery errors, and thus smears as a downrange ellipse. On Titan or Venus, this delivery locus is further convolved with displacement by the (uncertain) zonal and meridional winds. The resultant dispersion may or may not be formally ellipsoidal, but it is conventional to fit it with an ellipse and to call it such. Usually such an ellipse is defined as one in which 99% of the random cases in a Monte-Carlo simulation fall.

[28] 'The Manner of Computing the Weight or Force of the Winds", an unprinted paper read to the Royal Society on November 18, 1691, see [565] and also [566].

[29] http://dragonfly.jhuapl.edu

11

Worlds Beyond: Exoplanet Climate

1992 – Wolszczan and Frail discover pulsar planets

1993 – Kasting, Whitmire and Reynolds explore the habitable zone around different star types

1995 – Mayor and Queloz discover 51 Peg b; Marcy and Butler report a second exoplanet soon after

1997 – First GCM model of a synchronously rotating exoplanet

2001 – Sodium atmosphere discovered around HD 209458b ("Osiris")

2005 – Direct detection of photons from an exoplanet (TrES-1)

2007 – Day/night temperature contrast measured on HD 189733b

2009 – Carbon dioxide, water, methane indicated on HD 209458b

2009 – Kepler launched

2010 – Transmission spectrum through atmosphere of Earth-sized GJ 1214b

2012 – Kepler 1520 (KIC 1255748b) observed to be possibly disintegrating

2013 – Blue color determined for HD 189733b

2016 – Planet discovered in habitable zone around Proxima Centauri, only 4.2 light years from Earth

2017 – TRAPPIST-1 system discovered with seven planets; some possibly habitable, some runaway greenhouse

2018 – Almost 4000 confirmed planets in over 2700 systems, with at least 600 systems having more than one planet

This book has confined itself up to this point primarily with the terrestrial planets of our solar system, plus Titan. Yet even the earliest thinkers about the conditions on other worlds also imagined there might be other planets around other stars. The search for planets around other stars, like any other scientific endeavor, had a number of negative searches and a few false alarms. The first firm discovery was of planets around a pulsar, observed by radio astronomy, but the existence of solid bodies around the neutronium remnant of a supernova was such an exotic astrophysical scenario, with little prospect for atmospheres or life, that it aroused only mild interest among planetary scientists.

The situation was quite different in 1995, when Mayor and Queloz discovered evidence of an object, not too different in mass from the planet Jupiter, around a normal star. This was followed up with several comparable discoveries – the enabling technology had been gas cells used as precision wavelength references in astronomical spectroscopy. These allowed rather precise measurements of the Doppler shift of stars, a periodic signal which could be interpreted as the reflex motion of the star back and forth along the line of sight due to the gravity of a planet in orbit around it. This spectroscopic technique was just what had been anticipated by Struve in 1952 [110].

This technique, like the transit technique which rapidly discovered many other planets soon afterwards, most easily detects those planets that are most massive, and closest to their star.[1] And indeed, the 51 Peg b discovered in 1995 was at least half the mass of Jupiter, but, unlike the location and period of the giant planets in our solar system, swept around its star in only four Earth days, placing it at a distance of only 0.05 AU from its star!

A gas giant like Jupiter isn't of obvious interest as an abode for life (although cannot be excluded entirely), even in a more clement location, but it was quickly noted that moons of giant planets around other stars might be habitable. Darren Williams and Jim Kasting at Penn State University explored this possibility [569].

In fact, the general question of habitable zones around arbitrary stars had been explored only shortly before by Kasting, working with Whitmire and Reynolds. They used Kasting's radiative–convective model for atmospheres with CO_2–H_2O greenhouses (used previously in Venus studies and elsewhere) but took into account the different spectra from different types of star. The portrayal in

---

[1] Actually, the Doppler method favors massive planets, while the transit method favors planets with large diameters, not quite the same thing. The Doppler method only places a lower limit on mass M, depending on the inclination i of the planet's orbit to the line of sight: $M \sin i = 1.5 M_e$ means M is $1.5 M_e$ (1.5 Earth masses) if the observation is edge-on ($i = 90°$) to the orbit, but $M \sim 2 M_e$ if $i \sim 45$ degrees, for example.

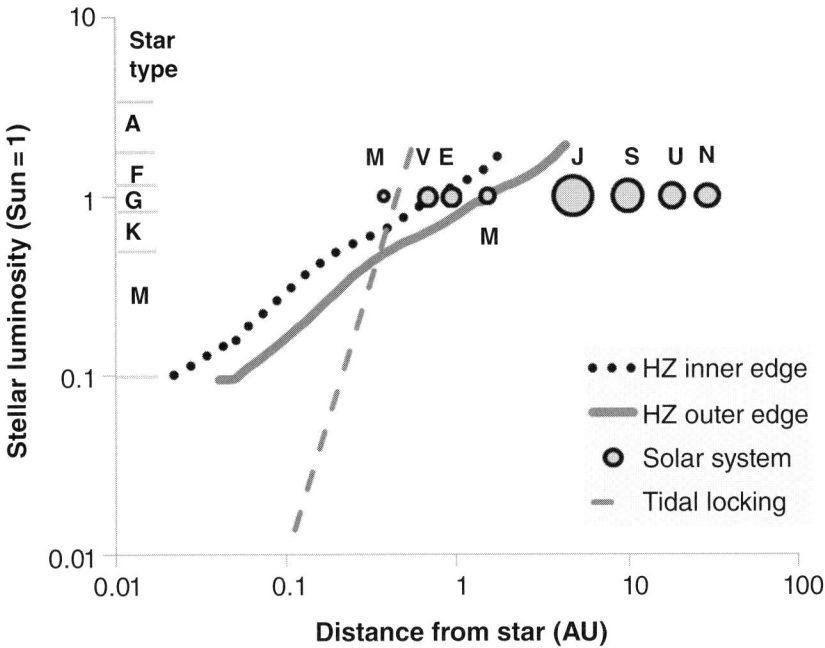

Figure 11.1 The inner and outer edge of the habitable zone (HZ) as defined by Kasting and colleagues in 1993. Mars (lower M) is on the hairy outer edge, Venus (V) a little beyond the inner edge, and Earth (E) nicely in between for our Sun, a G-type star. All are beyond the radius (dashed line) inside which a planet might be expected to become tidally locked so that it rotated synchronously (a zone in which many exoplanets would subsequently be found!). Note that the HZ does not extend up to the right – A-type stars evolve too quickly to be considered habitable environments. (Figure by author.)

this paper [570][2] of the boundary of the habitable zone (Figure 11.1) in our solar system became a fundamental one in exoplanet studies.[3]

The definition of habitability for life as we know it is fundamentally determined by water – sometimes popularized as "The Goldilocks Problem" – making a planet not too hot, not too cold (for a nice review of the question at this point see [571]). Often a global average temperature at the freezing point is considered the outer boundary, but the insolation conditions for this depend on how effective the greenhouse effect is. Similarly, the upper limit of temperatures that life

---

[2] At the time of writing, this paper has an impressive 1600 citations.
[3] Recall that the term "habitable zone" had been used by Maunder a century earlier, and some general assessments of planetary conditions were discussed in a book (*Habitable Planets for Man*) by Stephen Dole in 1964.

can tolerate is not well defined, there being some remarkably hardy organisms on Earth that live in deep-sea hydrothermal vents.

In Kasting's study, the outer edge of the habitable zone is defined by the planet being so cold that CO_2 clouds could form, these causing the albedo of the planet to rise and causing the planet surface to be too cold, regardless of how much CO_2 the planet had in its atmosphere. The inner edge was defined by the Venus-like loss of water through a "leaky" tropopause. These two boundaries define a slightly wavy band in star-distance space – a brighter star allows a slightly more distant planetary orbit.

Importantly, Kasting et al. noted that planets close enough to their star could become tidally locked, as our Moon is to Earth, or Titan is to Saturn. Thus one side will be permanently warmed by the star, and the other continuously exposed to the cold black of space.

This issue is of special interest because, contrary to the idea of our Sun being a "typical" G-type star, most stars actually are "M-dwarfs" – small, cool (and thus red) stars.[4] Now, if the atmosphere can transport enough heat from the dayside to the nightside, all is well, and the average temperature (depending on the greenhouse, albedo etc.) may allow liquid water on both sides. But this is the crux of the matter – how much heat will the atmosphere (and perhaps oceans) carry? If it isn't enough, then the cold side of the planet is so cold that all the water vapor, or possibly even some of the CO_2 as well, will get frozen out and be unable to support a greenhouse. Thus Kasting et al. argued that planets to the left of the dotted tidal-locking line in Figure 11.1 might not be habitable.

The simplest assessment of this day/night temperature contrast mirrors studies of the climate variation with latitude some decades before, with an energy balance model. But the choice of heat transport coefficient in these exotic conditions is not obvious. Thus, it did not take long before the growing capability of global circulation models began to be applied to the problem. Manoj Joshi at NASA Ames (where the group led by Bob Haberle had been growing Mars GCM capability) simulated a synchronous planet around an M-dwarf [572]. In fact, as Joshi noted, a typical rotation rate of a tidally locked M-dwarf in inertial space is about the same as that of Titan, and so the time-history of results in this paper were plotted in terms of "Titan days"!

---

[4] This classification is based on the spectrum of the star, the ordering from blue (O, hot) to red (M, cool) can be remembered as "Only Boring Astronomers Find Gratification Knowing Mnemonics". Intrinsic stellar brightness and spectrum have, since about 1910, been mapped on the famous Hertzsprung–Russell diagram, on which most stars form a track called the main sequence.

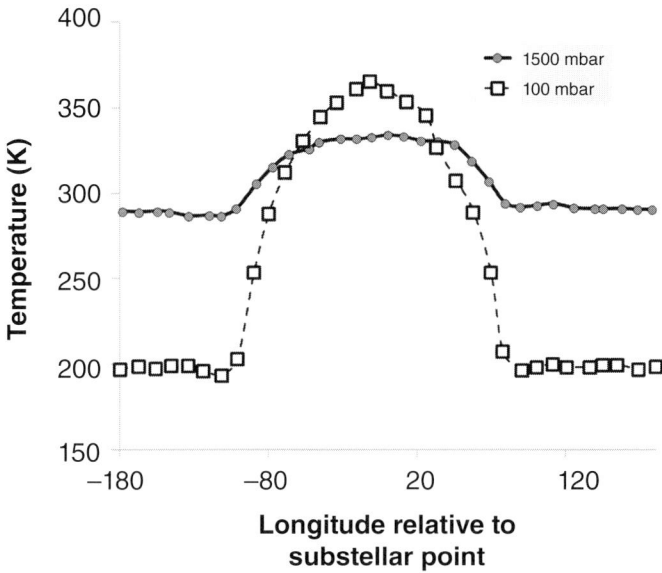

Figure 11.2 Joshi's global circulation model of a tidally locked planet around an M-dwarf star, with a CO_2 greenhouse of 100 mbar or 1500 mbar. The thin-atmosphere case has a cold nightside which would trap any water as ice, making this world's habitability unpromising. (Figure by author.)

The substellar point (where it is always high noon) is rather hot for the thin-atmosphere case, but a zone nearby would be quite pleasant in temperature terms. However, the long-term habitability of this situation is not favorable, as the permanently frozen nightside would trap out any water vapor as permanent ice deposits. On the other hand, the 1500 mbar case is warmer overall (due to the stronger greenhouse effect) and much more even in temperature (due to day–night transport of heat by the thicker atmosphere) – see Figure 11.2.

In fact, this situation is much the same as the problem of Mars' climate, except here the temperature is determined by distance from the substellar point rather than latitude. The same volatile-trapping effect occurs, and the system has the potential for hysteresis – if the atmosphere is thin to begin with, the nightside is cold, and squirting in some water or CO_2 doesn't save the planet – the volatiles just freeze out on the nightside. But if the planet is warm enough to begin with, the volatiles stay in the vapor phase and can contribute to heat transport and greenhouse warming.

Several new exoplanet discoveries were made in the 1990s, but all were "Hot Jupiters" – hot because they orbited close to their stars (how they ended up being close to their stars is another story... likely they didn't form there). Because the radial velocity (Doppler) detection technique looks for a periodic signal, the longer the time series of data to work with, the weaker a signal that can be picked out of the noise. Thus, even with no improvement in instrumentation, over time the sensitivity to ever-less-massive planets should improve. The same, in fact, is true of the transit technique (also anticipated by Struve in 1952), which bagged its first exoplanet, HD 209458b (a 1.39 Jupiter radii planet, also at about 0.05 AU) in 1999. There have also been dramatic improvements in the instrumentation over time: while the first exoplanet discoveries were in fact made with somewhat modest telescopes on which time could be spared for a speculative endeavor, once exoplanet searches became legitimate (i.e. offered the reasonable prospect of positive – or even exciting – results) front-line telescopes at major observatories got into the act, and dedicated space observatories have been launched. These efforts have yielded an impressively consistent march of progress – like some Moore's law of astrophysics, the frontier of the smallest planetary mass has been rolled back by about a factor of 2 every year for the past two decades (see Figure 11.3) and indeed the trend was obvious enough that

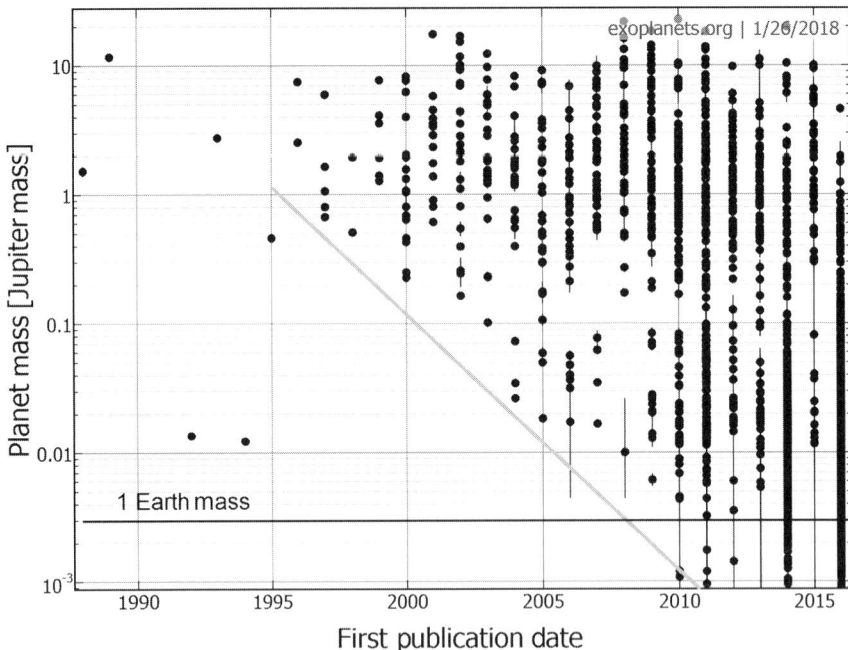

Figure 11.3 The masses of exoplanets as a function of date of discovery. The burgeoning number in total, and the progressive recession of the low-mass frontier is evident. (Figure by author: plotted using data from the Exoplanet Orbit Database [573] and the Exoplanet Data Explorer at exoplanets.org.)

some even predicted in the mid 2000s that the discovery of an Earth-mass (or 0.003 Jupiter-mass) planet was only a few years away, as it proved to be.

The exercise of planet detection picked up its pace significantly in 2009 with the launch of the Kepler mission by NASA. This is a space telescope, rather smaller than Hubble (it has a 1.4-m mirror vs. Hubble's 2.4 m), launched into an orbit around the Sun that slowly trails the Earth. This vantage point allows the telescope to stare at a target region of space, without disturbances from eclipses or obscuration by the Earth that Hubble's low orbit experiences. Kepler's task, armed with a giant camera (42 detectors, each of 2 megapixels!) that gazes at a 12-degree-diameter patch of sky, is to watch for planetary transits, seen as little dips in the brightness of the 150,000 stars it monitors. And Kepler actually "watches" – the images acquired every minute or half-hour are far too voluminous to send back entirely, so the pixels around each star are extracted onboard (in fact the telescope is deliberately slightly out of focus, so that the light from each star is spread over several pixels, improving the quality of the measurement). Kepler is able to measure light changes of the order of 20 parts per million (0.002%) – a factor of 3 or 4 better than the expected dip due to an Earth-sized planet.

Whereas a world hunter's success once relied upon stamina at the telescope (e.g. Herschel) and/or skill in optics (q.v. Huygens), the modern planet hunters distinguish themselves by their ability to sift through the deluge of data from observatories such as Kepler. Actual planetary transits must be sifted from noise glitches, including the uncooperatively variable nature of many stars (e.g. sunspots cause brightness variations too) and formal statistical methods such as Bayesian tests have become the stock-in-trade of the exoplanet literature. In essence, does the improved fit due to introducing a hypothetical planet into a model of the lightcurve data justify the introduction of the new parameters? Efficient means to search the option space (phase, period etc.) of the problem have been developed – Markov chain Monte Carlo (MCMC) being probably the most popular, and computer clusters hum in search of yet more detections.

This enterprise has been enormously successful – Kepler data have yielded thousands (Figure 11.4) of candidate planets, of which over a thousand have been confirmed. Many groundbased observatories have been set up to perform transit searches too. Since these rely on staring at many stars, a large telescope is not necessarily needed, just lots of pixels. Many such observatories are built around camera clusters with quite small telescopes or even just large telephoto lenses. It has become the convention in the exoplanet business that planets discovered around stars that do not already have a prominent catalog entry (e.g. the first such exoplanet was discovered around Star 51 in Flamsteed's catalog of the constellation Pegasus, so the star is termed 51 Pegasi and the planet 51 Pegasi b, or 51 Peg b for short) become named after the observing project that studied

New Kepler Planet Candidates
As of July 23, 2015

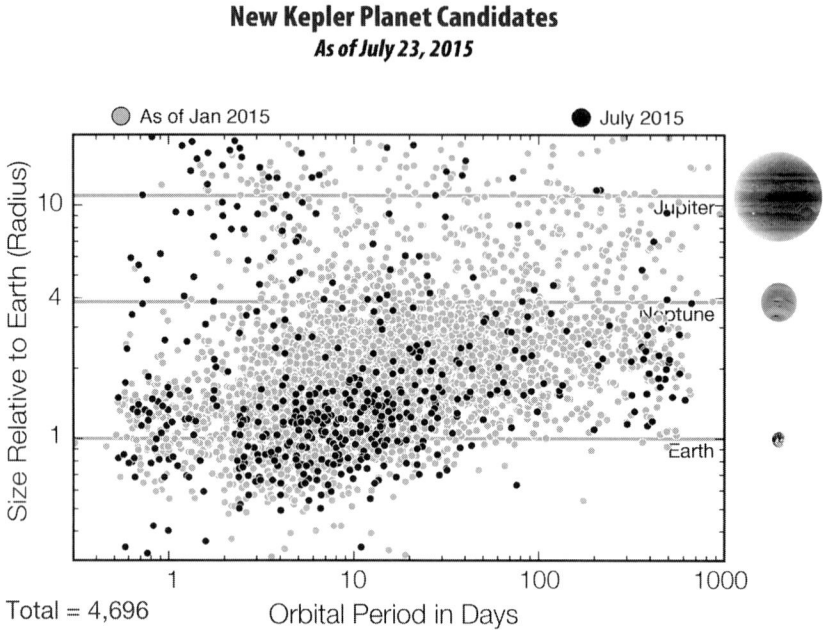

Figure 11.4 The basic properties of thousands of exoplanet candidates have now been determined. (Source: NASA.)

them. Thus, the planet from which photons were first isolated (by a much larger telescope) is TrES-1b: first discovered by the Trans-Atlantic Exoplanet Survey (TrES) which uses three 4-inch telescopes sited at Lowell Observatory, Palomar and the Canary Islands. Thus, exoplanet names include other surveys such as OGLE, TRAPPIST and Kepler, although many of the prominent exoplanets are around brighter stars in the Henry Draper star catalog (e.g. HD 409258b). A few stars have been discovered to have many planets, in which case letters c, d, e, f, g etc. have to be invoked. There have been proposals to offer more informal (and less ungainly) names for the most prominent exoplanets, but these may take a long time to accept, if ever.

Merely discovering that a planet exists is no longer of wide interest, however. The orbital period gives a sense of how far a planet is from its star, but that is only part of the equation (as the debates by Lowell, Poynting and others about Mars' habitability in the early 1900s demonstrate). How reflective is the planet, and what kind of atmosphere might it have? From tens to hundreds of light years away, these sorts of detail are not easy to come by.

In a few precious instances with very careful observations with the most advanced (spaceborne) telescopes it has become possible to isolate light (or heat) from the planet and deduce some hints of temperatures. The first successes in

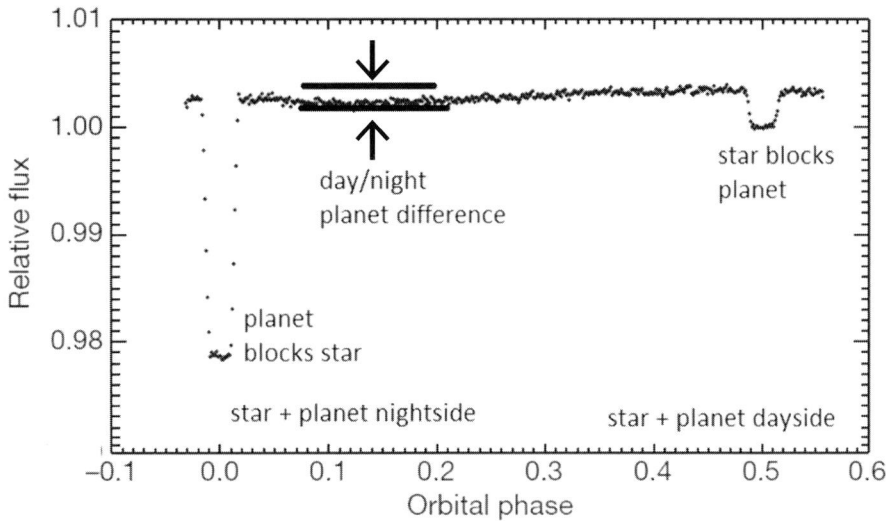

Figure 11.5 Infrared lightcurve of HD189733 and the planet HD 189733b, plotted against fraction of an orbit (i.e. year, or day[5]). When the planet transits across the stellar disk, it obscures about 2% of the star's light, whereas when the star blocks the line-of-sight to the planet (a "secondary transit") the total light drops by about 0.3%; thus at this wavelength the planet is about 300 times less bright than the star. Nonetheless, the measurements are made with sufficient precision that this contribution can be seen not to be constant and the reflected light variation with phase can be isolated from a variation of planetary temperature with longitude. (Figure by author, with data from [575].)

doing so were led by Joe Harrington [574] of the University of Central Florida, and Heather Knutson [575] of the Harvard-Smithsonian, both using the Spitzer telescope – essentially the infrared counterpart to Hubble. Observing light from the star HD 189733 at 8 μm with exquisite precision (Figure 11.5) yielded several pieces of information. The most prominent feature of such light histories is the primary transit, where the planet blocks light from the star, indicating the relative size of planet and star. Also visible is a secondary transit, where the star blocks the radiation from the planet (which includes both reflected starlight and the thermal glow of the planet's dayside). By careful fitting of the data, Knutson

---

[5] A sidereal day is the rotation period in inertial space, i.e. relative to a fixed reference frame such as the stars. A solar day on Earth, the time between successive noons, is a little shorter because the Earth goes around the Sun once a year. On Mars and Titan, with longer years, the solar and sidereal day are even closer in value. On a synchronously rotating planet, however, the solar day is infinitely long, and the sidereal day is equal to the year. Venus is an unusual case since it rotates backwards – its solar day is 120 Earth days long.

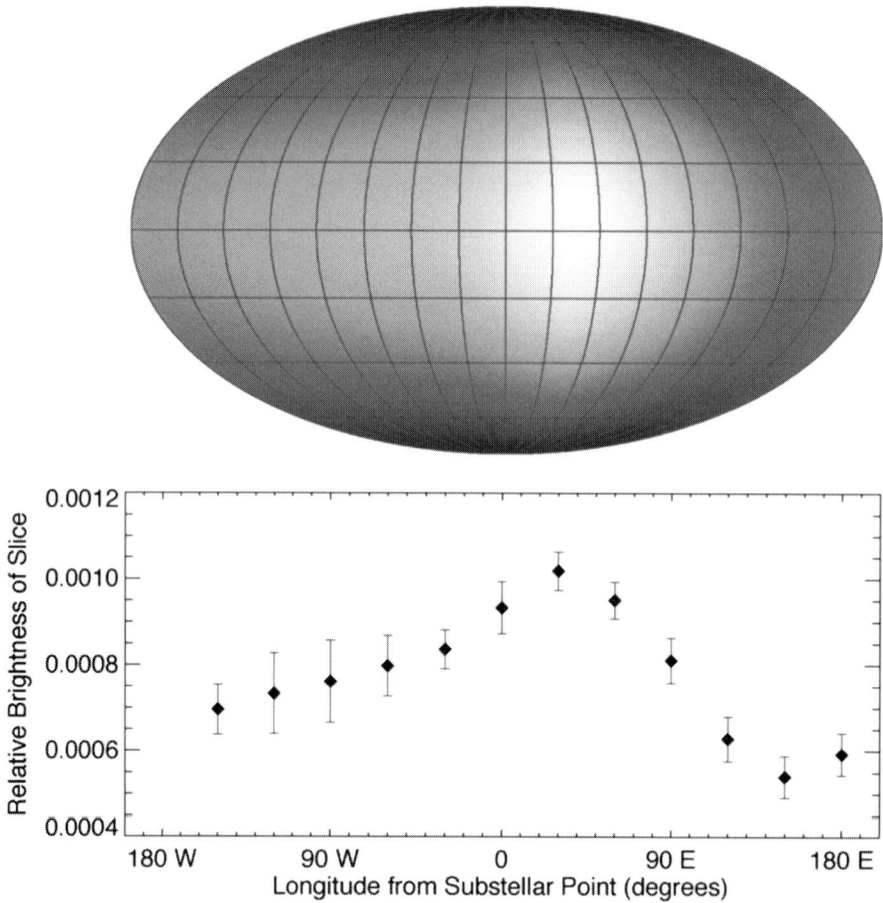

Figure 11.6 The brightness of HD 189733b derived from the transit lightcurve in Figure 11.5, portrayed as a temperature "map". Actually, the term map is misleading as the data have no latitude information whatsoever – the two-dimensional map is more of a guess, but useful for press purposes. The displacement of the brightest (hottest) part of the planet from the substellar point is real and interesting, however. (Knutson et al., 2007 [575].)

and colleagues could determine a minimum brightness temperature of 973 ± 33 K and a maximum of 1212 ± 11 K, indicating that energy from the irradiated dayside is efficiently redistributed throughout the atmosphere;[6] interestingly, the hottest point was displaced 16 degrees eastward of the substellar point (Figure 11.6). On

---

[6] This result was of some interest to Jonathan Lunine and myself who had applied the simple maximum entropy production (MaxEP/MEP) model to a synchronously rotating exoplanet, before such measurements became available. The MEP result, published in a Lunar and Planetary Science Conference abstract in 2002, is that the day/night temperature contrast should be one-third of the effective temperature, not too different from the

the other hand, Harrington's earlier observation, at 24 μm, of the star Upsilon Andromeda suggested a qualitatively weak transport of heat on that planet, but had rather less data to work with.

The Knutson result in particular gave modelers something to aim at, and despite the near-total absence of information to constrain such models (not least, the atmospheric composition, which affects the depths at which solar energy is absorbed and where it is emitted[7]) many numerical circulation models were fired up to attempt to reproduce the observed numbers. Further observations, fiercely competed for, are bringing some composition measurements to at least a handful of exoplanets.

Some of this zoo of worlds have exotic sizes and compositions indeed [577] – as well as Hot Jupiters there are Black Jupiters and Puffy Jupiters. The tidal locking of some worlds may mean some "Super Earths" are "Eyeball Earths", frozen snowballs with a pool of open water around the subsolar point (a term Ray Pierrehumbert coined for the planet Gleise 581g [578]). Note that the term "Super Earths", used by some to denote exoplanets larger than Earth but smaller than Neptune, is a misleading term that might be best avoided as implying we know more about them than we do [579, 580] – such a world might just as well be a "Super Venus", which implies a rather different environment! Indeed, because Venus is closer to the Sun than the Earth, an astronomer on some distant exoplanet observing transits of the Sun would be more likely to discover Venus before the Earth because our sister planet's transits would be deeper and would occur more often.

Statistics of temperatures and temperature contrasts are also starting to emerge [581], which demand a more general (or simpler) approach to understanding planetary heat transport at a more fundamental level and its dependence on various parameters, which yields some fundamental understanding that can be more generally applied. A recent GCM parameter-sweep exercise by Yohai Kaspi and Adam Showman is notably useful in this respect [582]. Ultimately, as in the solar system, thermodynamic limits (such as the maximum work output from a Carnot engine driven by the day–night heat flow) may be a useful constraint on the circulation dynamics [583].

From being a field that had essentially no information to work with at all a decade and a half ago, exoplanet climate is a busy field with dedicated conferences

---

Knutson value. However, perhaps that was just lucky – a range of contrasts is observed, perhaps indicating different pressure levels of absorption and emission.

[7] In fact, there is indication, in the lack of sodium or potassium features in the optical transmission spectrum measured by Hubble, that HD 189733b has atmospheric haze (e.g. [576]).

("Exoclimes", "Extreme Planetary Systems" etc.) with hundreds of participants. While the amount of data per world that this community has to work with is not too dissimilar from that available to astronomers trying to decipher conditions on our solar system's planets at the beginning of the twentieth century, there are more worlds to choose from, and the modeling tools are far superior.

A cottage industry of global circulation model studies has spun up, despite the fact that there are almost no data to constrain them. However, these studies – benefitting from fast computer clusters and the internet for easy data and code-sharing – are usefully exposing fundamental dependencies of day–night and equator–pole heat transport on such parameters as diameter, rotation rate, friction and so on. However, the exotic setting of many exoplanets means that unfamiliar processes may be significant, such as magnetic drag from the star. The exoplanet community appears to have adopted their own convention in expressing day/night temperature contrast in terms of a "redistribution efficiency", the ratio of heat advection time to heat radiation time, although fundamentally this is related to the heat transport parameter in Budyko–Sellers-type models.

One particularly hot exoplanet, 55 Cancri e, has been measured to have a brightness temperature (recorded by the Spitzer telescope at 4.5 μm) of some 2700 K, while the nightside temperature was still some 1300 K [584].[8] Although it was speculated that rivers of lava might somehow transport heat on this torrid world to its nightside, modeling seems to favor transport by an atmosphere [585], which would have to be substantial – perhaps 1.4 bar.

Among the thousands of exoplanets, a few have been found with extreme orbital characteristics – a notable example is HD 80606 b. This planet has a 111.4-day orbit, with a comet-like eccentricity of some 0.927! Even Saturn's modest eccentricity of 0.05 (and Mars of 0.09) is enough to cause a profound asymmetry in the seasons, and Pluto's 0.25 drives a partial atmospheric collapse. But HD 80606 b sees a variation in its stellar irradiation by a factor of some 828. During a 6-hour observation with Spitzer (this time at 8 μm), Greg Laughlin and others [586, 587] observed the brightness temperature rise from ~800 K to some ~1300 K!

Some planets are strong candidates to be in their star's habitable zone – Kepler 62e and f being cases in point. Studies of tidally locked planets with GCMs suggest that some might have climates that are at least stable in the instantaneous sense (e.g. Robin Wordsworth at LMD found that the exoplanet Gleise 581d, which receives 35% less stellar flux than Mars, could be maintained habitable with 10 bars of CO_2, even allowing for cloud formation [588]). Whether these planets

---

[8] The 55 Cancri system has several exoplanets.

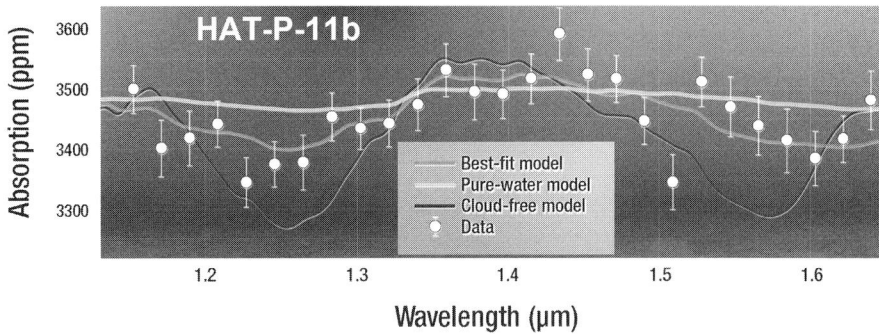

Figure 11.7 A spectrum of the planet HAT-P-11b (around a star observed by the Hungarian Automated Telescope network; the planet is also Kepler 3b) showing probable indication of water absorption bands (which imply, unlike Venus, some extensive water vapor above the clouds). Deriving this spectrum required data from both the Hubble and Spitzer space telescopes. (Source: NASA/JPL-Caltech.)

are habitable in the sustained sense, when silicate weathering or volcanic accumulation may alter the CO_2 abundance, is of course another question.

In a few precious cases, some indications of atmospheric composition have been gleaned, usually won by the expenditure of lots of observing time with the largest observatories or spaceborne telescopes. These spectra are necessarily of low resolution (there just not being that many photons to go around!) and so in some respects the state of knowledge of these bodies (HAT-P-11b being one example – see Figure 11.7; CH_4, CO_2 and water vapor have been indicated around HD 189733b, e.g. [589]) resembles that of Titan around 1970, or of Mars or Venus in the early twentieth century. The modeling machinery is now well developed (in many cases adapted from the models used on solar system objects) to attempt to interpret these spectra. In some cases, as was once the case for the solar system planets, laboratory data are lacking to fully interpret what are often exotic compounds under extreme conditions (titanium oxide being a prominently suspected cloud-forming material in Hot Jupiters, for example!).

Just like the stochastic resonance of record-breaking heatwaves seeming to be recent on our planet, which is slowly warming, two of the most exciting developments in exoplanet studies came about just as work on this book was being completed.

A Belgian team detected a system of seven planets around the star TRAPPIST-1, a red dwarf 40 light years away [590].[9] These planets are in resonant orbits with each other (as are the Galilean moons of Jupiter), which may have kept these

---

[9] The name of the survey instrument (Transiting Planet and Planetesimals Small Telescope, at La Silla Observatory) is a nod to a famous Belgian beer style.

orbits relatively stable for an extended period. The planets are all likely tidally locked, and indeed tidal heating is probably very significant for the nearest ones, enough to maintain magma oceans. The equilibrium temperatures (assuming zero albedo) range from 400 K to 168 K, so some are likely habitable in the classic sense, and some are probably runaway greenhouse candidates, if they have retained atmospheres (see later).

More exciting still was when a roughly Earth-mass planet ($M \sin i = 1.3 M_e$) was discovered in late 2016 [591] orbiting around Proxima Centauri, the closest star to Earth! (Actually, the star we see in the sky 4.2 light years away and call Alpha Centauri is itself a double star, and a third member of the system, Proxima Centauri, orbits these at a distance of 0.2 light years, so is sometimes the closest to Earth, hence its name.[10]) Proxima Cen is a red dwarf star, only about 12 percent the mass of our Sun.

The keen interest in modeling the climate of this world is evidenced by the release, essentially simultaneous with the paper announcing the discovery of the planet itself, of several papers assessing its habitability [592, 593].[11] A thicket of papers have followed since, considering how Proxima Centauri b might be observed, and how stable an atmosphere might be to the strong X-ray flux and solar wind from the star – while red dwarfs are not especially luminous, they can be highly variable and active (flaring), which may make their habitability challenging in the long term.

Undeterred, however, GCM models were quickly run with what few pieces of information did exist, and of course a range of answers resulted, depending on what is assumed about how much atmosphere and how much water, and whether the planet is synchronous or not (see Figure 11.8). This body will surely be the object of much forthcoming study, not only by models and by future telescopes, but (just) conceivably by visit by spacecraft. As an example, New Horizons is tearing away from Pluto at about 14 km/s, or 3 AU/yr. At this speed, it would take about 80,000 years to fly the 4.2 light years (quarter of a million AU) to Proxima Cen. Even Huygens, who calculated that a bullet, the fastest thing that he knew, would take 250 years to reach Saturn, might be daunted by this timescale. But who knows what progress may bring, and in fact exploring

---

[10] Proxima Cen was discovered in 1915 by the Scots astronomer Robert Innes, then director of the Union Observatory in South Africa. It is too faint to be seen by the naked eye.

[11] Fast-moving astrophysicists are in the habit of putting new manuscripts on a preprint server for all to see immediately, without the screening of peer review – indeed, the same day another paper [594] with a title confusingly similar to the first of the two above hit the Arxiv, followed six days later by yet another [595]. In contrast to this frenzy, planetary scientists for the most part let journal publication run its course before distributing work.

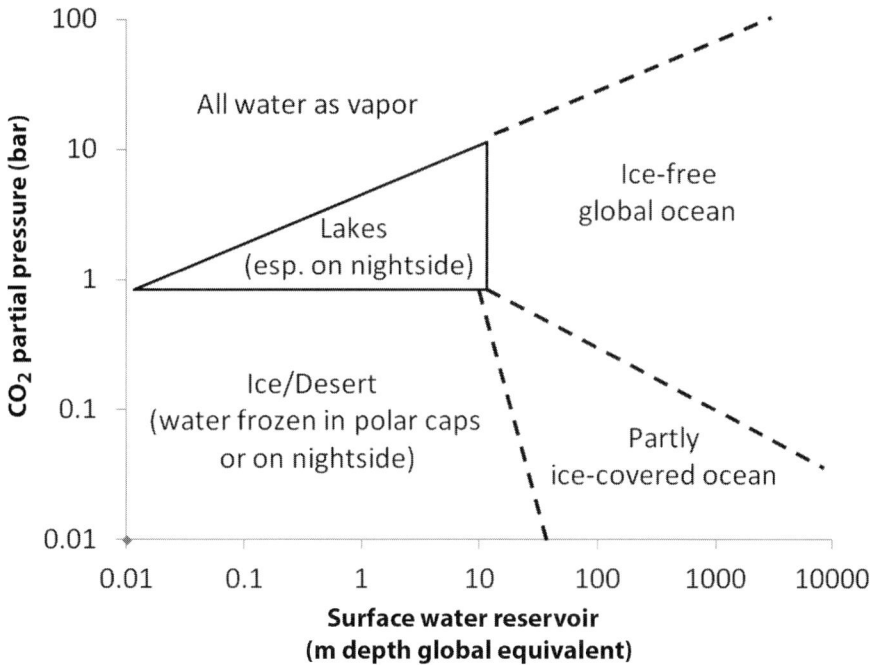

Figure 11.8 The characteristic climates of Proxima Centauri b, as simulated by Turbet et al. [593]. The lines should be considered schematic only – they of course depend on such unknowns as whether there is much background gas such as nitrogen. The boundaries and their interpretation also depend on rotation rate (e.g. in the lower left box, the ice deposits form at the poles if the planet rotates non-synchronously, but on the nightside in the case of tidal locking). (Figure by author, based on [593].)

Proxima Cen b will get easier as time goes on – the Alpha Centauri system is in fact moving closer to Earth, and will be only 3 light years away about 30,000 years from now!

New astronomy missions are under development to make further exoplanet detections and characterization. The Transiting Exoplanet Survey Satellite (TESS) launched in April 2018, and is expected to find exoplanets around the nearest and brightest stars – thus the ones from which follow-up observations might characterize atmospheres and temperature distributions. The long-awaited James Webb Space Telescope (JWST) will be a prime tool for exoplanet characterization, its powerful instruments and giant 18-segment 6.5-m mirror going far beyond Spitzer's capabilities.

In the long term it may be possible to observe planetary lightcurves with sufficient fidelity to detect surface features (as was once done for Titan and Pluto). There has even been modeling to assess whether the specular reflection of the Sun on the (relatively) smooth ocean surface might be used to detect seas (and thus water and habitability) on exoplanets [596–599]. Cassini's observation of sunglint on Titan (and indeed its nondetection at low latitudes) often serves as a prototype for this consideration, although it is often forgotten that the idea was invoked in the nineteenth century in studies of Mars.

The study of exoplanets can be seen as the natural extension, a third realm perhaps, of the study of the climate. First, from the Earth, where we have huge, diverse daily datasets (terabits per day) from dozens of satellites, balloon flights and surface stations, together with records of the past from deep sea sediments and ice cores, all digested with elaborate but often empirically tuned supercomputer models. Second, our neighboring planetary climates of Mars, Venus and Titan, where a few isolated *in-situ* investigations serve as ground truth for a bigger picture emerging from meteorological and geological observations from orbit (perhaps gigabits per day) and where the more exotic conditions and materials demand a more fundamental approach to how climate processes are modeled. Exoplanet observations, by contrast, can be described by quantifications of perhaps only megabits per month, but reveal a much wider parameter space of even more extreme environments, challenging our physical understanding of how climates work. It is harder for an individual to span all three domains than it was in past centuries, when the pace was slower, but scientists working in these three domains today can still all learn from each other, and it is hoped that this book may help in this respect.

12

Conclusions

The goal of this work has been to lay out some aspects of the solar system's four principal planetary climates, and in particular to describe how, over several centuries but especially in the past five decades, we have learned about them. The threads of planetary exploration and of climatology, it seems, are closely intertwined. There are some stray loops here and there, for example the divergence of climate study and terrestrial weather prediction in the early twentieth century, but more often than not the two disciplines have closely inspired and informed each other. While global circulation models initially developed for terrestrial use have been widely adapted for planetary applications, many techniques deployed initially in planetary exploration have been employed to advantage in the observation of the Earth – the microwave radiometer being a case in point. Many prominent terrestrial climate scientists have also "cut their teeth" on planetary climate problems – Hansen, Watson, Henderson-Sellers, Lovelock and others.

On Earth, our intimate viewpoint confronts us with bewildering complexity. The basics of greenhouse warming, understood over a century ago, and with anthropogenic climate change becoming evident even in the 1930s, are not in serious scientific dispute. And yet the detailed effects of certain fundamentals such as cloud feedbacks remain poorly understood. It is not obvious to this author that ever-higher-resolution simulations are the answer: perhaps taking a step back and contemplating the other experiments in the solar system's climate laboratory may help. Of course, other worlds lack – as far as we know – the biological systems that introduce many of the confounding complications in our climate evolution, but it is worth recalling that much of the recognition of the role of biota in regulating climate itself emerged in part from contemplation of extraterrestrial biospheres.

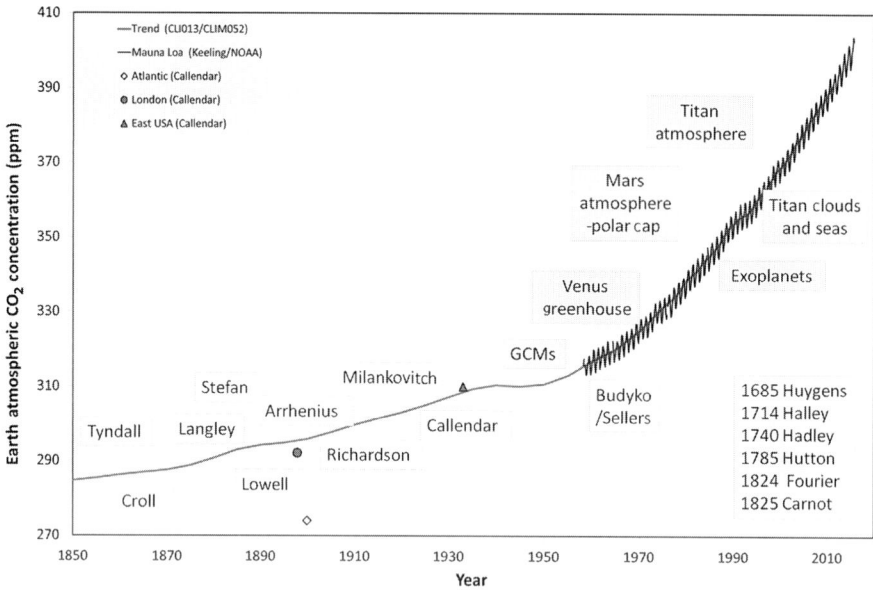

Figure 12.1 Historical landmarks described in this book, set against the measured and reconstructed abundance of carbon dioxide in the Earth's atmosphere.

The Earth's atmospheric evolution over the period covered by this book (Figure 12.1) is striking: while a temperature plot has many distracting bumps and wiggles, notably due to sulfate aerosol injections from large volcanic eruptions, the CO_2 growth in the past centuries is quite relentless. The curve turns notably upwards just at the start of the space age. This is perhaps no coincidence, as the relative prosperity that permits such indulgences as planetary exploration came in part as a result of economic activity associated with energy consumption and cement production.

Under the barrage of space missions sent there in the past decades, many of Mars' most fundamental mysteries have been solved. The holy grail of discovering evidence of past or present life will continue, doubtless, to motivate further exploration, but it seems that the returns from the enterprise of simply landing and looking at rocks may be diminishing – we can only usefully discover so many times that Mars had liquid water. On the other hand, Mars is overdue for systematic long-term network observations of surface meteorology, and its polar regions beckon as sites both of exotic processes in the present epoch, and as an archive of Mars' past. The relationship of the layered polar deposits to orbital forcing seems compelling, but is not yet fully elucidated – as is also true for the Earth, amplifying feedbacks and intrinsic variability distort the history.

Perhaps, as at Earth, close dissection of the polar layers will shed light on at least the most recent Mars climate history.

But it seems doubtful that the question of how warm and wet Mars was, and what greenhouse or other mechanism allowed that to happen, will be conclusively resolved – most likely because there is probably not a single answer. As one recent review put it, the requirements of a solution to the faint young Sun paradox at Mars are demanding, and have yet to be met [600].[1] Modelers will doubtless continue to fiddle with gas mixtures, cloud schemes and various model permutations, but it seems that if one wishes a given result (a habitable Mars, or a frigid, dead one) plausible atmospheric conditions can be invoked and a model constructed to support that result's existence, either way. Only once we have good paleoclimate records, perhaps lake sediment cores, at many different locations will we develop a robust sense of how long and how clement conditions may have been at the planetary scale. That said, the explanatory capability of mesoscale and large-eddy simulations to model the formation and evolution of specific geological structures and atmospheric features has become impressive, and most of the basics of how Mars works are now becoming reasonably well understood.

Venus is a problematic world. Our understanding of the cloud structure can be improved, and perhaps some new probe measurements can refine models of the greenhouse effect under the rather extreme conditions that pertain on that world. But the basics are already understood: a thick, hot greenhouse resulting from and leading to the loss of water. The disastrous divergence of Venus and Earth's climates has become a more poignant question now that so many exoplanets have been found at and beyond the hot end of the habitable zone.

Perhaps when we more closely interrogate Venus' surface globally we can learn more – Venus today is mapped no better than was Mars by Mariner 9 – there is over 40 years and two orders of magnitude of resolution to catch up on. It is clear, however, that our understanding of the near-surface atmosphere is woefully inadequate. Composition measurements – both of the surface, and the near-surface atmosphere – that help inform whether chemical equilibria regulate the Venusian climate are urgently needed, and the planetary boundary layer is essentially unknown. We had better measurements of the temperature structure of the lowest 10 km of our atmosphere from balloons in the nineteenth century than we do of Venus even now. The challenge is of course that the lowest

---

[1] The authors suggest "Any greenhouse theory must (a) produce the warming and rainfall needed, (b) have a plausible source for the gases required, (c) be sustainable, and (d) explain how the atmosphere evolved to its present state." These are challenging requirements and, judging from the literature, they have yet to be met.

part of the atmosphere is the most hostile, and present technology can only operate for a few hours before it is cooked. Progress in these measurements will likely be expensive and slow.

Venus' atmosphere is challenging to model, both from the standpoint of having very limited knowledge of the atmosphere beneath the cloud deck, and from the practical issue that the massive atmosphere has a long memory, taking many Venus days and computational cycles to settle down to a steady state. The wide range of temperatures also means that some atmospheric properties, such as the heat capacity, vary enough to be awkward in some numerical models (a useful summary of the challenges in modeling different planetary atmospheres is Forget and Lebonnois [601]). On the other hand, Venus' lack of strong seasons simplifies matters somewhat.

Titan's distance from the Sun has kept its orange atmosphere in a somewhat primordial state. And its distance from Earth has similarly helped it retain its mysteries. This sweetest fruit in planetary climate is a billion miles away, not exactly low-hanging. But while distant, Titan's environment with a thick atmosphere, albeit cold, is rather benign and inviting for exploration systems of all sorts – balloons, airplanes, landers, even boats.

The fact that Titan's hydrological cycle, with century-long droughts punctuated by torrential downpours, seems like the terrestrial one taken to a greenhouse extreme, suggests that there is much to be learned there, of direct import to understanding at a more fundamental level processes that affect us on Earth. The additional phenomenological richness of Titan's seas promises a new laboratory for air–sea exchange of heat, moisture and momentum, the processes that affect El Niño and hurricanes. On Titan, however, the seas are neither extensive nor deep enough to drive the global climate as powerfully as they do on Earth – Titan's seas are mostly about effects, not causes. This may make things easier to understand. Titan's relatively warm tropopause, allowing methane moisture to leak upwards into stratospheric loss, also serves as an instructive proxy for how the Earth may ultimately lose its oceans, and perhaps how Venus lost its water in the past.

With Cassini's studies of Titan now over, Titan's mapping is less complete even than Venus' is today, and yet we already know of abundant dunes and their relation to climate, of ubiquitous river channels, and that surface changes due to rainfall and sea surface roughness changes can be observed even from flyby measurements. An orbiter would be able to map Titan much better, and to watch storm clouds puff up and dissipate as rain and hail lash the surface beneath. Inspection of the shorelines from a boat or aircraft, or even coring of seafloor sediments by a submersible, might one day reveal layers of evaporites and muds recording climate change. There seems little reason to imagine that, when we can study Titan in anything like the detail afforded to Mars, there will not be many wonderful processes and a rich history to discover.

While Pluto's remarkable conditions have attracted a flurry of modeling effort, there will be little new data for the forseeable future, and scientific interest will likely move on to other worlds. Of these, however, there appears to be no shortage – the thousands of exoplanets now discovered offering a wide canvas on which to paint new models. It is striking how studies of Titan in the 1970s parallel the state of the art of exoplanets, with just a handful of datapoints from spectra and lightcurves.

As a historical narrative, the understanding our own climate began easily enough, with it being straightforward to identify a few key realizations and protagonists. Even in the 1950s and 1960s there were perhaps only dozens of scientists pursuing questions of climate, but by the 1980s and 1990s this had grown to hundreds. An individual scientist publishing a paper is difficult to identify as a landmark in progress when there are hundreds or thousands of papers per year, many of which are mutually contradictory.

It has been easier to sustain a coherent tale in planetary science, where progress is by and large punctuated by the more-or-less discrete events of space-craft encounters, especially in the first couple of decades. But the story of our evolving understanding of our own climate and those of our planetary neighbors is unfinished – science is after all a journey, not a destination. A classic storyline is that of the hopeful quest, and it seems evident where the planets are concerned that hope has been a major factor in shaping scientific opinion, rather more than most scientists might tend to admit. This seems the only explanation for the grudgingly slow decrease in the estimated pressure of the Martian atmosphere, until scientists were finally confronted with Mariner 4's results showing the atmosphere to be a hundred times thinner than ours. Similarly, it is only by the progressive bludgeoning of results from one space mission after another that we admitted that Venus' atmosphere is a hundred times more massive than ours, and the temperatures nothing short of hellish. Even expectations about Titan were first conditioned by notions that it might be somehow habitable; and then after its bracing surface climate was learned, scientists sought surface oceans (albeit hydrocarbon ones) as much for aesthetic reasons as because there was reason to think scientifically that they should exist.

Any story worth telling is one about people, and many climate scientists fit the bill of scientific heroes, having to bravely articulate novel ideas, often to a hostile reception, to bring progress. Callendar's greenhouse, Lovelock's Gaia Hypothesis and Hoffman's Snowball Earth are just examples. The sometimes capricious nature of scientific history and attribution – of who did something first – is also evident in this tale. Publication is the gold standard – it isn't enough for a scientist to have figured something out in their own head. But publication where? Many nineteenth- and twentieth-century discoveries languished in obscure Danish or Irish or Japanese journals – little better in a way than Huygens or da Vinci writing their discoveries in anagrams or mirror-writing centuries before – only to be rediscovered by someone else in a more popular or accessible form. That said, many climate scientists have shown prominent talent for communication – Tyndall, Hansen, Sagan and others come to mind. A surprising number have played roles in setting up or editing journals.[2] The struggle of ideas aside, many climate scientists have been (and are!) jolly interesting and adventurous people, with a striking appetite for climbing mountains, ascending to great heights in balloons or depths in diving bells. Many have sailed the oceans or struck out across deserts or icy wastes in search of clues about climate.

The recent withdrawal of the United States from the Paris Accord on climate change may be greeted by those of a rational disposition with dismay. Yet seen with the broader historical perspective laid out in this book, this will be but a temporary setback. Indeed, political pushback is nothing new – long battles over the wording of the 1990s reports of the IPCC took place, with some countries whose economies depended almost completely on fossil fuel production pushing to minimize the asserted certainty of the causes and effects of climate change [602]. But the truth will out, and mitigation policies will eventually be seen as enlightened self-interest.

As in politics, so it is in the scientific sphere. At the cutting edge, where ingenious theorists scramble to explain observations at the limit of detection (which may be "wrong" anyway), in a myriad of different (often incompatible) ways, the truth is difficult to see through the turbulence of the scientific process. But with the long view, the trade winds of progress bring us reliably ever closer to understanding how climates work.

The study of planetary climates is intimately connected with the issue of habitability, and ultimately the origins of life and the question: Are we alone in the dark of the universe? We may fear a positive answer, but, as we enter the Anthropocene contemplating the stars and planets through an atmosphere thickening with greenhouse gases, it seems we will not be shivering.

---

[2] I was also surprised on researching this book how many scientists seem to be related to other scientists.

Glossary

Albedo – The reflectivity of a surface to optical (solar) light. Although difficult to measure in practice, since observations are usually only made from one direction, the climatologically relevant albedo indicates radiation reflected in all directions (the Bond albedo).

Antigreenhouse effect – The effect of atmospheric opacity at optical (solar) wavelengths, leading to a warm stratosphere and reducing surface temperatures. Relevant for "nuclear winter" or "impact winter" scenarios.

Black body radiation – The electromagnetic radiation (usually heat) from a warm body that is a perfect emitter, proportional to the fourth power of absolute temperature. This is an important reference case in radiative transfer. Typical surfaces have an emissivity less than unity, so are warmer than a black body would be for the same heat loss.

Climate feedback – The property of a climate system to amplify or attenuate the effects of external changes, such as the increase of sunlight. Positive feedbacks include the ice–albedo feedback where polar caps are more reflective. Negative (stabilizing) feedbacks include the hypothetical Daisyworld.

Convective adjustment – A procedure wherein a radiative equilibrium model is corrected for upward energy flow by convection by not allowing the lapse rate to exceed a critical value.

Croll–Milankovitch cycles – Periodic changes in climate caused by variation in astronomical parameters, i.e. orbit and spin, which influence the distribution of sunlight with latitude and season. Responsible for the Earth's ice ages, as well as the Martian polar layered terrain, and the apparent redistribution of methane seas on Titan.

Daisyworld – A conceptual model of climate feedback invoking albedo changes due to growth of colored flowers at different temperatures.

Energy balance model – A relatively simple climate model where a planetary body is divided into a number of zones and the heat coming in and out of each zone is evaluated. Typically a one-dimensional model where the zones are bands of latitude: examples include Budyko–Sellers models of the Earth's climate and the Leighton–Murray model for Mars.

Faint young Sun problem – A fundamental concern in planetary climate, that conventional solar evolution models predict that the solar flux was ~30% lower in the early history of the planets, ~4 billion years ago, but life was still somehow supported on Earth, and Mars appears to have had water on its surface. An enhanced greenhouse effect is usually invoked to explain the paradox.

Gaia Hypothesis – At its core, merely the idea that living things can profoundly influence their planetary environment.

General circulation model (GCM) – A computer model of atmospheric (or sometimes atmosphere and ocean) dynamics. When such a model is also modified to include appropriate heat flows, it may also be called a global climate model.

Greenhouse effect – The effect of atmospheric opacity in the thermal infrared, which impedes the outward radiation of heat from a planet's surface and thereby increases the temperature.

Habitable zone – Originally the region around a star that permits liquid water to exist on a planetary surface. Various elaborations now exist.

Isotopes – Atoms or compounds chemically identical but of different atomic weight. Slightly different rates of processes such as escape or evaporation mean ratios of different isotopes can be used to trace past climate conditions. Notably the D/H ratio is an indicator of loss of water from Venus, and the O-18 to O-16 ratio in ice or seafloor deposits on Earth can indicate past climate conditions.

Lapse rate – The change of temperature with altitude. In the deep (convective) parts of a planetary atmosphere in dry air, the lapse rate usually has a near-constant ("adiabatic") value, determined by the gas properties and local gravity. The presence of a condensible (water vapor, methane etc.) can modify the lapse rate significantly.

Mean Meridional Circulation – The overall pattern of winds projected along lines of longitude. On Earth, this pattern (sometimes called "Hadley circulation", which should be understood to be a large-scale average, rather than a portrayal at any instant) is largely symmetric about the equator, with three cells in each hemisphere (Hadley, Ferrel and Polar). On Titan, due to the long year and slow rotation, and on Mars, due to its thin quick-responding atmosphere, the circulation is asymmetric – from one hemisphere to the other – for much of the year.

Mesoscale model – A numerical model of atmospheric dynamics on a less-than-global scale.

Polar vortex – A circulation pattern, sometimes purely circular, sometimes wavy, around a planet's polar region. This may confine a body of air during polar night, leading to anomalous chemistry (as in Titan's polar hood, and Earth's ozone hole).

Proxy – An indirect climate record, such as flower pollen.

Radiative transfer – The science of heat transport by the transmission and absorption of electromagnetic energy, fundamental to climate as well as to stellar structure. Much of this science is concerned with how much radiation is absorbed at particular wavelengths (lines or bands) and in the relatively clear gaps ("windows") between them.

Radio occultation – An observation where a radio signal between a spacecraft and the ground passes through a planetary atmosphere. The refraction by the atmosphere bends the signal, and this change in path length can be recovered by precise frequency measurements: analysis yields the refractivity profile of the atmosphere, which corresponds roughly to the density variation with height. With some assumptions, this can then be translated into a pressure and temperature profile. It is with this method that the surface conditions on Mars and Titan were first measured, as well as the structure of the atmospheres of the outer planets.

Runaway greenhouse – A positive climate feedback invoking a condensible greenhouse gas (typically water) wherein increased temperatures lead to more greenhouse vapor, and hence higher temperatures, etc., until the condensible reservoir has been evaporated completely.

Solar constant – A flux density measuring mean solar electromagnetic radiation (solar irradiance) per unit area. It is measured on a surface perpendicular to the rays, one astronomical unit (AU) from the Sun (roughly the distance from the Sun to the Earth).

Spectroscopy – The science of measuring different wavelengths of electromagnetic radiation. Most often in climate applications this is of visible or infrared light, dispersed with a prism or diffraction grating but relevant ("photometer", "radiometer" etc.) measurements may be made, especially early in the study of a distant world, by simply filtering radiation. Infrared or microwave measurements may use interferometric or other techniques. Spectroscopy can also be applied to nonelectromagnetic radiation, as in the energy distribution of neutrons, used to measure the hydrogen content of planetary surfaces such as Mars.

Stellar occultation – An observation wherein light from a star is observed through a planetary atmosphere (from a spacecraft or another planetary body). This can reveal the presence and amount of haze and/or gases by their absorption, and also (via refraction – see also Radio occultation) yields the density profile of the atmosphere. This technique yielded estimates of Triton's and Pluto's atmospheres.

Teleconnection – A correlation between climate signals at different locations, which may be the result of a common cause rather than a cause and effect per se. The most famous is the El Niño/Southern Oscillation.

Further Reading

The references to this book draw attention to sources on specific topics. More general directions for further reading are given here. The two essential books on planetary climate are the CCTP volume and Pierrehumbert's text.

S. J. Mackwell, A. A. Simon-Miller, J. W. Harder and M. A. Bullock, (Eds.), *Comparative Climatology of Terrestrial Planets*, University of Arizona Press, 2014. This massive yet affordable volume (700+ pages, $35) is an authoritative academic review of what is known about the climates of Mars, Venus and Titan, and also discusses principles of various dynamical, greenhouse, photochemical and other models applied to different worlds, including the Earth. The various review chapters were stimulated by a workshop ("CCTP", the first of three so far) attempting to bring together terrestrial and planetary climate scientists, as well as those studying exoplanets. (In part, the motivation for the workshop was about enhancing collaboration across NASA endeavors – planetary science, solar physics, Earth science and astrophysics being run by different divisions within the organization. While the same administrative separations may not be as wide in other countries, the intellectual bridge-building exercise is well worthwhile.)

The magisterial yet readable textbook by Raymond Pierrehumbert, *Principles of Planetary Climate*, Cambridge University Press, 2010, explains in engaging detail how to do climate calculations yourself, complete with Python computer code. Pierrehumbert's book serves a formidable banquet of topics in rich depth and may take years to fully digest. For a less heavyweight yet still authoritative summary of the key physics, with crisp graphics and simple equations, I recommend Frederick Taylor, *Elementary Climate Physics*, Oxford University Press, 2005. Both books have exercises and directions for further reading.

A small but useful supplement to these is Andew Ingersoll's *Planetary Climates*, Princeton University Press, 2013. Although this is as much about dynamics as climate per se, it somehow retains an informal tone while confronting such topics as potential vorticity. It also fits in a pocket. Other volumes in the Primers in Climate series include books on paleoclimate and on the roles of the Sun and the oceans.

Planets and Climate

Climate is one of many topics that can be considered at a planetary level, and several books that discuss the planetary sciences in general (including e.g. planetary formation and orbital mechanics) are worth studying. Perhaps one of the best texts is I. de Pater and J. Lissauer, *Planetary Science*, Cambridge University Press, 2015.

As for weather and climate on Earth specifically, there are a host of books available. The two books I've found perennially useful are A. McIlveen, *An Introduction to Weather and Climate*, Oxford University Press, 2010, and J. Peixoto and A. Oort, *Physics of Climate*, American Institute of Physics, 1992.

The Reports of the Intergovernmental Panel on Climate Change (IPCC) are the places to go for the latest on the current evolution of our climate and its impacts, and are free online. A succinct but authoritative review is J. Haughton, *Global Warming: The Complete Briefing*, Cambridge University Press, 1994. Haughton was lead editor of the first three IPCC reports.

Planetary Exploration

There are many books to choose from. A good set of overall accounts is the four-volume series *Robotic Exploration of the Solar System* by P. Ulivi and D. Harland, published by Praxis/Springer.

The best book on Russian exploration specifically is W. T. Huntress and M. Ya. Marov, *Soviet Robots in the Solar System: Mission Technologies and Discoveries*, Praxis/Springer, 2011.

Venus

(In addition to the more comprehensive volumes below, there are a number of books focussed on Venus' geology.)

D. M. Hunten, L. Colin, T. M. Donahue and V. I. Moroz (Eds.), *Venus*, University of Arizona Press, 1983. Although a quarter of a century old, this is still a good summary of the state of the art after the initial Venera and Pioneer Venus missions, with an emphasis on the atmosphere.

D. H. Grinspoon, *Venus Revealed: A New Look Below The Clouds Of Our Mysterious Twin Planet*, Perseus, 1997. This is a nice, popular-level account of the exploration and understanding of Venus, by one of the major players in studies of Venus' climate history, with many illuminating anecdotes.

M. Ya. Marov and D. H. Grinspoon, *The Planet Venus*, Yale University Press, 1998. This is a well-presented and readable book summarizing the results of the twentieth-century missions, translated from Marov's Russian authoritative original. I personally find it easier to find facts and graphs in this book than in the Arizona book above.

S. W. Bougher, et al., *Venus II*, University of Arizona Press,1997. Although more up-to-date, as its name and year of publication would imply, than its predecessor, the tome *Venus II* contains more on geology and perhaps less on climate-related topics, since its production was driven by the Magellan mission, which was focussed on radar mapping of the surface. At 1362 pages (plus a CD-ROM) it is nothing if not formidable.

L. W. Esposito, E. R. Stofan and T. E. Cravens (Eds.), *Exploring Venus as a Terrestrial Planet*, American Geophysical Union (AGU Geophysical Monograph 176), 2007, is an authoritative and up-to-date review of Venus research and would be an excellent place to start if embarking on serious Venus research.

L. Bengtsson et al. (Eds.), *Towards Understanding the Climate of Venus*, ISSI Scientific Report 11, Springer, 2013, is a more recent, rather compact volume with authoritative chapters on circulation and radiation modeling, but rather shallow treatment of other areas.

F. W. Taylor, *The Scientific Exploration of Venus*, Cambridge University Press, 2014, is a good overview and is up-to-date with Venus Express findings. The anecdotes about various proposed missions and their nonselection are also illuminating.

There is a *Venus III* book presently in the making, to more broadly cover Venus Express findings.

Mars

There are many books on Mars, and especially about the history of its exploration – just a few of the best are listed here.

P. Read and S. Lewis, *The Martian Climate Revisited*, Springer, 2004, is an excellent treatment of the dynamics of the Martian atmosphere. It is less comprehensive on studies of past climate, and has an odd title, since it is not obvious to me that the Martian climate was visited a first time in book form, but this had been until now the best single volume on its topic.

The new "one-stop-shop" is the massive (570 two-column pages) volume: R. M. Haberle, R. T. Clandy, F. Forget, M. D. Smith and R. W. Zurek, *The Atmosphere and Climate of Mars*, Cambridge University Press, 2017. This captures the now more than two decades of continuous Mars observation by spacecraft, and the tremendous amount of modeling work that has been going on.

H. H. Kieffer, B. M. Jakosky, C. W. Snyder and M. S. Matthews (Eds.) *Mars*, University of Arizona Press, 1993. This, if you can get hold of it, is still one of the most comprehensive volumes about the red planet, consolidating most of the knowledge available post-Viking.

Michael Carr, *The Surface of Mars*, Cambridge, 2007, is a good primer on Martian geology, with a specific focus on the role of water (his earlier book, *Water on Mars*, is good too, but out of print)

F. Forget, F. Costard and P. Lognonne, *La planete Mars: Histoire d'un autre monde*, Editions Belin, Paris, 2003, is a delightful and well-illustrated introduction to Mars (albeit in French).

Titan

A readable (I am told) account of what we knew about Titan before the Cassini mission is given by R. Lorenz and J. Mitton, *Lifting Titan's Veil*, Cambridge University Press, 2002. The story is continued, capturing with first-hand anecdotes of many of Cassini's first few years of discoveries, in R. Lorenz and J. Mitton, *Titan Unveiled*, Princeton University Press, 2008. A slightly higher-level book is *Titan: Exploring an Earth-Like World* by A. Coustenis and F. Taylor, World Scientific, 2008 (do not confuse this with the earlier book, now superseded, by the same authors and publisher, *Titan: The Earth-Like Moon*, 1999).

An authoritative academic summary of the findings at Titan of Cassini's nominal mission (the first four years), written by dozens of scientists on the mission, is R. Brown, J.-P. Lebreton and H. Waite, *Titan from Cassini-Huygens*, Springer, 2009.

The most recent book, edited by I. Mueller-Wodarg, C. Griffith and E. Lellouch et al., is *Titan: Interior, Surface, Atmosphere, and Space Environment*, Cambridge University Press, 2014, and is similarly authoritative. It captures more recent Cassini data, and explains what is going on rather well: its comparisons of different climate/circulation models for Titan are very useful.

Pluto

The books about Pluto are being rewritten. The Pre-New-Horizons view is in S. Stern and D. Tholen (Eds.), *Pluto and Charon*, University of Arizona Press, 1997.

Exoplanets

This is a fast-moving field that I confess I have not followed closely. A couple of starting points are the books by Sara Seager: S. Seager (Ed.), *Exoplanets*, University of Arizona Press, 2010. Like CCTP, this Arizona Press volume has heft, with chapters contributed by various authors. It is also affordably priced.

S. Seager, *Exoplanet Atmospheres: Physical Processes*, Princeton University Press, 2010, is an introductory text on the relevant physics (including atmospheric escape). It serves as a more recent replacement for the classic J. Chamberlain and D. Hunten, *Theory of Planetary Atmospheres*, Academic Press 1989.

Alternative Planets

Extrasolar planetary climate has been considered by science fiction writers far longer than there has been evidence for astronomers that such planets might exist. Guides for science fiction authors on making plausible worlds are thus interesting and relevant reading, one example being *Worldbuilding* by Steve Gillett, 1996.

Similarly, planetary climate in the context of how it might be modified has been the subject of fiction and some speculative papers over the years. Most work on that topic is found in *Journal of the British Interplanetary Society*. But a comprehensive text on the subject is *Terraforming* by Martyn Fogg (SAE Press, 1996). A more recent and succinct (but less detailed) work is Martin Beech's, *Terraforming: The Making of Habitable Worlds*, Springer, 2009.

Bibliography

[1] M. Beech (2010). Atmospheric height by twilight's glow. *Journal of the Royal Astronomical Society of Canada*, 104, 147–148.

[2] C. Powell (2014). Did Cosmos pick the wrong hero. *Discover Magazine* blog.

[3] G. Tierie (1932). Cornelis Drebbel (1572–1633). Ph.D. Thesis, University of Leiden (available online).

[4] C. Huygens, *The Celestial Worlds Discover'd: Or, Conjectures Concerning the Inhabitants, Plants and Productions of the Worlds in the Planets*, London: Timothy Childe, 1698.

[5] J. Wilkins, *A Discovery of a New World, or A Discourse tending to prove, that 'tis Probable there may be another Habitable World in the Moon, with a discourse on the probability of a passage thither*, London: J. Gillibrand, 1684.

[6] A. MacDonald, *The Long Space Age*, New Haven, CT: Yale University Press, 2017.

[7] A. Sobester, *Stratospheric Flight: Aeronautics at the Limit*, Springer, 2011.

[8] E. Halley (1693). A discourse concerning the proportional heat of the Sun in all latitudes, with the method of collecting the same, as it was read before the Royal Society in one of their Late Meetings. *Philosphical Transactions of the Royal Society*, 17, 878–885.

[9] E. Halley (1686). An estimate of the quantity of vapour raised out of the sea by the warmth of the Sun; derived from an experiment shown before the Royal Society, at one of Their Late Meetings. *Philosphical Transactions of the Royal Society, 1686–1692*, 16, 366–370, doi:10.1098/rstl.1686.0067

[10] R.D. Lorenz (2014). The flushing of Ligeia: composition variations across Titan's seas in a simple hydrological model. *Geophysical Research Letters*, 41, 5764–5770, doi:10.1002/2014GL061133.

[11] E. Halley (1714). A short account of the cause of the saltness of the ocean, and of the several lakes that emit no rivers; with a proposal, by help thereof, to discover the age of the World. *Philosphical Transactions of the Royal Society*, 29, 296–300, doi:10.1098/rstl.1714.0031.

[12] E. Halley (1686). An historical account of the trade winds, and monsoons, observable in the seas between and near the tropicks, with an attempt to assign the phisical cause of the said winds. *Philosphical Transactions of the Royal Society*, 16(179–191), 153–168.

[13] G. Hadley (1735). VI. Concerning the cause of the general trade-winds. *Philosphical Transactions of the Royal Society*, 39(437), 58–62.

[14] A.O. Persson (2006). Hadley's principle: understanding and misunderstanding the trade winds. *History of Meteorology*, 3, 17–42.

[15] G.B. Dalrymple (1994). *The Age of the Earth*, Stanford, CA: Stanford University Press.

[16] B. Franklin (1755). Letter to Peter Collinson, dated Philadelphia, August 25, 1755, quoted in R.D. Lorenz, M.R. Balme, Z. Gu, et al. (2016), History and Applications of Dust Devil Studies. *Space Science Reviews*, 203, 5–37.

[17] T. Jefferson (1787). *Notes on the State of Virginia*. London: John Stockdale.

[18] J.R. Fleming, *Historical Perspectives on Climate Change*, New York: Oxford University Press, 1998.

[19] K. Thompson (1980). Forests and climate change in America: some early views. *Climatic Change*, 3, 47–64.

[20] W. Herschel (1801). Observations tending to investigate the nature of the Sun, in order to find the causes or symptoms of its variable emission of light and heat; with remarks on the use that may possibly be drawn from solar observations. *Philosophical Transactions of the Royal Society of London*, 91, 265–318.

[21] D.V. Hoyt & K.H. Schatten (1997). *The Role of the Sun in Climate Change*, Oxford: Oxford University Press.

[22] A. Wulf (2015). *The Invention of Nature: The Adventures of Alexander von Humboldt, the Lost Hero of Science*. London: John Murray.

[23] R. Hamblyn (2011). *The Invention of Clouds: How an Amateur Meteorologist Forged the Language of the Skies*, London: Pan Macmillan.

[24] E. Burgess (1837). General remarks on the temperature of the terrestrial globe and the planetary spaces by Baron Fourier. *American Journal of Science and Arts*, 32, 1–20.

[25] J.R. Fleming (1999). Joseph Fourier, the 'greenhouse effect', and the quest for a universal theory of terrestrial temperatures. *Endeavour*, 23, 72–75.

[26] E. Bard (2004). Greenhouse effect and ice ages: historical perspective. *Comptes Rendus Geoscience*, 336, 603–638.

[27] C.P. Smyth (1856). Note on the constancy of solar radiation. *Monthly Notices of the Royal Astronomical Society*, 16, 220.

[28] C. Piazzi-Smyth (1858). *Tenerife, An Astronomer's Experiment: or, Specialities of a Residence above the Clouds*, London: Lovell Reeve.

[29] P.F.H. Baddeley (1860). *Whirlwinds and Dust Storms of India*, London: Bell.

[30] R.D. Lorenz, M.R. Balme, Z. Gu, et al. (2016). History and Applications of Dust Devil Studies. *Space Science Reviews*, 203, 5–37.

[31] M. Monmonier (1999). *Air Apparent: How Meteorologists Learned to Map, Predict and Dramatize Weather*, Chicago: University of Chicago Press,

[32] P. Moore (2015). *The Weather Experiment*, New York: Farrar.

[33] R. Holmes (2013). *Falling Upwards: How We Took to the Air*, New York: Pantheon.

[34] J. Lequex (2009). Early infrared astronomy. *Journal of Astronomical History and Heritage*, 12(2), 125–140.

[35] J. Tyndall (1861). The Bakerian Lecture: On the absorption and radiation of heat by gases and vapours, and on the physical connexion of radiation, absorption, and conduction. *Philosophical Transactions of the Royal Society of London*, 151, 1–36.

[36] Eunice Foote (1856). Article XXL – Circumstances affecting the Heat of the Sun's Rays, read before the American Association, August 23, 1856. *The American Journal of Science and Arts*, XXII, 382–383.

[37] J.J. Ebelmen (1847). *Recherches sur la décomposition des roches*, Paris: Carilian-Goeury et Vor. Dalmont.

[38] R. Berner (2012). Jacques-Joseph Ébelmen, the founder of earth system science. *Comptes Rendus Geoscience*, 344(11), 544–548.

[39] M.E. Galvez & J. Gaillardet (2012). Historical constraints on the origins of the carbon cycle concept. *Comptes Rendus Geoscience*, 344(11), 549–567.

[40] J. Murray et al. (1885). *Report on the Scientific Results of the Voyage of H.M.S. Challenger during the years 1873–1876*, London: Her Majesty's Stationery Office.

[41] D. Roemmich et al. (2012). 135 years of global ocean warming between the Challenger Expedition and the Argo Programme. *Nature Climate Change*, 2, 425–428.

[42] W. Leitch (1867). *God's Glory in the Heavens*, London: Alexander Strahan.

[43] H.D. Taylor (1895). Mars, A negative optical proof of the absence of seas in Mars. *Monthly Notices of the Royal Astronomical Society*, 55, 462–474.

[44] J.H. Poynting (1904). Radiation in the Solar System: its effect on temperature and its pressure on small bodies. *Philosophical Transactions of the Royal Society of London, Series A*, 202, 525–552.

[45] Earl of Rosse (1871). On the Radiation of Heat from the Moon. No. II. *Proceedings of the Royal Society of London*, 19, 9–14.

[46] S.P. Langley (1886). The temperature of the Moon. *Science*, 7(158) 8–9.

[47] S.P. Langley (1889). *The Temperature of the Moon, Third Memoir*, National Academy of Sciences Vol. 4 part 2.

[48] S.P. Langley (1884). *Researches on Solar Heat and Its Absorption by the Earth's Atmosphere: A Report of the Mount Whitney Expedition*. US Government Printing Office.

[49] S. Arrhenius (1896). On the influence of carbonic acid in the air upon the temperature of the ground. *The London, Edinburgh, and Dublin Philosophical Magazine and Journal of Science*, 41(251), 237–276.

[50] T. Mellard Reade (1879). *Chemical Denudation in Relation to Geological Time*, London: D. Bogue.

[51] T.C. Chamberlin (1898). The influence of great epochs of limestone formation upon the constitution of the atmosphere. *The Journal of Geology*, 6(6), 609–621.

[52] J. R. Fleming (2000). TC Chamberlin, climate change, and cosmogony. *Studies in History and Philosophy of Science Part B: Studies in History and Philosophy of Modern Physics*, 31(3), 293–308.

[53] T.C. Chamberlin (1897). A Group of Hypotheses Bearing on Climatic Changes. *Journal of Geology*, 5, 653–683.

[54] T.C. Chamberlin (1906). On a possible reversal of deep-sea circulation and its influence on geological climates. *Journal of Geology*, 14, 363–373.

[55] N. Ekholm (1901). On the variations of the climate of the geological and historical past and their causes. *Quarterly Journal of the Royal Meteorological Society*, 27, 11–62.

[56] G.T. Walker (1923). Correlation in seasonal variations of weather. VIII. A preliminary study of world-weather. *Memoirs of the Indian Meteorological Department*, 24(Part 4), 75–131.

[57] R.W. Katz (2002). Sir Gilbert Walker and a connection between El Niño and statistics. *Statistical Science*, 17, 97–112.

[58] G.J. Stoney (1898). Of atmospheres upon planets and satellites. *The Astrophysical Journal*, 7, 25.

[59] K.P. Hoinka (1997). The tropopause: discovery, definition and demarcation. *Meteorologische Zeitschrift, N.F.*, 6, 281–303.

[60] W. Dines (1908). The Registering Balloon Ascents in England of July 22–27, 1907, Preliminary Account. *Quarterly Journal of the Royal Meteorological Society*, 34, 1–5.

[61] A.L. Rotch (1897). On obtaining meterological records in the upper air by means of kites and balloons. *Proceedings of the American Academy of Arts and Sciences*, 32, 245–251.

[62] R.H. Goddard (1919). *A Method of Reaching Extreme Altitudes*, Washington, DC: Smithsonian Institution.

[63] W. Sheehan (1996). The Planet Mars, A History of Observation and Discovery, Tucson, AZ: University of Arizona Press.

[64] R.D. Lorenz (1997). Did Comas Sola discover Titan's atmosphere? *Astronomy and Geophysics*, 38(3) 16–18.

[65] P. Lowell (1907). A general method for evaluating the surface-temperature of the planets: with special reference to the temperature of Mars. *The London, Edinburgh, and Dublin Philosophical Magazine and Journal of Science*, 14(79), 161–176.

[66] J. Poynting (1907). On Prof. Lowell's method for evaluating the surface-temperatures of the planets; with an attempt to represent the effect of day and night on the temperature of the Earth. *The London, Edinburgh, and Dublin Philosophical Magazine and Journal of Science*, 14(84), 749–760.

[67] A.R. Wallace (1907). Is Mars Habitable, London: Macmillan.

[68] S. Arrhenius (1918). *The Destinies of Stars*, London: Putnams (translated by J. Fries from the 1915 Swedish publication).

[69] E.W. Maunder (1913). *Are the Planets Inhabited?* London: Harper and Brothers.

[70] C. Pekeris (1933). *The Development and Present Status of the Theory of the Heat Balance in the Atmosphere*. D.Sc. Thesis, Massachusetts Institute of Technology.

[71] R. Goody & Y. Yung (1961). *Atmospheric Radiation: Theoretical Basis*, Oxford: Oxford University Press.

[72] G. Simpson (1927). Some studies in terrestrial radiation. *Memoirs of the Royal Meteorological Society*, 2(16) 69–95.

[73] G. Simpson (1928). Further studies in terrestrial radiation. *Memoirs of the Royal Meteorological Society*, 3(21) 1–26.

[74] G. Simpson (1929). The distribution of terrestrial radiation. *Memoirs of the Royal Meteorological Society*, 3(23), 53–78.

[75] C. Abbe (1901). The physical basis of long-range weather forecasts. *Monthly Weather Review*, 29, 551–561.

[76] Milutin Milanković (1879–1958). From his autobiography with comments by his son, Vasko, and a preface by Andre Berger, European Geophysical Society, Kaltenburg-Lindau, 1995.

[77] M. Milankovitch (1941). *Canon of Insolation and the Ice-Age Problem*, Belgrade: Royal Serbian Academy.

[78] J. Imbrie & K. Imbrie, *Ice Ages, Solving the Mystery*, Cambridge, MA: Harvard University Press, 1979.

[79] L.F. Richardson, *Weather Prediction by Numerical Process*, Cambridge: Cambridge University Press, 1922.

[80] E. Pettit & S. Nicholson (1924). Radiation Measures on the Planet Mars. *Publications of the Astronomical Society of the Pacific*, 36, 269–272.

[81] W. Sinton (1986). Through the infrared with logbook and lantern slides, a history of infrared astronomy from 1868 to 1960. *Publications of the Astronomical Society of the Pacific*, 98, 246–251.

[82] D. Harland (2005). *Water and the Search for Life on Mars*, Springer.

[83] J. Fleming (2016). *Inventing Atmospheric Science: Bjerknes, Rossby, Wexler, and the Foundations of Modern Meteorology*, Cambridge, MA: MIT Press.

[84] W.D. Flower (1936). *Sand Devils.* Meteorological Office Professional Notes, 5, 1–16. London: His Majesty's Stationery Office.

[85] R. Lorenz (2013). The longevity and aspect ratio of dust devils: effects on detection efficiencies and comparison of landed and orbital imaging at Mars. *Icarus*, 226(1), 964–970.

[86] R. Courant, K. Friedrichs & H. Lewy (1928). Über die partiellen Differenzengleichungen der mathematischen Physik. *Mathematische Annalen* (in German), 100(1), 32–74.

[87] J. Harlen Bretz (1923). The channeled scabland of the Columbia Plateau. *Journal of Geology*, 31, 617–649.

[88] J. Soennichsen (2008). *Bretz's Flood: The Remarkable Story of a Rebel Geologist and the World's Greatest Flood*, Seattle, WA: Sasquatch Books.

[89] V.R. Baker & D. Nummedal (1978). The channeled scabland: a guide to the geomorphology of the Columbia Basin, Washington. Prepared for the Comparative Planetary Geology Field Conference held in the Columbia Basin, June 5–8, 1978, sponsored by Planetary Geology Program, NASA Office of Space Science.

[90] W. Köppen & A. Wegener (1924). *Die Klimate der Geologischen Vorzeit*, Berlin: Borntraeger. An English translation (*Climates of the Geological Past*) was published in 2015.

[91] V. Vernadsky (1998). *The Biosphere*, Annotated Edition, Springer.

[92] G.S. Callendar (1938). The artificial production of carbon dioxide and its influence on temperature. *Quarterly Journal of the Royal Meteorological Society*, 64(275), 223–240.

[93] J.R. Fleming (2007). The Callendar Effect, The Life and Work of Guy Stewart Callendar (1898–1964), Boston, MA: American Meteorological Society.

[94] E. Hawkins & P. Jones (2013). On Global Temperatures: 75 years after Callendar. *Quarterly Journal of the Royal Meteorological Society*, 139, 1961–1963.

[95] W.S. Adams & T. Dunham (1932). Absorption bands in the infra-red spectrum of Venus. *Publications of the Astronomical Society of the Pacific*, 44, 243–245.

[96] R. Wildt (1940). Note on the surface temperature of Venus. *The Astrophysical Journal*, 91, 266–268.

[97] H. Spencer Jones (1940). *Life on Other Worlds*, London: English Universities Press.

[98] A. Adel & V.M. Slipher (1934). Concerning the carbon dioxide content of the atmosphere of the planet Venus. *Physical Review*, 46(3), 240.

[99] A. Adel & V.M. Slipher (1934). The constitution of the atmospheres of the giant planets. *Physical Review*, 46, 902–906.

[100] M. Livio (2017). Winston Churchill's essay on alien life found. *Nature*, 542, 289–291.

[101] K. Harper, *Weather by the Numbers: The Genesis of Modern Meteorology*, Boston, MA: MIT Press, 2012.

[102] J.M. Lewis (2003). Ooishi's observation viewed in the context of jet stream discovery. *Bulletin of the American Meteorological Society*, 84, 357–369.

[103] R. Coen (2014). *Fu-go: The Curious History of Japan's Balloon Bomb Attack on America*, Lincoln, NE: University of Nebraska Press.

[104] R.E. Drapeau (2011). Operation Outward: Britain's World War II offensive balloons. *IEEE Power & Energy Magazine*, 9, 94–105.

[105] H.U. Sverdrup & W.H. Munk (1947). *Wind, Sea and Swell: Theory of Relations for Forecasting*, US Navy Hydrographic Office Publication No. 601, 44 pp.

[106] R.D. Lorenz & A.G. Hayes (2012). The growth of wind-waves in Titan's hydrocarbon seas. *Icarus*, 219(1), 468–475.

[107] R.D. Lorenz, J.M. Dooley, J.D. West & M. Fujii (2003). Backyard spectroscopy and photometry of Titan, Uranus and Neptune. *Planetary and Space Science*, 51(2), 113–125.

[108] H. Keiffer et al. The planet Mars: from antiquity to the present, in H. Keiffer et al. (eds.), *Mars*, Tucson, AZ: University of Arizona Press, 1986.

[109] D.M. Harland, *Water and the Search for Life on Mars*, Springer, 2005.

[110] O. Struve (1952). Proposal for a project of high-precision stellar radial velocity work. *The Observatory*, 72, 199–200.

[111] C.H. Mayer, T.P. McCullough & R.M. Sloanaker (1958). Observations of Venus at 3.15-cm wave length. *The Astrophysical Journal*, 127, 1.

[112] W.M. Sinton & J. Strong (1960). Radiometric observations of Venus. *The Astrophysical Journal*, 131, 470.

[113] E. Pettit & S.B. Nicholson (1955). Temperatures on the bright and dark sides of Venus. *Publications of the Astronomical Society of the Pacific*, 67, 293–303.

[114] K.L. Franklin (1959). An account of the discovery of Jupiter as a radio source. *The Astronomical Journal*, 64, 37–39.

[115] P. Lynch (2008). The origins of computer weather prediction and climate modeling. *Journal of Computational Physics*, 227, 3431–3444.

[116] J.G. Charney, R. Fjörtoft & J. Von Neumann (1950). Numerical integration of the barotropic vorticity equation. *Tellus A*, 2, 237–254.

[117] N.A. Phillips (1956). The general circulation of the atmosphere: a numerical experiment. *Quarterly Journal of the Royal Meteorological Society*, 82(352), 123–164.

[118] D. Fultz (1949). A preliminary report on experiments with thermally produced lateral mixing in a rotating hemispherical shell of liquid. *Journal of Meteorology*, 6(1), 17–33.

[119] H. Riehl & D. Fultz (1957). Jet stream and long waves in a steady rotating-dishpan experiment: structure of the circulation. *Quarterly Journal of the Royal Meteorological Society*, 83(356), 215–231.

[120] M. Ghil, P. Read & L. Smith (2010). Geophysical flows as dynamical systems: the influence of Hide's experiments. *Astronomy & Geophysics*, 51(4), 4–28.

[121] H. Stommel (1961). Thermohaline circulation with two stable regimes of flow. *Tellus*, 13, 224–241.

[122] C.C. Langway, *The History of Early Polar Ice Cores*, US Army Cold Regions Research and Engineering Laboratory, ERDC/CRREL TR-08–1, January 2008.

[123] C. Sterken (2011). Ernst Julius Opik: Solar variability and climate change. *Baltic Astronomy*, 20 195–203,

[124] E. Opik (1952). Ice ages. *Irish Astronomical Journal*, 2, 71–84.

[125] E. Opik (1965). Climatic change in cosmic perspective. *Icarus*, 4, 289–307.

[126] H. Wexler (1956). Variations in insolation, general circulation and climate. *Tellus A*, 8, 480–494.

[127] H. Shapley, *Climatic Change, Evidence, Causes and Effects*, Cambridge, MA: Harvard University Press, 1953.

[128] G.N. Plass (1956). The influence of the 15μ carbon-dioxide band on the atmospheric infra-red cooling rate. *Quarterly Journal of the Royal Meteorological Society*, 82(353), 310–324.

[129] G.N. Plass (1956). Effect of carbon dioxide variations on climate. *American Journal of Physics*, 24(5), 376–387.

[130] S. Weart, *The Discovery of Global Warming*, Cambridge, MA: Harvard University Press, 2008.

[131] R. Revelle & H.E. Suess (1957). Carbon dioxide exchange between atmosphere and ocean, and the question of an increase of atmospheric CO_2 during the past decades. *Tellus*, 9, 18–27.

[132] C.D. Keeling (1960). The concentration and isotopic abundances of carbon dioxide in the atmosphere. *Tellus*, 12(2), 200–203.

[133] H. Strughold, *The Green and Red Planet*, Albuquerque, NM: University of New Mexico Press, 1953.

[134] C. Cockell (2001). 'Astrobiology' and the ethics of new science. *Interdisciplinary Science Reviews*, 26, 90–96.

[135] H. Spinrad, G. Münch & L.D. Kaplan (1963). Letter to the Editor: The detection of water vapor on Mars. *The Astrophysical Journal*, 137, 1319.

[136] L.D. Kaplan, G. Münch & H. Spinrad (1964). An analysis of the spectrum of Mars. *The Astrophysical Journal*, 139, 1–15.

[137] P. Latil & T. Mar (1959). Planetary observations by the multi-balloon technique. *New Scientist*, 7 May, 1005–1007.

[138] A. Dollfus, Observations of water vapor on Mars and Venus. In *The Origin and Evolution of Atmospheres and Oceans*, Proceedings of a Conference, held at the Goddard Institute for Space Studies, NASA, New York, April 8–9, 1963. Edited by P.J. Brancazio & A.G.W. Cameron, New York: Wiley, 1964, p. 257.

[139] F.B. House, A. Gruber, G.E. Hunt & A.T. Mecherikunnel (1986). History of satellite missions and measurements of the Earth radiation budget (1957–1984). *Reviews of Geophysics*, 24(2), 357–377.

[140] H. Wexler (1954). Observing the weather from a satellite vehicle. *Journal of the British Interplanetary Society*, 13, 269–276.

[141] C. Sagan, *The Radiation Balance of Venus*. JPL Technical Report 32–34, September 1960.

[142] A. Kuzmin & B. Clark (1965). The polarization measurement and the brightness temperature of Venus at 10.6 cm wavelength. *The Astronomical Journal*, 43, 595–617.

[143] E.J. Öpik (1961). The aeolosphere and atmosphere of Venus. *Journal of Geophysical Research*, 66(9), 2807–2819.

[144] J.E. Hansen & S. Matsushima (1967). The atmosphere and surface temperature of Venus: a dust insulation model. *The Astrophysical Journal*, 150, 1139.

[145] R.E. Samuelson (1967). Greenhouse effect in semi-infinite scattering atmospheres: application to Venus. *The Astrophysical Journal*, 147, 782.

[146] R.E. Danielson, J.J. Caldwell & D.R. Larach (1973). An inversion in the atmosphere of Titan. *Icarus*, 20(4), 437–443.

[147] B. Zellner (1973). The polarization of Titan. *Icarus*, 18(4), 661–664.

[148] D. Hunten (Ed.), *The Atmosphere of Titan: A Workshop held at Ames Research Center, July 1973*, NASA SP-340.

[149] A. Kliore, D.L. Cain, G.S. Levy, V.R. Eshleman, G. Fjeldbo & F.D. Drake (1965). Occultation experiment: results of the first direct measurement of Mars's atmosphere and ionosphere. *Science*, 149(3689), 1243–1248.

[150] G. Fjeldbo, W.C. Fjeldbo & V.R. Eshleman (1966). Models for the atmosphere of Mars based on the Mariner 4 occultation experiment. *Journal of Geophysical Research*, 71(9), 2307–2316.

[151] R.B. Leighton & B.C. Murray (1966). Behavior of carbon dioxide and other volatiles on Mars. *Science*, 153(3732), 136–144.

[152] S. Manabe & R.F. Strickler (1964). Thermal equilibrium of the atmosphere with a convective adjustment. *Journal of the Atmospheric Sciences*, 21(4), 361–385.

[153] *Restoring the Quality of Our Environment*. Report of the Environment Pollution Panel, President's Science Advisory Committee, The White House, November 1965.

[154] F. Möller (1963). On the influence of changes in the CO_2 concentration in air on the radiation balance of the Earth's surface and on the climate. *Journal of Geophysical Research*, 68(13), 3877–3886.

[155] S. Manabe & R.T. Wetherald (1967). Thermal equilibrium of the atmosphere with a given distribution of relative humidity. *Journal of Atmospheric Sciences*, 24, 241–259.

[156] E.N. Lorenz (1963). Deterministic nonperiodic flow. *Journal of Atmospheric Sciences*, 20(2), 130–148.

[157] J. Gleick, *Chaos: Making a New Science*, New York: Viking, 1987.

[158] S. Manabe & K. Bryan (1969). Climate calculations with a combined ocean-atmosphere model. *Journal of Atmospheric Science*, 26, 786–789.

[159] C. Leovy & Y. Mintz (1969). Numerical simulation of the atmospheric circulation and climate of Mars. *Journal of the Atmospheric Sciences*, 26(6), 1167–1190.

[160] E. Opik (1965). Climatic change in cosmic perspective. *Icarus*, 4, 289–307.

[161] E. Eriksson, Air-ocean-icecap interactions in relation to climatic fluctuations and glaciation cycles. In J.M. Mitchell (Ed.), *Causes of Climatic Change*. Meteorological Monographs, 8(30), Boston, MA: American Meteorological Society, 1968, pp. 68–92.

[162] S. Fritz, The heating distribution in the atmosphere and climatic change. In R. Pfeffer (Ed.), *Dynamics of Climate: The Proceedings of a Conference on the Application of Numerical*

Integration Techniques to the Problem of the General Circulation, October 26–28, 1955, New York: Pergamon Press, 1955, pp. 96–102.

[163] W.E. Cobb & H.J. Wells (1970). The electrical conductivity of oceanic air and its correlation to global atmospheric pollution. *Journal of the Atmospheric Sciences*, 27(5), 814–819.

[164] S.I. Rasool & S.H. Schneider (1971). Atmospheric carbon dioxide and aerosols: effects of large increases on global climate. *Science*, 173(3992), 138–141.

[165] T.C. Peterson, W.M Connolley & J. Fleck (2008). The myth of the 1970s global cooling scientific consensus. *Bulletin of the American Meteorological Society*, 89(9), 1325–1337.

[166] C. Goldblatt & A.J. Watson (2012). The runaway greenhouse: implications for future climate change, geoengineering and planetary atmospheres. *Philosophical Transactions of the Royal Society of London A: Mathematical, Physical and Engineering Sciences*, 370(1974), 4197–4216.

[167] M. Komabayashi (1967). Discrete equilibrium temperatures of a hypothetical planet with the atmosphere and the hydrosphere of a one component–two phase system under constant solar radiation. *Journal of the Meteorological Society of Japan*, 45, 137–139.

[168] A.P. Ingersoll (1969). The runaway greenhouse: a history of water on Venus. *Journal of the Atmospheric Sciences*, 26(6), 1191–1198.

[169] S.I. Rasool & C. de Bergh (1970). The runaway greenhouse and the accumulation of CO_2 in the Venus atmosphere. *Nature*, 226, 1037–1039.

[170] E. Lorenz (1970). Climatic change as a mathematical problem, *Journal of Applied Meteorology*, 9, 325–329.

[171] A. Feagre (1972). An intransitive model of the Earth–atmosphere–ocean system, *Journal of Applied Meteorology*, 11, 4–6.

[172] J.B. Pollack (1969). A nongray CO_2-H_2O greenhouse model of Venus. *Icarus*, 10(2), 314–341.

[173] D. Morrison, D.P. Cruikshank & R.E. Murphy (1972). Temperatures of Titan and the Galilean satellites at 20 microns. *The Astrophysical Journal*, 173, L143.

[174] F.C. Gillett, W.J. Forrest & K.M. Merrill (1973). 8–13 micron observations of Titan. *The Astrophysical Journal*, 184, L93.

[175] F.C. Gillett (1975). Further observations of the 8–13 micron spectrum of Titan. *The Astrophysical Journal*, 201, L41–L43.

[176] C. Sagan (1973). The greenhouse of Titan. *Icarus*, 18(4), 649–656.

[177] F.J. Low & G.H. Rieke (1974). Infrared photometry of Titan. *The Astrophysical Journal*, 190, L143.

[178] B.C. Murray, W.R. Ward & S.C. Yeung (1973). Periodic insolation variations on Mars. *Science*, 180(4086), 638–640.

[179] W.R. Ward (1974). Climatic variations on Mars: 1. Astronomical theory of insolation. *Journal of Geophysical Research*, 79(24), 3375–3386.

[180] S. Byrne (2009) The polar deposits of Mars. *Annual Reviews of Earth and Planetary Science*, 37, 8.1–8.26.

[181] C. Sagan and G. Mullen (1972). Earth and Mars: evolution of atmospheres and surface temperatures. *Science*, 177(4043), 52–56.

[182] A. Henderson-Sellers & A.J. Meadows (1979). A simplified model for deriving planetary surface temperatures as a function of atmospheric chemical composition. *Planetary and Space Science*, 27(8), 1095–1099.

[183] A. Henderson-Sellers & A.J. Meadows (1976). The evolution of the surface temperature of Mars. *Planetary and Space Science*, 24(1), 41–44.

[184] R.E. Dickinson & R.J. Cicerone (1986). Future global warming from atmospheric trace gases. *Nature*, 319, 109–115.

[185] J.E. Lovelock (1965). A physical basis for life detection experiments. *Nature*, 207(997), 568–570.

[186] D.R. Hitchcock & J.E. Lovelock (1967). Life detection by atmospheric analysis. *Icarus*, 7(1), 149–159.

[187] J.E. Lovelock (1972). Gaia as seen through the atmosphere. *Atmospheric Environment*, 6(8), 579–580.

[188] J.E. Lovelock & L. Margulis (1974). Atmospheric homeostasis by and for the biosphere: the Gaia hypothesis. *Tellus: Series A* (Stockholm: International Meteorological Institute), 26(1–2), 2–10.

[189] L. Margulis & J.E. Lovelock (1974). Biological modulation of the Earth's atmosphere. *Icarus*, 21(4), 471–489.

[190] D. Schwartzman, *Life, Temperature and The Earth*, New York: Columbia University Press, 1999.

[191] J.E. Lovelock, *The Ages of Gaia*, New York: Norton, 1988.

[192] J.E. Lovelock, *Homage to Gaia*, Oxford: Oxford University Press, 2001.

[193] J.E. Hansen & J. Hovenier (1974). Interpretation of the polarization of Venus. *Journal of the Atmospheric Sciences*, 31(4), 1137–1160.

[194] P.J. Gierasch (1975). Meridional circulation and the maintenance of the Venus atmospheric rotation. *Journal of the Atmospheric Sciences*, 32(6), 1038–1044.

[195] S.B. Fels & R.S. Lindzen (1974). The interaction of thermally excited gravity waves with mean flows. *Geophysical and Astrophysical Fluid Dynamics*, 6(2), 149–191.

[196] G. Schubert & J.A. Whitehead (1969). Moving flame experiment with liquid mercury: possible implications for the Venus atmosphere. *Science*, 163(3862), 71–72.

[197] E. Kálnay De Rivas (1973). Numerical models of the circulation of the atmosphere of Venus. *Journal of the Atmospheric Sciences*, 30(5), 763–779.

[198] J.B. Pollack, O.B. Toon & R. Boese (1980). Greenhouse models of Venus' high surface temperature, as constrained by Pioneer Venus measurements. *Journal of Geophysical Research*, 85, A13, 8223–8231.

[199] W. McCrea (1975). Ice ages and the Galaxy. *Nature*, 255, 607–609.

[200] H. Shapley (1921). Note on a possible factor in changes of geological climate. *Journal of Geology*, 29, 502–204.

[201] J.D. Hays, J. Imbrie & N.J. Shackleton (1976). Variations in the Earth's orbit: pacemaker of the ice ages. *Science*, 194(4270), 1121–1132.

[202] A. Henderson-Sellers (1979). Clouds and the long-term stability of the Earth's atmosphere and climate. *Nature*, 279, 786–788.

[203] W.B. Rossow, A. Henderson-Sellers & S.K. Weinreich (1982). Cloud feedback: a stabilizing effect for the early Earth? *Science*, 217(4566), 1245–1247.

[204] S.L. Hess, R.M. Henry, C.B. Leovy, J.A. Ryan & J.E. Tillman (1977). Meteorological results from the surface of Mars: Viking 1 and 2. *Journal of Geophysical Research*, 82(28), 4559–4574.

[205] Y. Nakamura & D.L. Anderson (1979). Martian wind activity detected by a seismometer at Viking lander 2 site. *Geophysical Research Letters*, 6(6), 499–502.

[206] R.D. Lorenz (1996). Martian surface wind speeds described by the Weibull distribution. *Journal of Spacecraft and Rockets*, 33(5), 754–756.

[207] J.B. Pollack, C.B. Leovy, Y.H. Mintz & W. Van Camp (1976). Winds on Mars during the Viking season: predictions based on a general circulation model with topography. *Geophysical Research Letters*, 3(8), 479–482.

[208] S.D. Wall (1981). Analysis of condensates formed at the Viking 2 Lander site: the first winter. *Icarus*, 47(2), 173–183.

[209] J. Hansen, D. Johnson, A. Lacis, et al. (1981). Climate impact of increasing atmospheric carbon dioxide. *Science*, 213(4511), 957–966.

[210] R.S. Lindzen, A.Y. Hou & B.F. Farrell (1982). The role of convective model choice in calculating the climate impact of doubling CO_2. *Journal of the Atmospheric Sciences*, 39(6), 1189–1205.

[211] S.H. Schneider & S.L. Thompson (1980). Cosmic conclusions from climatic models: can they be justified? *Icarus*, 41(3), 456–469.

[212] M.I. Hoffert, A.J. Callegari, C.T. Hsieh & W. Ziegler (1981). Liquid water on Mars: an energy balance climate model for CO_2/H_2O atmospheres. *Icarus*, 47(1), 112–129.

[213] P.B. James & G.R. North (1982). The seasonal CO_2 cycle on Mars: an application of an energy balance climate model. *Journal of Geophysical Research: Solid Earth*, 87(B12), 10271–10283.

[214] G.W. Paltridge (1975). Global dynamics and climate: a system of minimum entropy exchange. *Royal Meteorological Society, Quarterly Journal*, 101, 475–484.

[215] G.W. Paltridge (1978). The steady-state format of global climate. *Quarterly Journal of the Royal Meteorological Society*, 104(442), 927–945.

[216] E.N. Lorenz, Generation of available potential energy and the intensity of the general circulation. In H. Pfeffer (ed.) *Dynamics of Climate: Proceedings of a Conference*. Symposium Publications Division, Pergamon Press, 1960.

[217] M.I. Hoffert, A.J. Callegari, C.T. Hsieh & W. Ziegler (1981). Liquid water on Mars: an energy balance climate model for CO_2/H_2O atmospheres. *Icarus*, 47(1), 112–129.

[218] P.B. James & G.R. North (1982). The seasonal CO_2 cycle on Mars: an application of an energy balance climate model. *Journal of Geophysical Research: Solid Earth*, 87(B12), 10271–10283.

[219] L.M. François, J.C.G. Walker & W.R. Kuhn (1990). A numerical simulation of climate changes during the obliquity cycle on Mars. *Journal of Geophysical Research: Solid Earth*, 95 (B9), 14761–14778.

[220] W. Jaffe, J. Caldwell & T. Owen (1980). Radius and brightness temperature observations of Titan at centimeter wavelengths by the Very Large Array. *The Astrophysical Journal*, 242, 806–811.

[221] G.F. Lindal, G.E. Wood, H.B. Hotz, et al. (1983). The atmosphere of Titan: an analysis of the Voyager 1 radio occultation measurements. *Icarus*, 53(2), 348–363. (Gunnar Fjeldbo

at Stanford University, who had been part of the team executing the Mariner 4 experiment at Mars, had changed his name to Gunnar Lindal.)

[222] D.M. Hunten. In T. Gehrels and M. Shapley Matthews (eds.), *Saturn*, Tucson, AZ: University of Arizona Press, 1984.

[223] R.E. Samuelson (1983). Radiative equilibrium model of Titan's atmosphere. *Icarus*, 53 (2), 364–387.

[224] J.B. Pollack, J.F. Kasting, S.M. Richardson & K. Poliakoff (1987). The case for a wet, warm climate on early Mars. *Icarus*, 71(2), 203–224.

[225] S.W. Squyres & M.H. Carr (1986). Geomorphic evidence for the distribution of ground ice on Mars. *Science*, 231(4735), 249–252.

[226] G.D. Clow (1987). Generation of liquid water on Mars through the melting of a dusty snowpack. *Icarus*, 72(1), 95–127.

[227] M.H. Carr (1979). Formation of Martian flood features by release of water from confined aquifers. *Journal of Geophysical Research*, 84(13), 2995–3007.

[228] S.M. Clifford (1993). A model for the hydrologic and climatic behavior of water on Mars. *Journal of Geophysical Research*, 98(E6), 10–973.

[229] T.J. Parker, R.S. Saunders & D.M. Schneeberger (1989). Transitional morphology in West Deuteronilus Mensae, Mars: implications for modification of the lowland/upland boundary. *Icarus*, 82(1), 111–145.

[230] V.R. Baker, R.G. Strom, V.C. Gulick, J.S. Kargel & G. Komatsu (1991). Ancient oceans, ice sheets and the hydrological cycle on Mars. *Nature*, 352, 589–594.

[231] M. Carr (1987). Water on Mars. *Nature*, 326(6108), 30–35.

[232] M. Carr, *Water on Mars*, Oxford: Oxford University Press, 1996.

[233] L.W. Esposito (1984). Sulfur dioxide: episodic injection shows evidence for active Venus volcanism. *Science*, 223(4640), 1072–1074.

[234] M. Marov & D. Grinspoon, *The Planet Venus*, New Haven, CT: Yale University Press, 1998.

[235] V. Moroz (2001). Spectra and spacecraft. *Planetary and Space Science*, 49, 173–190.

[236] V.A. Krasnopolsky (2006). Chemical composition of Venus atmosphere and clouds: some unsolved problems. *Planetary and Space Science*, 54(13), 1352–1359.

[237] R.Z. Sagdeev, et al. (1986). Overview of VEGA Venus balloon in situ meteorological measurements. *Science*, 231, 1411–1414.

[238] D. Crisp, A.P. Ingersoll, C.E. Hildebrand & R.A. Preston (1990). VEGA balloon meteorological measurements. *Advances in Space Research*, 10(5), 109–124.

[239] J.F. Kasting, J.B. Pollack & T.P. Ackerman (1984). Response of Earth's atmosphere to increases in solar flux and implications for loss of water from Venus. *Icarus*, 57(3), 335–355.

[240] A.J. Watson, T.M. Donahue & W.R. Kuhn (1984). Temperatures in a runaway greenhouse on the evolving Venus: implications for water loss. *Earth and Planetary Science Letters*, 68(1), 1–6.

[241] T.M. Donahue, J.H Hoffman, R.R. Hodges & A.J. Watson (1982). Venus was wet: a measurement of the ratio of deuterium to hydrogen. *Science*, 216(4546), 630–633.

[242] J.F. Kasting & J.B. Pollack (1983). Loss of water from Venus. I. Hydrodynamic escape of hydrogen. *Icarus*, 53(3), 479–508.

[243] D.H. Grinspoon (1987). Was Venus wet? Deuterium reconsidered. *Science*, 238(4834), 1702–1704.

[244] A.J. Watson & J.E. Lovelock (1983). Biological homeostasis of the global environment: the parable of Daisyworld. *Tellus B*, 35(4), 284–289.

[245] A.J. Wood, G.J. Ackland, J.G. Dyke, H.T. Williams & T. M. Lenton (2008). Daisyworld: a review. *Reviews of Geophysics*, 46(1) RG1001.

[246] J.H. Koeslag, P.T. Saunders & J.A. Wessels (1997). Glucose homeostasis with infinite gain: further lessons from the Daisyworld parable? *Journal of Endocrinology*, 154(2), 187–192.

[247] J.C.G. Walker, P.B. Hays & J.F. Kasting (1981). A negative feedback mechanism for the long-term stabilization of Earth's surface temperature. *Journal of Geophysical Research: Oceans*, 86(C10), 9776–9782.

[248] R.P. Turco, O.B. Toon, T.P. Ackerman, J.B. Pollack & C. Sagan (1983). Nuclear winter: global consequences of multiple nuclear explosions. *Science*, 222(4630), 1283–1292.

[249] C.P. McKay, J. B. Pollack & R. Courtin (1989). The thermal structure of Titan's atmosphere. *Icarus*, 80(1), 23–53.

[250] M.T. Lemmon, E. Karkoschka & M. Tomasko (1993). Titan's rotation: surface feature observed. *Icarus*, 103, 329–332.

[251] C.P. McKay, J.B. Pollack & R. Courtin (1991). The greenhouse and antigreenhouse effects on Titan. *Science*, 253(5024), 1118–1121.

[252] M. Thatcher, Speech to United Nations General Assembly (Global Environment), November 8, 1989.

[253] T.D. Robinson & D.C. Catling (2014). Common 0.1-bar tropopause in thick atmospheres set by pressure-dependent infrared transparency. *Nature Geoscience*, 7(1), 12–15.

[254] J. Hillier, P. Helfenstein, A. Verbiscer, et al. (1990). Voyager disk-integrated photometry of Triton. *Science*, 250(4979), 419–421.

[255] G.L. Tyler, D.N. Sweetnam, J.D. Anderson, et al. (1989). Voyager radio science observations of Neptune and Triton. *Science*, 246(4936), 1466–1473.

[256] C.J. Hansen & D.A. Paige (1992). A thermal model for the seasonal nitrogen cycle on Triton. *Icarus*, 99(2), 273–288.

[257] A.M. Zalucha & T.I. Michaels (2013). A 3D general circulation model for Pluto and Triton with fixed volatile abundance and simplified surface forcing. *Icarus*, 223(2), 819–831.

[258] J.A. Stansberry, D.J. Pisano & R.V. Yelle (1996). The emissivity of volatile ices on Triton and Pluto. *Planetary and Space Science*, 44(9), 945–955.

[259] J.A. Stansberry & R.V. Yelle (1999). Emissivity and the fate of Pluto's atmosphere. *Icarus*, 141(2), 299–306.

[260] S.J. Johnsen, H.B. Clausen, W. Dansgaard, et al. (1992). Irregular glacial interstadials recorded in a new Greenland ice core. *Nature*, 359(6393), 311–313.

[261] J.R. Petit, J. Jouzel, D. Raynaud, et al. (1999). Climate and atmospheric history of the past 420,000 years from the Vostok ice core, Antarctica. *Nature*, 399(6735), 429–436.

[262] S. Hong, J.P. Candelone, C.C. Patterson & C.F. Boutron (1994). Greenland ice evidence of hemispheric lead pollution two millennia ago by Greek and Roman civilizations. *Science*, 265(5180), 1841–1843.

[263] W. Dansgaard, S.J. Johnsen, H.B. Clausen, et al. (1993). Evidence for general instability of past climate from a 250-kyr ice-core record. *Nature*, 364(6434), 218–220.

[264] R.B. Alley & D.R. MacAyeal (1994). Ice-rafted debris associated with binge/purge oscillations of the Laurentide Ice Sheet. *Paleoceanography*, 9(4), 503–511.

[265] H. Heinrich (1988). Origin and consequences of cyclic ice rafting in the northeast Atlantic Ocean during the past 130,000 years. *Quaternary Research*, 29(2), 142–152.

[266] S. Rahmstorf (1994). Rapid climate transitions in a coupled ocean–atmosphere model. *Nature*, 372(6501), 82–85.

[267] A. Ganopolski, S. Rahmstorf, V. Petoukhov & M. Claussen (1998). Simulation of modern and glacial climates with a coupled global model of intermediate complexity. *Nature*, 391(6665), 351–356.

[268] A. Ganopolski & S. Rahmstorf (2001). Rapid changes of glacial climate simulated in a coupled climate model. *Nature*, 409(6817), 153–158.

[269] Y.L. Yung, M. Allen & J.P. Pinto (1984). Photochemistry of the atmosphere of Titan: comparison between model and observations. *The Astrophysical Journal Supplement Series*, 55, 465–506.

[270] Y.L. Yung & J.P. Pinto (1978). Primitive atmosphere and implications for the formation of channels on Mars. *Nature*, 273(5665), 730–732.

[271] F.M. Flasar (1983). Oceans on Titan? *Science*, 221(4605), 55–57.

[272] J.I. Lunine, D.J. Stevenson & Y.L. Yung (1983). Ethane ocean on Titan. *Science*, 222(4629), 1229–1230.

[273] D.J. Stevenson & B.E. Potter (1986). Titan's latitudinal temperature distribution and seasonal cycle. *Geophysical Research Letters*, 13(2), 93–96.

[274] C.P. McKay, J.B Pollack., J.I. Lunine & R. Courtin (1993). Coupled atmosphere–ocean models of Titan's past. *Icarus*, 102(1), 88–98.

[275] J.I. Lunine & B. Rizk (1989). Thermal evolution of Titan's atmosphere. *Icarus*, 80(2), 370–389.

[276] C.P. McKay, O.B. Toon & J.F. Kasting (1991). Making Mars habitable. *Nature*, 352(6335), 489–496.

[277] C.P. McKay (1982). Terraforming Mars. *Journal of the British Interplanetary Society*, 35, 427–433.

[278] R.E. Dickinson & R.J. Cicerone (1986). Future global warming from atmospheric trace gases. *Nature*, 319, 109–115.

[279] S.W. Bougher, D.M. Hunten & R.J. Phillips (Eds.) (1997). *Venus II: Geology, Geophysics, Atmosphere, and Solar Wind Environment*. Tucson, AZ: University of Arizona Press.

[280] L.W. Esposito, E.R. Stofan & T.E. Cravens (Eds.) (2013). *Exploring Venus as a Terrestrial Planet*, AGU Geophysical Monograph 176. John Wiley & Sons.

[281] N.H. Sleep, K.J. Zahnle, J.F. Kasting & H.J. Morowitz (1989). Annihilation of ecosystems by large asteroid impacts on the early Earth. *Nature*, 342(6246), 139.

[282] Y. Abe & T. Matsui (1988). Evolution of an impact-generated H_2O–CO_2 atmosphere and formation of a hot proto-ocean on Earth. *Journal of the Atmospheric Sciences*, 45(21), 3081–3101.

[283] K.J. Zahnle, J.F. Kasting & J.B. Pollack (1988). Evolution of a steam atmosphere during Earth's accretion. *Icarus*, 74(1), 62–97.

[284] K. Hamano, Y. Abe & H. Genda (2013). Emergence of two types of terrestrial planet on solidification of magma ocean. *Nature*, 497(7451), 607.

[285] J.F. Kasting, D.P. Whitmire & R.T. Reynolds (1993). Habitable zones around main sequence stars. *Icarus*, 101(1), 108–128.

[286] E. Pierazzo, D.A. Kring & H.J. Melosh (1998). Hydrocode simulation of the Chicxulub impact event and the production of climatically active gases, *Journal of Geophysical Research*, 103, 28607–28625.

[287] M.A. Bullock & D.H. Grinspoon (1996). The stability of climate on Venus. *Journal of Geophysical Research: Planets*, 101(E3), 7521–7529.

[288] F. Taylor & D. Grinspoon (2009). Climate evolution of Venus. *Journal of Geophysical Research: Planets*, 114(E9).

[289] G.L. Hashimoto, Y. Abe & S. Sasaki (1997). CO_2 amount on Venus constrained by a criterion of topographic-greenhouse instability. *Geophysical Research Letters*, 24(3), 289–292.

[290] K.B. Klose, J.A. Wood & A. Hashimoto (1992). Mineral equilibria and the high radar reflectivity of Venus mountaintops. *Journal of Geophysical Research: Planets*, 97(E10), 16353–16369.

[291] M.A. Bullock & D.H. Grinspoon (2001). The recent evolution of climate on Venus. *Icarus*, 150(1), 19–37.

[292] S.C. Solomon, M.A. Bullock & D.H. Grinspoon (1999). Climate change as a regulator of tectonics on Venus. *Science*, 286(5437), 87–90.

[293] F.S. Anderson & S.E. Smrekar (1999). Tectonic effects of climate change on Venus. *Journal of Geophysical Research: Planets*, 104(E12), 30743–30756.

[294] R. J. Phillips, M.A. Bullock & S.A. Hauck (2001). Climate and interior coupled evolution on Venus. *Geophysical Research Letters*, 28(9), 1779–1782.

[295] P.F. Hoffman, A.J. Kaufman, G.P. Halverson & D.P. Schrag (1998). A Neoproterozoic snowball Earth. *Science*, 281(5381), 1342–1346.

[296] P.F. Hoffman & D.P. Schrag (2000). Snowball Earth. *Scientific American*, 282(1), 68–75.

[297] J.L. Kirschvink, Late Proterozoic low-latitude global glaciation: the snowball Earth. In J. Schopf et al. (Eds.), *The Proterozoid Biosphere: A Multidisciplinary Study*, Cambridge: Cambridge University Press, 1992.

[298] R.D. Lorenz, C.P. McKay & J.I. Lunine (1997). Photochemically driven collapse of Titan's atmosphere. *Science*, 275(5300), 642–644.

[299] R.D. Lorenz, C.P. McKay & J.I. Lunine (1999). Analytic investigation of climate stability on Titan: sensitivity to volatile inventory. *Planetary and Space Science*, 47(12), 1503–1515.

[300] E.M. Conway, *Exploration and Engineering: The Jet Propulsion Laboratory and the Quest for Mars*, Baltimore, MD: Johns Hopkins University Press, 2015.

[301] O. Morton, *Mapping Mars: Science, Imagination and the Birth of a World*, New York: Picador, 2002.

[302] M.C. Malin & K.S. Edgett (2000). Evidence for recent groundwater seepage and surface runoff on Mars. *Science*, 288(5475), 2330–2335.

[303] A.D. Del Genio, W. Zhou & T.P. Eichler (1993). Equatorial superrotation in a slowly rotating GCM: implications for Titan and Venus. *Icarus*, 101(1), 1–17.

[304] F. Hourdin, O. Talagrand, R. Sadourny, et al. (1995). Numerical simulation of the general circulation of the atmosphere of Titan. *Icarus*, 117(2), 358–374.

[305] F. Forget & R.T. Pierrehumbert (1997). Warming early Mars with carbon dioxide clouds that scatter infrared radiation. *Science*, 278(5341), 1273–1276.

[306] Y.L. Yung, H. Nair & M.F. Gerstell (1997). CO_2 greenhouse in the early Martian atmosphere: SO_2 inhibits condensation. *Icarus*, 130(1), 222–224.

[307] J.W. Head, H. Hiesinger, M.A. Ivanov, et al. (1999). Possible ancient oceans on Mars: evidence from Mars Orbiter Laser Altimeter data. *Science*, 286(5447), 2134–2137.

[308] D.E. Smith, M.T. Zuber & G.A. Neumann (2001). Seasonal variations of snow depth on Mars. *Science*, 294(5549), 2141–2146.

[309] A.B. Ivanov & D.O. Muhleman (2001). Cloud reflection observations: results from the Mars Orbiter Laser Altimeter. *Icarus*, 154(1), 190–206.

[310] D.S. McKay, E.K. Gibson, K.L. Thomas-Keprta, et al. (1996). Search for past life on Mars: possible relic biogenic activity in Martian meteorite ALH84001. *Science*, 273(5277), 924–930.

[311] K.L. Thomas-Keprta, D.A. Bazylinski, J.L. Kirschvink, et al. (2000). Elongated prismatic magnetite crystals in ALH84001 carbonate globules: potential Martian magnetofossils. *Geochimica et Cosmochimica Acta*, 64(23), 4049–4081.

[312] J.E.P. Connerney, M.H. Acuna, P.J. Wasilewski, et al. (1999). Magnetic lineations in the ancient crust of Mars. *Science*, 284(5415), 794–798.

[313] D.A. Allen & J.W. Crawford (1984). Cloud structure on the dark side of Venus. *Nature*, 307, 222–224.

[314] D. Crisp, D.A. Allen, D.H. Grinspoon & J.B. Pollack (1991). The dark side of Venus: near-infrared images and spectra from the Anglo-Australian Observatory. *Science*, 253(5025), 1263–1266.

[315] M.T. Lemmon, E. Karkoschka & M. Tomasko (1993). Titan's rotation: surface feature observed. *Icarus*, 103(2), 329–332.

[316] P.H. Smith, M.T. Lemmon, R.D. Lorenz, et al. (1996). Titan's surface, revealed by HST imaging. *Icarus*, 119(2), 336–349.

[317] R.D. Lorenz, M.T. Lemmon, P.H. Smith & G.W. Lockwood (1999). Seasonal change on Titan observed with the Hubble Space Telescope WFPC-2. *Icarus*, 142(2), 391–401.

[318] C.A. Griffith, T. Owen, G.A. Miller & T. Geballe (1998). Transient clouds in Titan's lower atmosphere. *Nature*, 395(6702), 575–578.

[319] C.A. Griffith, J.L. Hall & T.R. Geballe (2000). Detection of daily clouds on Titan. *Science*, 290(5491), 509–513.

[320] R. Lorenz & J. Mitton, *Lifting Titan's Veil*, Cambridge: Cambridge University Press, 2002.

[321] M.E. Brown, A.H. Bouchez & C.A. Griffith (2002). Direct detection of variable tropospheric clouds near Titan's south pole. *Nature*, 420(6917), 795–797.

[322] H.G. Roe, I. De Pater, B.A. Macintosh & C.P. McKay (2002). Titan's clouds from Gemini and Keck adaptive optics imaging. *The Astrophysical Journal*, 581(2), 1399.

[323] R.D. Lorenz, C.A. Griffith, J.I. Lunine, C.P. McKay & N.O. Rennò (2005). Convective plumes and the scarcity of Titan's clouds. *Geophysical Research Letters*, 32(1) L01201.

[324] H.G. Roe, M.E. Brown, E.L. Schaller, A.H. Bouchez & C.A. Trujillo (2005). Geographic control of Titan's mid-latitude clouds. *Science*, 310(5747), 477–479.

[325] D.M. Harland & R.D. Lorenz, *Space Systems Failures*, Springer, 2006.

[326] W.V. Boynton, S.H. Bailey, D.K Hamara, et al. (2001). Thermal and evolved gas analyzer: part of the Mars volatile and climate surveyor integrated payload. *Journal of Geophysical Research: Planets*, 106(E8), 17683–17698.

[327] J.F. Nye, W.B. Durham, P.M. Schenk & J.M. Moore (2000). The instability of a south polar cap on Mars composed of carbon dioxide. *Icarus*, 144(2), 449–455.

[328] R.D. Lorenz, E.F. Young & M.T. Lemmon (2001). Titan's smile and collar: HST observations of seasonal change 1994–2000. *Geophysical Research Letters*, 28(23), 4453–4456.

[329] R.D. Lorenz, M.T. Lemmon & P.H. Smith (2006). Seasonal evolution of Titan's dark polar hood: midsummer disappearance observed by the Hubble Space Telescope. *Monthly Notices of the Royal Astronomical Society*, 369(4), 1683–1687.

[330] Y.L. Yung (1987). An update of nitrile photochemistry on Titan. *Icarus*, 72(2), 468–472.

[331] R.E. Samuelson, N.R. Nath & A. Borysow (1997). Gaseous abundances and methane supersaturation in Titan's troposphere. *Planetary and Space Science*, 45(8), 959–980.

[332] R.E. Samuelson & L.A. Mayo (1997). Steady-state model for methane condensation in Titan's troposphere. *Planetary and Space Science*, 45(8), 949–958.

[333] R.D. Lorenz, J.I. Lunine, P.G. Withers & C.P. McKay (2001). Titan, Mars and Earth: entropy production by latitudinal heat transport. *Geophysical Research Letters*, 28(3), 415–418.

[334] A. Kleidon & R.D. Lorenz. *Non-equilibrium Thermodynamics and the Production of Entropy: Life, Earth, and Beyond*. Springer Science & Business Media, 2005.

[335] R. Dewar (2003). Information theory explanation of the fluctuation theorem, maximum entropy production and self-organized criticality in non-equilibrium stationary states. *Journal of Physics A: Mathematical and General*, 36(3), 631.

[336] R. Goody (2007). Maximum entropy production in climate theory. *Journal of the Atmospheric Sciences*, 64(7), 2735–2739.

[337] T.E. Jupp & P.M. Cox (2010). MEP and planetary climates: insights from a two-box climate model containing atmospheric dynamics. *Philosophical Transactions of the Royal Society B: Biological Sciences*, 365(1545), 1355–1365.

[338] S. Shimokawa & H. Ozawa (2001). On the thermodynamics of the oceanic general circulation: entropy increase rate of an open dissipative system and its surroundings. *Tellus A*, 53(2), 266–277.

[339] N. Hoffman (2000). White Mars: a new model for Mars' surface and atmosphere based on CO_2. *Icarus*, 146(2), 326–342.

[340] P.D. Lanagan, A.S. McEwen, L.P. Keszthelyi & T. Thordarson (2001). Rootless cones on Mars indicating the presence of shallow equatorial ground ice in recent times. *Geophysical Research Letters*, 28(12), 2365–2367.

[341] M.C. Malin, M.A. Caplinger & S.D. Davis (2001). Observational evidence for an active surface reservoir of solid carbon dioxide on Mars. *Science*, 294(5549), 2146–2148.

[342] S. Byrne & A.P. Ingersoll (2003). A sublimation model for Martian south polar ice features. *Science*, 299(5609), 1051–1053.

[343] S. Byrne (2009). The polar deposits of Mars. *Annual Review of Earth and Planetary Sciences*, 37, 535–560.

[344] W.V. Boynton, et al. (2002). Distribution of hydrogen in the near surface of Mars: evidence for subsurface ice deposits. *Science*, 297(5578), 81–85.

[345] W.C. Feldman, W.V. Boynton, R.L. Tokar, et al. (2002). Global distribution of neutrons from Mars: results from Mars Odyssey. *Science*, 297(5578), 75–78.

[346] C.T. Pillinger, with M.R. Sims & S. Clemmet, *The Guide to Beagle 2*. London: Faber and Faber, 2003.

[347] J.W. Head, G. Neukum, R. Jaumann, et al. (2005). Tropical to mid-latitude snow and ice accumulation, flow and glaciation on Mars. *Nature*, 434(7031), 346–351.

[348] V. Formisano, S. Atreya, T. Encrenaz, N. Ignatiev & M. Giuranna (2004). Detection of methane in the atmosphere of Mars. *Science*, 306(5702), 1758–1761.

[349] M.J. Mumma, G.L. Villanueva, R.E. Novak, et al. (2009). Strong release of methane on Mars in northern summer 2003. *Science*, 323(5917), 1041–1045.

[350] V.A. Krasnopolsky, J.P. Maillard & T.C. Owen (2004). Detection of methane in the Martian atmosphere: evidence for life? *Icarus*, 172(2), 537–547.

[351] K. Zahnle, R.S. Freedman & D.C. Catling (2011). Is there methane on Mars? *Icarus*, 212 (2), 493–503.

[352] F. Lefevre & F. Forget (2009). Observed variations of methane on Mars unexplained by known atmospheric chemistry and physics. *Nature*, 460(7256), 720–723.

[353] A.A. Pavlov, J.F. Kasting, L.L. Brown, K.A. Rages & R. Freedman (2000). Greenhouse warming by CH_4 in the atmosphere of early Earth. *Journal of Geophysical Research: Planets*, 105(E5), 11981–11990.

[354] M.G. Trainer, A.A. Pavlov, H.L. DeWitt, et al. (2006). Organic haze on Titan and the early Earth. *Proceedings of the National Academy of Sciences*, 103(48), 18035–18042.

[355] D.M. Kass, J.T. Schofield, T.I. Michaels, et al. (2003). Analysis of atmospheric mesoscale models for entry, descent, and landing. *Journal of Geophysical Research: Planets*, 108(E12).

[356] S.W. Squyres, R.E. Arvidson, J.F. Bell, et al. (2004). The Opportunity Rover's Athena science investigation at Meridiani Planum, Mars. *Science*, 306(5702), 1698–1703.

[357] R.V. Morris, S.W. Ruff, R. Gellert, et al. (2010). Identification of carbonate-rich outcrops on Mars by the Spirit rover. *Science*, 329(5990), 421–424.

[358] S.W. Ruff, J.D. Farmer, W.M. Calvin, et al. (2011). Characteristics, distribution, origin, and significance of opaline silica observed by the Spirit rover in Gusev crater, Mars. *Journal of Geophysical Research: Planets*, 116(E7), doi:10.1029/2010JE003767.

[359] R.D. Lorenz (2009). Power law of dust devil diameters on Earth and Mars. *Icarus*, 203, 683–684.

[360] R.D. Lorenz & D. Reiss (2014). Solar panel clearing events, dust devil tracks, and in-situ vortex detections on Mars. *Icarus*, 248, 162–164.

[361] M.T. Lemmon, M.J. Wolff, J.F. Bell, et al. (2015). Dust aerosol, clouds, and the atmospheric optical depth record over 5 Mars years of the Mars Exploration Rover mission. *Icarus*, 251, 96–111.

[362] R.D. Lorenz, M.T. Lemmon & P.H. Smith (2006). Seasonal evolution of Titan's dark polar hood: midsummer disappearance observed by the Hubble Space Telescope. *Monthly Notices of the Royal Astronomical Society*, 369(4), 1683–1687.

[363] R.D. Lorenz, P.H. Smith & M.T. Lemmon (2004). Seasonal change in Titan's haze 1992–2002 from Hubble Space Telescope observations. *Geophysical Research Letters*, 31(10).

[364] A. Marten, T. Hidayat, Y. Biraud & R. Moreno (2002). New millimeter heterodyne observations of Titan: vertical distributions of nitriles HCN, HC_3N, CH_3CN, and the isotopic ratio $^{15}N/^{14}N$ in its atmosphere. *Icarus*, 158(2), 532–544.

[365] P. Rannou, F. Hourdin & C.P. McKay (2002). A wind origin for Titan's haze structure. *Nature*, 418(6900), 853–856.

[366] T. Kostiuk, T.A. Livengood, T. Hewagama, et al. (2005). Titan's stratospheric zonal wind, temperature, and ethane abundance a year prior to Huygens insertion. *Geophysical Research Letters*, 32(22), doi:10.1029/2005GL023897.

[367] T. Tokano & F.M. Neubauer (2002). Tidal winds on Titan caused by Saturn. *Icarus*, 158 (2), 499–515.

[368] T. Tokano, G.J. Molina-Cuberos, H. Lammer & W. Stumptner (2001). Modelling of thunderclouds and lightning generation on Titan. *Planetary and Space Science*, 49(6), 539–560.

[369] R.D. Lorenz (2000). The weather on Titan. *Science*, 290(5491), 467–468.

[370] R.D. Lorenz, C.A. Griffith, J.I. Lunine, C.P. McKay & N.O. Rennò (2005). Convective plumes and the scarcity of Titan's clouds. *Geophysical Research Letters*, 32(1), doi:10.1029/2004GL021415/.

[371] M. Awal & J.I. Lunine (1994). Moist convective clouds in Titan's atmosphere. *Geophysical Research Letters*, 21(23), 2491–2494.

[372] H.W. Ou (2001). Possible bounds on the Earth's surface temperature: from the perspective of a conceptual global-mean model. *Journal of Climate*, 14(13), 2976–2988.

[373] M.T. Rosing, D.K. Bird, N.H. Sleep & C.J. Bjerrum (2010). No climate paradox under the faint early Sun. *Nature*, 464(7289), 744–747.

[374] C. Goldblatt & K. Zahnle (2011). Faint young Sun paradox remains. *Nature*, 474(7349), E1.

[375] C. Goldblatt & K. Zahnle (2011). Clouds and the faint young Sun paradox. *Climate of the Past*, 7, 203–220.

[376] R.T. Pierrehumbert (1995). Thermostats, radiator fins, and the local runaway greenhouse. *Journal of the Atmospheric Sciences*, 52(10), 1784–1806.

[377] R.S. Lindzen, M.D. Chou & A.Y. Hou (2001). Does the Earth have an adaptive infrared iris? *Bulletin of the American Meteorological Society*, 82(3), 417–432.

[378] R. Rondanelli & R.S. Lindzen (2010). Can thin cirrus clouds in the tropics provide a solution to the faint young Sun paradox? *Journal of Geophysical Research: Atmospheres*, 115(D2).

[379] D.L. Hartmann & M.L. Michelsen (2002). No evidence for iris. *Bulletin of the American Meteorological Society*, 83(2), 249–254.

[380] M.L. Roderick & G.D. Farquhar (2002). The cause of decreased pan evaporation over the past 50 years. *Science*, 298(5597), 1410–1411.

[381] A. Ohmura & M. Wild (2002). Is the hydrological cycle accelerating? *Science*, 298, 1345–1346.

[382] W. Brutsaert & M.B. Parlange (1998). Hydrologic cycle explains the evaporation paradox. *Nature*, 396(6706), 30.

[383] D.J. Travis, A.M. Carleton & R.G. Lauritsen (2002). Climatology: contrails reduce daily temperature range. *Nature*, 418(6898), 601–601.

[384] A.J. Kalkstein & R.C. Balling (2004). Impact of unusually clear weather on United States daily temperature range following 9/11/2001. *Climate Research*, 26(1), 1–4.

[385] S. Dietmüller, M. Ponater, R. Sausen, K.P. Hoinka & S. Pechtl (2008). Contrails, natural clouds, and diurnal temperature range. *Journal of Climate*, 21(19), 5061–5075.

[386] C.C. Porco, E. Baker, J. Barbara, et al. (2005). Imaging of Titan from the Cassini spacecraft. *Nature*, 434(7030), 159–168.

[387] E.P. Turtle, J.E. Perry, A.S. McEwen, et al. (2009). Cassini imaging of Titan's high-latitude lakes, clouds, and south-polar surface changes. *Geophysical Research Letters*, 36 (2), doi/10.1029/2008GL036186.

[388] C.A. Griffith, P. Penteado, K. Baines, et al. (2005). The evolution of Titan's mid-latitude clouds. *Science*, 310(5747), 474–477.

[389] C. Elachi, et al. (2005). Cassini radar views the surface of Titan. *Science*, 308(5724), 970–974.

[390] R.D. Lorenz, R.M. Lopes, F. Paganelli, et al. (2008). Fluvial channels on Titan: initial Cassini RADAR observations. *Planetary and Space Science*, 56(8), 1132–1144.

[391] R. Lorenz and J. Mitton, *Titan Unveiled*, Princeton University Press, 2008 (revised edition 2010).

[392] M.G. Tomasko, B. Archinal & T. Becker (2005). Rain, winds and haze during the Huygens probe's descent to Titan's surface. *Nature*, 438(7069), 765–778.

[393] H.B. Niemann et al. (2010). Composition of Titan's lower atmosphere and simple surface volatiles as measured by the Cassini-Huygens probe gas chromatograph mass spectrometer experiment. *Journal of Geophysical Research*, 115, E12006, doi:10.1029/ 2010JE003659.

[394] R.D. Lorenz, H. Niemann, D. Harpold & J. Zarnecki (2006). Titan's damp ground: constraints on Titan surface thermal properties from the temperature evolution of the Huygens GCMS inlet. *Meteoritics and Planetary Science*, 41, 1405–1414.

[395] R.D. Lorenz et al. (2014). Silence on Shangri-La: detection of Titan surface volatiles by acoustic absorption. *Planetary and Space Science*, 90, 72–80.

[396] E. Karkoschka & M.G. Tomasko (2009). Rain and dewdrops on Titan based on in situ imaging. *Icarus*, 199(2), 442–448.

[397] L.C. Kouvaris & F.M. Flasar (1991). Phase equilibrium of methane and nitrogen at low temperatures: Application to Titan. *Icarus*, 91(1), 112–124.

[398] R.D. Lorenz & J.I. Lunine (2002). Titan's snowline. *Icarus*, 158(2), 557–559.

[399] R.D. Lorenz, S. Wall, J. Radebaugh, et al. (2006). The sand seas of Titan: Cassini RADAR observations of longitudinal dunes. *Science*, 312(5774), 724–727.

[400] P. Rannou, F. Montmessin, F. Hourdin & S. Lebonnois (2006). The latitudinal distribution of clouds on Titan. *Science*, 311 (5758), 201–205.

[401] J. Mitchell (2008). The drying of Titan's dunes: Titan's methane hydrology and its impact on atmospheric circulation. *Journal of Geophysical Research: Planets*, 113(E8).

[402] T. Tokano & F. Neubauer (2002). Tidal winds on Titan caused by Saturn. *Icarus*,158, 499–515.

[403] J. Radebaugh et al. (2008). Dunes on Titan observed by Cassini radar. *Icarus*, 194, 690–703.

[404] J. Radebaugh et al. (2010). Linear dunes on Titan and Earth: initial remote sensing comparisons. *Geomorphology*, 121, 122–132.

[405] R.D. Lorenz & J. Radebaugh (2009). Global pattern of Titan's dunes: radar survey from the Cassini Prime Mission. *Geophysical Research Letters*, 36, L03202, doi:10.1029/2008GL036850, 2009.

[406] T. Tokano (2008). Dune-forming winds on Titan and the influence of topography. *Icarus*, 194, 243–262.

[407] C. Wald (2009). In dune map, Titan's winds seem to blow backward. *Science*, 323, 1418.

[408] S.E. Smrekar, E.R. Stofan, N. Mueller, et al. (2010). Recent hotspot volcanism on Venus from VIRTIS emissivity data. *Science*, 328(5978), 605–608.

[409] E. Marcq, J.L. Bertaux, F. Montmessin & D. Belyaev (2013). Variations of sulfur dioxide at the cloud top of Venus's dynamic atmosphere. *Nature Geoscience*, 6(1), 25–28.

[410] L. Kaltenegger, W.G. Henning & D.D. Sasselov (2010). Detecting volcanism on extrasolar planets. *The Astronomical Journal*, 140(5), 1370.

[411] F. Tian, M.W. Claire, J.D. Haqq-Misra, et al. (2010). Photochemical and climate consequences of sulfur outgassing on early Mars. *Earth and Planetary Science Letters*, 295(3), 412–418.

[412] I. Halevy, M.T. Zuber & D.P. Schrag (2007). A sulfur dioxide climate feedback on early Mars. *Science*, 318(5858), 1903–1907.

[413] S.S. Johnson, M.A. Mischna, T.L. Grove & M.T. Zuber (2008). Sulfur-induced greenhouse warming on early Mars. *Journal of Geophysical Research: Planets*, 113(E8).

[414] L. Kerber, F. Forget & R. Wordsworth (2015). Sulfur in the early Martian atmosphere revisited: experiments with a 3-D global climate model. *Icarus*, 261, 133–148.

[415] Y. Ueno, M.S. Johnson, S.O. Danielache, et al. (2009). Geological sulfur isotopes indicate elevated OCS in the Archean atmosphere, solving faint young Sun paradox. *Proceedings of the National Academy of Sciences*, 106(35), 14784–14789.

[416] C. Goldblatt, M.W. Claire, T.M. Lenton, et al. (2009). Nitrogen-enhanced greenhouse warming on early Earth. *Nature Geoscience*, 2(12), 891–896.

[417] S.M. Som, D.C. Catling, J.P. Harnmeijer, P.M. Polivka & R. Buick (2012). Air density 2.7 billion years ago limited to less than twice modern levels by fossil raindrop imprints. *Nature*, 484(7394), 359–362.

[418] L. Kavenagh & C. Goldblatt (2015). Using raindrops to constrain past atmospheric density. *Earth and Planetary Science Letters*, 413, 51–58.

[419] S.M. Som, R. Buick, J.W. Hagadorn, et al. (2016). Earth's air pressure 2.7 billion years ago constrained to less than half of modern levels. *Nature Geoscience*, 9(6), 448–452.

[420] B. Charnay, F. Forget, G. Tobie, C. Sotin & R. Wordsworth (2014). Titan's past and future: 3D modeling of a pure nitrogen atmosphere and geological implications. *Icarus*, 241, 269–279.

[421] O. Pauluis, V. Balaji & I.M. Held (2000). Frictional dissipation in a precipitating atmosphere. *Journal of the Atmospheric Sciences*, 57(7), 989–994.

[422] O. Pauluis & J. Dias (2012). Satellite estimates of precipitation-induced dissipation in the atmosphere. *Science*, 335(6071), 953–956.

[423] J. Goodman (2009). Thermodynamics of atmospheric circulation on hot Jupiters. *The Astrophysical Journal*, 693(2), 1645.

[424] K.A. Emanuel (1999). Thermodynamic control of hurricane intensity. *Nature*, 401(6754), 665–669.

[425] K. Emanuel (2005). Increasing destructiveness of tropical cyclones over the past 30 years. *Nature*, 436(7051), 686–688.

[426] A.L. Sprague et al. (2007). Mars' atmospheric argon: tracer for understanding Martian atmospheric circulation and dynamics. *Journal of Geophysical Research: Planets*, 112(E3).

[427] Y. Lian, M.I. Richardson, C.E. Newman, et al. (2012). The Ashima/MIT Mars GCM and argon in the Martian atmosphere. *Icarus*, 218(2), 1043–1070.

[428] A. Colaprete, J.R. Barnes, R.M. Haberle, et al. (2005). Albedo of the south pole on Mars determined by topographic forcing of atmosphere dynamics. *Nature*, 435(7039), 184–188.

[429] F. Forget, R.M. Haberle, F. Montmessin, B. Levrard & J.W. Head (2006). Formation of glaciers on Mars by atmospheric precipitation at high obliquity. *Science*, 311(5759), 368–371.

[430] J.B. Madeleine, F. Forget, J.W. Head, et al. (2009). Amazonian northern mid-latitude glaciation on Mars: a proposed climate scenario. *Icarus*, 203(2), 390–405.

[431] J.W. Head, G. Neukum, R. Jaumann, et al. (2005). Tropical to mid-latitude snow and ice accumulation, flow and glaciation on Mars. *Nature*, 434(7031), 346–351.

[432] N.T. Bridges, F. Ayoub, J.P. Avouac, et al. (2012). Earth-like sand fluxes on Mars. *Nature*, 485(7398), 339–342.

[433] R.D. Lorenz & J. Zimbelman, *Dune Worlds: How Wind-Blown Sand Shapes Planetary Landscapes*, Praxis-Springer, 2014.

[434] S. Piqueux, S. Byrne & M.I. Richardson (2003). Sublimation of Mars's southern seasonal CO_2 ice cap and the formation of spiders. *Journal of Geophysical Research: Planets*, 108(E8).

[435] H.H. Kieffer, P.R. Christensen & T.N. Titus (2006). CO_2 jets formed by sublimation beneath translucent slab ice in Mars' seasonal south polar ice cap. *Nature*, 442(7104), 793–796.

[436] J. Laskar, B. Levrard & J.F. Mustard (2002). Orbital forcing of the Martian polar layered deposits. *Nature*, 419(6905), 375–377.

[437] K.E. Fishbaugh & C.S. Hvidberg (2006). Martian north polar layered deposits stratigraphy: implications for accumulation rates and flow. *Journal of Geophysical Research: Planets*, 111(E6).

[438] C.S. Hvidberg, K.E. Fishbaugh, M. Winstrup, et al. (2012). Reading the climate record of the Martian polar layered deposits. *Icarus*, 221(1), 405–419.

[439] S.M. Milkovich & J.W. Head (2005). North polar cap of Mars: polar layered deposit characterization and identification of a fundamental climate signal. *Journal of Geophysical Research: Planets*, 110(E1).

[440] N. Schorghofer (2008). Temperature response of Mars to Milankovitch cycles. *Geophysical Research Letters*, 35(18).

[441] N. Schorghofer (2007). Dynamics of ice ages on Mars. *Nature*, 449(7159), 192–194.

[442] B. Levrard, F. Forget, F. Montmessin & J. Laskar (2007). Recent formation and evolution of northern Martian polar layered deposits as inferred from a Global Climate Model. *Journal of Geophysical Research: Planets*, 112(E6).

[443] J.T. Perron & P. Huybers (2009). Is there an orbital signal in the polar layered deposits on Mars? *Geology*, 37(2), 155–158.

[444] L.E. Lisiecki & P.A. Lisiecki (2002). Application of dynamic programming to the correlation of paleoclimate records. *Paleoceanography*, 17(4), 1.

[445] L.E. Lisiecki & M.E. Raymo (2005). A Pliocene-Pleistocene stack of 57 globally distributed benthic δ^{18}O records. *Paleoceanography*, 20(1).

[446] J. Imbrie & J.Z. Imbrie (1980). Modeling the climatic response to orbital variations. *Science*, 207(4434), 943–953.

[447] T. Naish, R. Powell, R. Levy, et al. (2009). Obliquity-paced Pliocene West Antarctic ice sheet oscillations. *Nature*, 458(7236), 322–328.

[448] C.M. Brierley & A.V. Fedorov (2010). Relative importance of meridional and zonal sea surface temperature gradients for the onset of the ice ages and Pliocene-Pleistocene climate evolution. *Paleoceanography*, 25(2).

[449] W.V. Boynton, D.W. Ming, S.P. Kounaves, et al. (2009). Evidence for calcium carbonate at the Mars Phoenix landing site. *Science*, 325(5936), 61–64.

[450] P.H. Smith, L.K. Tamppari, R.E. Arvidson, et al. (2009). H_2O at the Phoenix landing site. *Science*, 325(5936), 58–61.

[451] A. Kessler, *Martian Summer*, Open Road Media, 2011.

[452] M.H. Hecht, S.P. Kounaves, R.C. Quinn, et al. (2009). Detection of perchlorate and the soluble chemistry of Martian soil at the Phoenix lander site. *Science*, 325(5936), 64–67.

[453] S.P. Kounaves, N.A. Chaniotakis, V.F. Chevrier, et al. (2014). Identification of the perchlorate parent salts at the Phoenix Mars landing site and possible implications. *Icarus*, 232, 226–231.

[454] A.P. Zent, M.H. Hecht, D.R. Cobos, et al. (2010). Initial results from the thermal and electrical conductivity probe (TECP) on Phoenix. *Journal of Geophysical Research: Planets*, 115(E3).

[455] C. Holstein-Rathlou, H.P. Gunnlaugsson, J.P. Merrison, et al. (2010). Winds at the Phoenix landing site. *Journal of Geophysical Research: Planets*, 115(E5).

[456] R. Davy, J.A. Davis, P.A. Taylor, et al. (2010). Initial analysis of air temperature and related data from the Phoenix MET station and their use in estimating turbulent heat fluxes. *Journal of Geophysical Research: Planets*, 115(E3).

[457] M.D. Ellehoj, H.P. Gunnlaugsson, P.A. Taylor, et al. (2010). Convective vortices and dust devils at the Phoenix Mars mission landing site. *Journal of Geophysical Research: Planets*, 115(E4).

[458] J.A. Whiteway, L. Komguem, C. Dickinson, et al. (2009). Mars water-ice clouds and precipitation. *Science*, 325(5936), 68–70.

[459] A. Spiga, D.P. Hinson, J.B. Madeleine, et al. (2017). Snow precipitation on Mars driven by cloud-induced night-time convection. *Nature Geoscience*, 10(9), 652–657.

[460] T. Imamura, T. Higuchi, Y. Maejima, et al. (2014). Inverse insolation dependence of Venus' cloud-level convection. *Icarus*, 228, 181–188.

[461] O. Aharonson, A. Hayes, J.I. Lunine, R.D. Lorenz & C. Elachi (2009). An asymmetric distribution of lakes on Titan as a possible consequence of orbital forcing. *Nature Geoscience*, 2, 851–854.

[462] J.W. Barnes, J. Bow, J. Schwartz, et al. (2011). Organic sedimentary deposits in Titan's dry lakebeds: Probable evaporite. *Icarus*, 216(1), 136–140.

[463] S.M. MacKenzie, J.W. Barnes, C. Sotin, et al. (2014). Evidence of Titan's climate history from evaporite distribution. *Icarus*, 243, 191–207.

[464] A.G. Hayes, O. Aharonson, J.I. Lunine, et al. and the Cassini RADAR Team (2011). Transient surface liquid in Titan's polar regions from Cassini. *Icarus*, 211, 655–671.

[465] T. Tokano & F.M. Neubauer (2005). Wind-induced seasonal angular momentum exchange at Titan's surface and its influence on Titan's length-of-day. *Geophysical Research Letters*, 32(24), doi:10.1029/2005GL024456.

[466] R.D. Lorenz, B.W. Stiles, R.L. Kirk, et al. (2008). Titan's rotation reveals an internal ocean and changing zonal winds. *Science*, 319(5870), 1649–1651.

[467] K. Lambeck, *The Earth's Variable Rotation: Geophysical Causes and Consequences*. Cambridge: Cambridge University Press, 2005.

[468] P. Defraigne, O.D. Viron, V. Dehant, T. Van Hoolst & F. Hourdin (2000). Mars rotation variations induced by atmosphere and ice caps. *Journal of Geophysical Research: Planets*, 105(E10), 24563–24570.

[469] W.M. Folkner, C.F. Yoder, D.N. Yuan, E.M. Standish & R.A. Preston (1997). Interior structure and seasonal mass redistribution of Mars from radio tracking of Mars Pathfinder. *Science*, 278(5344), 1749–1752.

[470] R.K. Achterberg, B.J. Conrath, P.J. Gierasch, F.M. Flasar & C.A. Nixon (2008). Observation of a tilt of Titan's middle-atmospheric superrotation. *Icarus*, 197(2), 549–555.

[471] T. Tokano (2010). Westward rotation of the atmospheric angular momentum vector of Titan by thermal tides. *Planetary and Space Science*, 58(5), 814–829.

[472] R.D. Lorenz, P. Claudin, B. Andreotti, J. Radebaugh & T. Tokano (2010). A 3km atmospheric boundary layer on Titan indicated by dune spacing and Huygens data. *Icarus*, 205(2), 719–721.

[473] T. Tokano, F. Ferri, G. Colombatti, T. Mäkinen & M. Fulchignoni (2006). Titan's planetary boundary layer structure at the Huygens landing site. *Journal of Geophysical Research: Planets*, 111(E8), doi:10.1029/2006JE002704.

[474] B. Charnay & S. Lebonnois (2012). Two boundary layers in Titan's lower troposphere inferred from a climate model. *Nature Geoscience*, 5(2), 106–109.

[475] R.C. Ewing, A.G. Hayes & A. Lucas (2015). Sand dune patterns on Titan controlled by long-term climate cycles. *Nature Geoscience*, 8(1), 15–19.

[476] J.M. Lora, J.I. Lunine, J.L. Russell & A.G. Hayes (2014). Simulations of Titan's paleoclimate. *Icarus*, 243, 264–273.

[477] J.M. Lora & J.L. Mitchell (2015). Titan's asymmetric lake distribution mediated by methane transport due to atmospheric eddies. *Geophysical Research Letters*, 42(15), 6213–6220.

[478] E.P. Turtle, J.E. Perry & A.G. Hayes (2011). Rapid and extensive surface changes near Titan's equator: evidence of April showers. *Science*, 331(6023), 1414–1417.

[479] J.L. Mitchell, M. Ádámkovics, R. Caballero & E.P. Turtle (2011). Locally enhanced precipitation organized by planetary-scale waves on Titan. *Nature Geoscience*, 4(9), 589–592.

[480] V. Cottini, C.A. Nixon, D.E. Jennings, et al. (2012). Spatial and temporal variations in Titan's surface temperatures from Cassini CIRS observations. *Planetary and Space Science*, 60(1), 62–71.

[481] D.E. Jennings, V. Cottini, C.A. Nixon, et al. (2016). Surface temperatures on Titan during northern winter and spring. *The Astrophysical Journal Letters*, L17(816), 2041–8205.

[482] R.A. West, J. Balloch, P. Dumont, et al. (2011). The evolution of Titan's detached haze layer near equinox in 2009. *Geophysical Research Letters*, 38(6), doi/10.1029/2011GL046843.

[483] R. A. West, A.D. Del Genio, J.M. Barbara, et al. (2015). Cassini Imaging Science Subsystem observations of Titan's south polar cloud. *Icarus*, 270, 399–408.

[484] R.J. de Kok, N.A. Teanby, L. Maltagliati, P.G. Irwin & S. Vinatier (2014). HCN ice in Titan's high-altitude southern polar cloud. *Nature*, 514(7520), 65–67.

[485] E.S. Kite, J.P. Williams, A. Lucas & O. Aharonson (2014). Low palaeopressure of the Martian atmosphere estimated from the size distribution of ancient craters. *Nature Geoscience*, 7(5), 335–339.

[486] S. Engel, J.I. Lunine & W.K. Hartmann (1995). Cratering on Titan and implications for Titan's atmospheric history. *Planetary and Space Science*, 43(9), 1059–1066.

[487] B.A. Ivanov, A.T. Basilevsky & G. Neukum (1997). Atmospheric entry of large meteoroids: implication to Titan. *Planetary and Space Science*, 45(8), 993–1007.

[488] R. Kahn (1982). Deducing the age of the dense Venus atmosphere. *Icarus*, 49(1), 71–85.

[489] M. Manga, A. Patel, J. Dufek & E.S. Kite (2012). Wet surface and dense atmosphere on early Mars suggested by the bomb sag at Home Plate, Mars. *Geophysical Research Letters*, 39(1), doi:10.1029/2011GL050192.

[490] R.M. Ramirez, R. Kopparapu, M.E. Zugger, et al. (2014). Warming early Mars with CO_2 and H_2. *Nature Geoscience*, 7(1), 59–63.

[491] R. Pierrehumbert & E. Gaidos (2011). Hydrogen greenhouse planets beyond the habitable zone. *The Astrophysical Journal Letters*, 734(1), L13.

[492] D.J. Stevenson (1999). Life-sustaining planets in interstellar space? *Nature*, 400(6739), 32–32.

[493] M. Allison, D.A. Godfrey & R.F. Beebe (1990). A wave dynamical interpretation of Saturn's polar hexagon. *Science*, 247(4946), 1061–1063.

[494] R. Morales-Juberías, K.M. Sayanagi, A.A. Simon, L.N. Fletcher & R.G. Cosentino (2015). Meandering shallow atmospheric jet as a model of Saturn's north-polar hexagon. *The Astrophysical Journal Letters*, 806(1), L18.

[495] L.N. Fletcher, B.E. Hesman, P.G Irwin, et al. (2011). Thermal structure and dynamics of Saturn's northern springtime disturbance. *Science*, 332(6036), 1413–1417.

[496] G. Fischer, W.S. Kurth, D.A. Gurnett, et al. (2011). A giant thunderstorm on Saturn. *Nature*, 475(7354), 75–77.

[497] J.R. Spencer & T. Denk (2010). Formation of Iapetus' extreme albedo dichotomy by exogenically triggered thermal ice migration. *Science*, 327(5964), 432–435.

[498] G.J. Ackland, M.A. Clark & T.M. Lenton (2003). Catastrophic desert formation in Daisyworld. *Journal of Theoretical Biology*, 223(1), 39–44.

[499] A.M. Turing (1952). The chemical basis of morphogenesis. *Philosophical Transactions of the Royal Society of London B: Biological Sciences*, 237(641), 37–72.

[500] A.M. Earle, R.P. Binzel, L.A. Young, et al. (2018). Albedo matters: understanding runaway albedo variations on Pluto. *Icarus*, 303, 1–9.

[501] J.L. Mitchell, R.T. Pierrehumbert, D.M. Frierson & R. Caballero (2006). The dynamics behind Titan's methane clouds. *Proceedings of the National Academy of Sciences*, 103(49), 18421–18426.

[502] J.M. Lora, P.J. Goodman, J.L. Russell & J.I. Lunine (2011). Insolation in Titan's troposphere. *Icarus*, 216(1), 116–119.

[503] D.J. Seidel, Q. Fu, W.J. Randel & T.J. Reichler (2008). Widening of the tropical belt in a changing climate. *Nature Geoscience*, 1(1), 21–24.

[504] R.J. Allen, S.C. Sherwood, J.R. Norris & C.S. Zender (2012). Recent Northern Hemisphere tropical expansion primarily driven by black carbon and tropospheric ozone. *Nature*, 485(7398), 350–354.

[505] J. Lu, C. Deser & T. Reichler (2009). Cause of the widening of the tropical belt since 1958. *Geophysical Research Letters*, 36(3), doi:10.1029/2008GL036076.

[506] C.A. Griffith, P. Penteado, P. Rannou, et al. (2006). Evidence for a polar ethane cloud on Titan. *Science*, 313(5793), 1620–1622.

[507] D.E. Jennings, C.M. Anderson, R.E. Samuelson, et al. (2012). Seasonal disappearance of far-infrared haze in Titan's stratosphere. *The Astrophysical Journal Letters*, 754(1), L3.

[508] R.D. Lorenz (1993). The life, death and afterlife of a raindrop on Titan. *Planetary and Space Science*, 41(9), 647–655.

[509] E.L. Barth & S.C. Rafkin (2007). TRAMS: a new dynamic cloud model for Titan's methane clouds. *Geophysical Research Letters*, 34(3).

[510] R. Hueso & A. Sánchez-Lavega (2006). Methane storms on Saturn's moon Titan. *Nature*, 442(7101), 428–431.

[511] S.P. Faulk, J.L. Mitchell, S. Moon & J.M. Lora (2017). Regional patterns of extreme precipitation on Titan consistent with observed alluvial fan distribution. *Nature Geoscience*, 10(11), 827.

[512] D.B. Curtis, C.D. Hatch, C.A. Hasenkopf, et al. (2008). Laboratory studies of methane and ethane adsorption and nucleation onto organic particles: application to Titan's clouds. *Icarus*, 195(2), 792–801.

[513] S.D.B. Graves, C.P. McKay, C.A. Griffith, F. Ferri & M. Fulchignoni (2008). Rain and hail can reach the surface of Titan. *Planetary and Space Science*, 56(3), 346–357.

[514] J.W. Barnes, B.J. Buratti, E.P. Turtle et al. (2013). Precipitation-induced surface brightenings seen on Titan by Cassini VIMS and ISS. *Planetary Science*, 2(1), 1–22.

[515] S.E. Schröder, E. Karkoschka & R.D. Lorenz (2012). Bouncing on Titan: motion of the Huygens probe in the seconds after landing. *Planetary and Space Science*, 73(1), 327–340.

[516] R.D. Lorenz (1993). Wake-induced dust cloud formation following impact of planetary landers. *Icarus*, 101(1), 165–167.

[517] B. Charnay, E. Barth, S. Rafkin, et al. (2015). Methane storms as a driver of Titan's dune orientation. *Nature Geoscience*, 8, 362–366.

[518] D. Sun, K.M. Lau & M. Kafatos (2008). Contrasting the 2007 and 2005 hurricane seasons: evidence of possible impacts of Saharan dry air and dust on tropical cyclone activity in the Atlantic basin. *Geophysical Research Letters*, 35(15).

[519] J.M. Prospero & T.N. Carlson (1972). Vertical and areal distribution of Saharan dust over the western equatorial North Atlantic Ocean. *Journal of Geophysical Research*, 77(27), 5255–5265.

[520] C.S. Bristow, K.A. Hudson-Edwards & A. Chappell (2010). Fertilizing the Amazon and equatorial Atlantic with West African dust. *Geophysical Research Letters*, 37(14).

[521] J. Gribbin (1988). Any old iron? *Nature*, 331, 570.

[522] A. J. Watson (1997). Volcanic Fe, CO_2, ocean productivity and climate. *Nature*, 385, 587–588.

[523] C.E. Newman, S.R. Lewis, P.L. Read & F. Forget (2002). Modeling the Martian dust cycle, 1. Representations of dust transport processes. *Journal of Geophysical Research: Planets*, 107(E12), 6–1.

[524] R.J. Phillips, B.J. Davis, K.L. Tanaka, et al. (2011). Massive CO_2 ice deposits sequestered in the south polar layered deposits of Mars. *Science*, 332(6031), 838–841.

[525] C.S. Edwards & B.L. Ehlmann (2015). Carbon sequestration on Mars. *Geology*, 43(10), 863–866.

[526] M. Mastrogiuseppe, V. Poggiali, A. Hayes, et al. (2014). The bathymetry of a Titan sea. *Geophysical Research Letters*, 41(5), 1432–1437.

[527] O. Mousis & B. Schmitt (2008). Sequestration of ethane in the cryovolcanic subsurface of Titan. *The Astrophysical Journal Letters*, 677(1), L67.

[528] M. Choukroun & C. Sotin (2012). Is Titan's shape caused by its meteorology and carbon cycle? *Geophysical Research Letters*, 39(4).

[529] E.H. Wilson & S.K. Atreya (2009). Titan's carbon budget and the case of the missing ethane. *The Journal of Physical Chemistry A*, 113(42), 11221–11226.

[530] D.M. Hunten (2006). The sequestration of ethane on Titan in smog particles. *Nature*, 443(7112), 669–670.

[531] I.B. Smith & J.W. Holt (2010). Onset and migration of spiral troughs on Mars revealed by orbital radar. *Nature*, 465(7297), 450–453.

[532] I.B. Smith, J.W. Holt, A. Spiga, A.D. Howard & G. Parker (2013). The spiral troughs of Mars as cyclic steps. *Journal of Geophysical Research: Planets*, 118, 1835–1857, doi:10.1002/jgre.20142.

[533] S. Diniega, S. Byrne, N.T. Bridges, C.M. Dundas & A.S. McEwen (2010). Seasonality of present-day Martian dune-gully activity. *Geology*, 38(11), 1047–1050.

[534] S. Diniega, C.J. Hansen, J.N. McElwaine, et al. (2013). A new dry hypothesis for the formation of Martian linear gullies. *Icarus*, 225(1), 526–537.

[535] L. Ojha, M.B. Wilhelm, S.L. Murchie, et al. (2015). Spectral evidence for hydrated salts in recurring slope lineae on Mars. *Nature Geoscience*, 8(11), 829–832.

[536] A.M. Palumbo, J.W. Head & R.D. Wordsworth (2018). Late Noachian Icy Highlands climate model: exploring the possibility of transient melting and fluvial/lacustrine activity through peak annual and seasonal temperatures. *Icarus*, 300, 261–286.

[537] N.E. Batalha, R.K. Kopparapu, J. Haqq-Misra & J.F. Kasting (2016). Climate cycling on early Mars caused by the carbonate–silicate cycle. *Earth and Planetary Science Letters*, 455, 7–13.

[538] J.L. Heldmann, W. Pollard, C.P. McKay, et al. (2013). The high elevation Dry Valleys in Antarctica as analog sites for subsurface ice on Mars. *Planetary and Space Science*, 85, 53–58.

[539] R.E. Arvidson (2016). Aqueous history of Mars as inferred from landed mission measurements of rocks, soils, and water ice. *Journal of Geophysical Research: Planets*, 121(9), 1602–1626.

[540] L. Montabone, S.R. Lewis, P.L. Read, et al. (2011). Mars Analysis Correction Data Assimilation (MACDA): MGS/TES v1.0. NCAS British Atmospheric Data Centre, 29 November 2011. doi:10.5285/78114093-E2BD-4601-8AE5-3551E62AEF2B.

[541] I.B. Smith, N.E. Putzig, J.W. Holt & R.J. Phillips (2016). An ice age recorded in the polar deposits of Mars. *Science*, 352(6289), 1075–1078.

[542] S. Piqueux, S. Byrne, H.H. Kieffer, T.N. Titus & C.J. Hansen (2015). Enumeration of Mars years and seasons since the beginning of telescopic exploration. *Icarus*, 251, 332–338.

[543] S.D. Guzewich, A.D. Toigo, L. Kulowski & H. Wang (2015). Mars Orbiter Camera climatology of textured dust storms. *Icarus*, 258, 1–13.

[544] L.K. Fenton, P.E. Geissler & R.M. Haberle (2007). Global warming and climate forcing by recent albedo changes on Mars. *Nature*, 446(7136), 646–649.

[545] M.D. Smith (2004). Interannual variability in TES atmospheric observations of Mars during 1999–2003. *Icarus*, 167(1), 148–165.

[546] A.A. Pankine & A.P. Ingersoll (2004). Interannual variability of Mars global dust storms: an example of self-organized criticality? *Icarus*, 170(2), 514–518.

[547] C.E. Newman & M.I. Richardson (2015). The impact of surface dust source exhaustion on the Martian dust cycle, dust storms and interannual variability, as simulated by the MarsWRF General Circulation Model. *Icarus*, 257, 47–87.

[548] J.H. Shirley (2015). Solar system dynamics and large-scale dust storms on Mars. *Icarus*, 251, 128–144.

[549] M.A. Mischna & J.H. Shirley (2017). Numerical modeling of orbit-spin coupling accelerations in a Mars general circulation model: implications for global dust storm activity. *Planetary and Space Science*, 141, 45–72.

[550] G.W. Lockwood & D.T. Thompson (2009). Seasonal photometric variability of Titan, 1972–2006. *Icarus*, 200(2), 616–626.

[551] G.W. Lockwood & D.T. Thompson (1979). A relationship between solar activity and planetary albedos. *Nature*, 280, 43–45.

[552] L.A. Sromovsky, V.E. Suomi, J.B. Pollack, et al. (1981). Implications of Titan's north–south brightness asymmetry. *Nature*, 292, 698–702.

[553] R.D. Lorenz, M.T. Lemmon, P.H. Smith & G.W. Lockwood (1999). Seasonal change on Titan observed with the Hubble Space Telescope WFPC-2. *Icarus*, 142(2), 391–401.

[554] K.L. Aplin & R.G. Harrison (2016). Determining solar effects in Neptune's atmosphere. *Nature Communications*, 7, 11976, doi:10.1038/ncomms11976.

[555] E.P. Ney (1959). Cosmic radiation and the weather. *Nature*, 183, 451–452.

[556] B. Sicardy, T. Widemann, E. Lellouch, et al. (2003). Large changes in Pluto's atmosphere as revealed by recent stellar occultations. *Nature*, 424(6945), 168–170.

[557] M. Way et al. (2016). Was Venus the first habitable world of our solar system? *Geophysical Research Letters*, 43, doi:10.1002/2016GL069790.

[558] Y. Abe, A. Abe-Ouchi, N. Sleep & K. Zahnle (2001). Habitable zone limits for dry planets. *Astrobiology*, 11, 443–460.

[559] R.D. Lorenz (2015). Meteorological insights from planetary heat flow measurements. *Icarus*, 250, 262–267.

[560] T. Tokano & R.D. Lorenz (2006). GCM simulation of balloon trajectories on Titan. *Planetary and Space Science*, 54, 685–694.

[561] R.D. Lorenz, C.E. Newman, T. Tokano, et al. (2012). Formulation of an engineering wind specification for Titan late summer polar exploration. *Planetary and Space Science*, 70, 73–83.

[562] R.D. Lorenz, T. Tokano & C.E. Newman (2012). Winds and tides of Ligeia Mare: application to the drift of the Titan Mare Explorer (TiME) Mission. *Planetary and Space Science*, 60(1), 72–85.

[563] E. Stofan, R. Lorenz, J. Lunine, et al. (2013). TiME – The Titan Mare Explorer. *Proceedings of IEEE Aerospace Conference*, Big Sky, MT, March 2013, paper 2434.

[564] R.D. Lorenz & C.E. Newman (2015). Twilight on Ligeia: implications of communications geometry and seasonal winds for exploring Titan's seas 2020–2040. *Advances in Space Research*, 56(1), 190–204.

[565] E.F. MacPike, *Correspondence and Papers of Edmond Halley*. New York: Arno Press, 1975.

[566] R.D. Lorenz (2015). Voyage across Ligeia Mare: mechanics of sailing on the hydrocarbon seas of Saturn's moon, Titan. *Ocean Engineering*, 104, 119–128.

[567] R.D. Lorenz & J. Mann (2015). Seakeeping on Ligeia Mare: dynamic response of a floating capsule to waves on the hydrocarbon seas of Saturn's moon Titan. *Johns Hopkins APL Technical Digest*, 33(2), 82–94.

[568] R.D. Lorenz, E.P. Turtle, J.W. Barnes, et al. (2019). Dragonfly: a rotorcraft lander concept for scientific exploration at Titan. *Johns Hopkins APL Technical Digest*, in press.

[569] D.M. Williams, J.F. Kasting & R.A. Wade (1997). Habitable moons around extrasolar giant planets. *Nature*, 385(6613), 234–236.

[570] J.F. Kasting, D.P. Whitmire & R.T. Reynolds (1993). Habitable zones around main sequence stars. *Icarus*, 101(1), 108–128.

[571] M.R. Rampino & K. Caldeira (1994). The Goldilocks Problem: climatic evolution and long-term habitability of terrestrial planets. *Annual Reviews of Astronomy and Astrophysics*, 32, 83–114.

[572] M.M. Joshi, R.M. Haberle & R.T. Reynolds (1997). Simulations of the atmospheres of synchronously rotating terrestrial planets orbiting M dwarfs: conditions for atmospheric collapse and the implications for habitability. *Icarus*, 129(2), 450–465.

[573] E. Han, S.X. Wang, J.T. Wright, et al. (2014). Exoplanet orbit database. II. Updates to exoplanets.org. *Publications of the Astronomical Society of the Pacific*, 126(943), 827.

[574] J. Harrington, B.M. Hansen, S.H. Luszcz, et al. (2006). The phase-dependent infrared brightness of the extrasolar planet υ Andromedae b. *Science*, 314(5799), 623–626.

[575] H.A. Knutson, D. Charbonneau, L.E. Allen, et al. (2007). A map of the day–night contrast of the extrasolar planet HD 189733b. *Nature*, 447(7141), 183.

[576] F. Pont, H.A. Knutson, R.L. Gilliland, C. Moutou & D. Charbonneau (2008). Detection of atmospheric haze on an extrasolar planet: the 0.55–1.05 μm transmission spectrum of H 189733b with the Hubble Space Telescope. *Monthly Notices of the Royal Astronomical Society*, 385, 109–118.

[577] R.T. Pierrehumbert (2013). Strange news from other stars. *Nature Geoscience*, 6(2), 81–83.

[578] R.T. Pierrehumbert (2011). A palette of climates for Gliese 581g. *The Astrophysical Journal Letters*, 726(1), L8.

[579] W.B. Moore, A. Lenardic, A.M. Jellinek, et al. (2017). How habitable zones and super-Earths lead us astray. *Nature Astronomy*, 1, 0043.

[580] E. Tasker, J. Tan, K. Heng, et al. (2017). The language of exoplanet ranking metrics needs to change. *Nature Astronomy*, 1, 0042.

[581] N. Cowan & E. Agol (2011). The statistics of albedo and heat recirculation on hot exoplanets. *The Astrophysical Journal*, 729(1), 54.

[582] Y. Kaspi & A. Showman (2015). Atmospheric dynamics of terrestrial exoplanets over a wide range of orbital and atmospheric parameters. *The Astrophysical Journal*, 804(1), 60.

[583] J. Goodman (2009). Thermodynamics of atmospheric circulation on hot Jupiters. *The Astrophysical Journal*, 693(2), 1645.

[584] B.O. Demory, M. Gillon, J. de Wit, et al. (2016). A map of the large day–night temperature gradient of a super-Earth exoplanet. *Nature*, 532(7598), 207.

[585] I. Angelo & R. Hu (2017). A case for an atmosphere on super-Earth 55 Cancri e. *The Astronomical Journal*, 154(6), 232.

[586] G. Laughlin, D. Deming, J. Langton, et al. (2009). Rapid heating of the atmosphere of an extrasolar planet. *Nature*, 457(7229), 562.

[587] D. Naef, D.W. Latham, M. Mayor, et al. (2001). HD 80606 b, a planet on an extremely elongated orbit. *Astronomy & Astrophysics*, 375(2), L27–L30.

[588] R.D. Wordsworth, F. Forget, F. Selsis, et al. (2011). Gliese 581d is the first discovered terrestrial-mass exoplanet in the habitable zone. *The Astrophysical Journal Letters*, *733*(2), L48.

[589] M.R. Swain, P. Deroo, C.A. Griffith, et al. (2010). A ground-based near-infrared emission spectrum of the exoplanet HD 189733b. *Nature*, 463(7281), 637–639.

[590] M., Gillon, A.H. Triaud, B.O. Demory, et al. (2017). Seven temperate terrestrial planets around the nearby ultracool dwarf star TRAPPIST-1. *Nature*, 542(7642), 456.

[591] G. Anglada-Escudé, P.J. Amado, J. Barnes, et al. (2016). A terrestrial planet candidate in a temperate orbit around Proxima Centauri, *Nature*. 536, 437–440.

[592] I. Ribas, E. Bolmont, F. Selsis, et al. (2016). The habitability of Proxima Centauri b. I: Irradiation, rotation and volatile inventory from formation to the present. *Astronomy & Astrophysics*, 596, A111.

[593] M. Turbet, J. Leconte, F. Selsis, et al. (2016). The habitability of Proxima Centauri b. II. Possible climates and observability. *Astronomy & Astrophysics*, 596, A112.

[594] R. Barnes, R. Deitrick, R. Luger, et al. (2016). The habitability of Proxima Centauri b. I: Evolutionary scenarios, arXiv:1608.06919.

[595] V.S. Meadows, G.N. Arney, E.W. Schwieterman, et al. (2016). The habitability of Proxima Centauri b: II: environmental states and observational discriminants, arXiv:1608.08620.

[596] D.M. Williams & E. Gaidos (2008). Detecting the glint of starlight on the oceans of distant planets. *Icarus*, 195(2), 927–937.

[597] T.D. Robinson, V.S. Meadows & D. Crisp (2010). Detecting oceans on extrasolar planets using the glint effect. *The Astrophysical Journal Letters*, 721(1), L67.

[598] A. Marshak, T. Várnai & A. Kostinski (2017). Terrestrial glint seen from deep space: Oriented ice crystals detected from the Lagrangian point. *Geophysical Research Letters*, 44, 5197–5202, doi:10.1002/2017GL073248.

[599] N.B. Cowan, D.S. Abbot & A. Voigt (2012). A false positive for ocean glint on exoplanets: the latitude-albedo effect. *The Astrophysical Journal Letters*, 752(1), L3.

[600] R.M. Haberle, D.C. Catling, M.H. Carr & K.J. Zahnle, The Early Mars Climate System. In R.M. Haberle et al. (eds.), *The Atmosphere and Climate of Mars*, Cambridge: Cambridge University Press, 2017, pp. 526–568.

[601] F. Forget & S. Lebonnois. Global climate models of the terrestrial planets. In S. J. Mackwell, A.A. Simon-Miller, J.W. Harder & M.A. Bullock (eds.), *Comparative Climatology of Terrestrial Planets*, Tucson, AZ: University of Arizona Press, 2013, pp. 213–229.

[602] J. Houghton, *In the Eye of the Storm*, Oxford: Lion Books, 2013.

Index